Biochar for Sustainable Soil Health and Crop Production

Owais Ali Wani • Munir Ahmad
Mohammad I. Al-Wabel
Editors

Biochar for Sustainable Soil Health and Crop Production

Springer

Editors
Owais Ali Wani
Division of Agronomy
Faculty of Agriculture
SKUAST-Kashmir, Wadura
Sopore, Jammu and Kashmir, India

Munir Ahmad
Soil Sciences Department, College of Food and Agricultural Sciences
King Saud University
Riyadh, Saudi Arabia

Mohammad I. Al-Wabel
Soil Sciences Department, College of Food and Agricultural Sciences
King Saud University
Riyadh, Saudi Arabia

ISBN 978-981-92-2229-2 ISBN 978-981-92-2230-8 (eBook)
https://doi.org/10.1007/978-981-92-2230-8

This Springer imprint is published by the registered company Springer Nature Singapore Pte Ltd.
The registered company address is: 152 Beach Road, #21-01/04 Gateway East, Singapore 189721, Singapore

Foreword

Soil health is the foundation upon which food systems, ecological stability, and human well-being firmly rest. In an era marked by accelerating climate change, declining natural resources, and mounting pressure on global food production, sustainable solutions that simultaneously enhance productivity and environmental resilience are no longer optional—they are imperative. Restoring the health of our soils, a resource central to civilization yet often taken for granted, must be a priority for science, policy, and practice alike.

Among the most promising nature-based innovations emerging in recent decades, *biochar* has captured global scientific and development attention for its transformative potential. Produced through the pyrolysis of organic biomass, biochar embodies both scientific innovation and ecological prudence. Its capacity to improve soil structure, increase water-holding ability, enhance nutrient use efficiency, immobilize contaminants, and sequester atmospheric carbon positions it as a powerful tool for advancing climate-smart and regenerative agriculture. Biochar exemplifies the true essence of a circular economy—converting organic waste streams into long-term soil wealth while closing carbon and nutrient loops in agro-ecosystems.

The edited volume *Biochar for Sustainable Soil Health and Crop Production* provides a comprehensive, evidence-driven, and interdisciplinary synthesis of the subject. The book brings together leading experts who unravel the biochemical, physical, and agronomic mechanisms through which biochar enriches soils and supports crops under abiotic stresses, particularly drought and nutrient-limited conditions. The chapters reflect not only depth of scientific inquiry but also clarity of practical relevance—linking laboratory advances with field-scale applicability, and academic insight with on-ground impact.

I commend the editors and authors for advancing a subject that harmonizes ancient wisdom with modern science. Their contributions span multiple geographies and scientific domains, underscoring the importance of cross-disciplinary collaboration to address soil degradation, climate vulnerability, and sustainable intensification of agriculture. This book will be valuable to researchers, students, extension specialists, industry stakeholders, and policymakers engaged in building carbon-efficient, input-smart, and resilient agricultural systems.

The future of agriculture must shift from extractive models toward *regenerative systems* that rebuild soil organic carbon, restore biological balance, optimize water and nutrient flows, and reduce environmental footprints. Biochar offers one such bridge—grounded in science, scalable in practice, and meaningful for climate mitigation and adaptation. This volume contributes significantly to that global transition.

I am confident that this book will serve as a key scientific reference and inspire further innovations for sustainable soil and crop management. I extend my best wishes for its success and impact across the world.

Vice-Chancellor
Sher-e-Kashmir University of Agricultural
Sciences & Technology of Kashmir
Shalimar, Srinagar, Jammu and Kashmir, India
7 January 2026

Nazir Ahmad Ganai

Acknowledgment

The authors gratefully acknowledge the financial support provided under the Scheme for Young Scientist & Technologist (SYST) Programme of the Department of Science and Technology (DST), Government of India, through Project No. SP/YO/2021/2204. The support and encouragement extended by DST is sincerely appreciated.

Contents

Editors and Contributors

About the Editors

Owais Ali Wani Presently working as DST Young Scientist at Division of Agronomy, Faculty of Agriculture, Sher-e-Kashmir University of Agricultural Sciences and Technology of Kashmir, Jammu and Kashmir, obtained Ph.D. from Sher-e-Kashmir University of Agricultural Sciences and Technology of Kashmir, Jammu and Kashmir. He is Postdoctoral research associate at DST India. His main research interests are sustainable residue management strategies, carbon sequestration, climate smart agriculture and biochar production technologies. Dr. Wani has actively contributed to the field of climate smart and sustainable residue management strategies. He has developed dozens of innovations in this field. He has developed pyrolyser which is acclaimed worldwide. He has also proved dominance of climate change in Himalayas and its influence on carbon pools and soil erosion rates. He has 3200 + citations at this young age and h- index of 24. He has published 30 + research papers in SCI indexed journals and editors of famous books on biochar. Dr received various prestigious awards, including DST Young Scientist award, NAAS Young Scientist award 2026, BRICS Young Scientist award Russia, AIT Thailand Visiting Fellow, Swactha Saarthi Fellowship and other national and internation innovation awards. He has received various research and innovation grants.

Munir Ahmad is currently working in Soil Science Department, College of Food and Agricultural Sciences, King Saud University Riyadh, Saudi Arabia. He is also serving as a Scientist in Saudi Biochar Research Group (SBRG). His research focuses on engineering/designing and fabricating biochar-based low-cost sorbents for soil and water remediation. He has been involved in synthesizing biochar-based nanocomposites for soil rehabilitation and carbon sequestration. His research interests include biochar production and applications, sustainable waste management, soil and water pollution remediation, climate change mitigation, and developing strategies for rehabilitating degraded lands and enhancing crop productivity. He has successfully completed eight research projects funded by various sources and is currently involved in two ongoing projects. He has published more than 65 research documents, including peer-reviewed international research articles, review articles, and book chapters. He has been serving as a Guest Editor in various journals including *Discover Sustainability*, *Toxics*, *Frontiers in Materials*, and *Sustainability*. He is also serving as reviewer for various high-ranked international peer-reviewed journals.

Mohammad I. Al-Wabel received his PhD in Soil Chemistry from Colorado State University, Fort Collins, USA, in 2001. He is serving as Professor of Environmental Soil Chemistry at the Department of Soil Sciences, College of Food and Agricultural Sciences, King Saud University Riyadh, Saudi Arabia (KSU). He has been serving as the Chairman of the Department of Soil Sciences at KSU from 2008 to 2012 and 2017–2018. He has also been serving as the Dean of the Deanship of Scientific Research, King Saud University Riyadh, Saudi Arabia, from 2018 to 2022. He is the founder and leader of the Saudi Biochar Research Group (SBRG), which is the pioneer biochar research group in Saudi Arabia. His area of research interests includes production and application of biochar-based low-cost adsorbents, remediation of contaminated soil and water, carbon sequestration and climate change mitigation, environment friendly utilization of waste resources, and revitalization of abandoned soils. He has published more than 180 research articles in peer-reviewed national and international journals, and more than 8 book chapters in various books. He has an *H-index* of 53 and *i-10* index of 114 with more than 10,000 citations. He has completed more than 20 research projects as PI and Co-PI in various interdisciplinary subjects. Due to his research contributions, he has been awarded with Research Excellency Award, High-Impact Publication Award, Lifetime Achievement Award, and various other awards.

Contributors

Vikas Abrol Division of Soil Science and Agricultural Chemistry, SKUAST Jammu, Jammu, Jammu and Kashmir, India

Jahangir Ahmad School of Earth, Atmosphere, and Environment, Monash University, Clayton, Victoria, Australia

Maher Iftekhar Ahmad Muhammad Nawaz Shareef University of Agriculture, Multan, Punjab, Pakistan

Munir Ahmad Soil Sciences Department, College of Food and Agricultural Sciences, King Saud University, Riyadh, Saudi Arabia

Mutair A. Akanji Soil Sciences Department, College of Food and Agricultural Sciences, King Saud University, Riyadh, Saudi Arabia

Abdullah S. F. Al-Farraj Soil Sciences Department, College of Food and Agricultural Sciences, King Saud University, Riyadh, Saudi Arabia

Hamed A. Al-Swadi Soil Sciences Department, College of Food and Agricultural Sciences, King Saud University, Riyadh, Saudi Arabia

Mohammad I. Al-Wabel Soil Sciences Department, College of Food and Agricultural Sciences, King Saud University, Riyadh, Saudi Arabia

Vivak M. Arya Division of Soil Science and Agricultural Chemistry, Sher-e-Kashmir University of Agricultural Sciences and Technology of Jammu, Jammu, Jammu and Kashmir, India

Rumisa Ashraf Division of Entomology, FoH, Shalimar, SKUAST-Kashmir, Srinagar, Jammu and Kashmir, India

Sabreena Ashraf Punjab Agricultural University, Ludhiana, Punjab, India

Department of Entomology, Punjab Agricultural University, Ludhiana, Punjab, India

Abid Shabir Bali Division of Agronomy, Sher-e-Kashmir University of Agricultural Sciences and Technology of Kashmir, Srinagar, Jammu and Kashmir, India

Javid Ahmad Bhat Division of Soil Science and Agricultural Chemistry, FoA, Wadura, SKUAST-Kashmir, Srinagar, Jammu and Kashmir, India

Ashima Choudhary Division of Fruit Science, Faculty of Horticulture & Forestry, Sher-e-Kashmir University of Agricultural Sciences and Technology of Jammu, Jammu, Jammu and Kashmir, India

K. Damodar Department of Agronomy, School of Agriculture, SR University, Warangal, Telangana, India

Tsering Dolker Division of Soil Science and Agricultural Chemistry, FoA, Wadura, SKUAST-Kashmir, Srinagar, Jammu and Kashmir, India

Adnan Fareed Muhammad Nawaz Shareef University of Agriculture, Multan, Punjab, Pakistan

College of Life Sciences and Agriculture, Southwest University of Science and Technology, Mianyang, Sichuan, China

Saurav Gupta Krishi Vigyan Kendra, Samba, Sher-e-Kashmir University of Agricultural Sciences and Technology,, Jammu, Jammu and Kashmir, India

Abid Hussain Muhammad Nawaz Shareef University of Agriculture, Multan, Punjab, Pakistan

Punjab Agricultural Research Board, Lahore, Punjab, Pakistan

Altaf Hussain Division of Soil Science and Agricultural Chemistry, SKUAST, Jammu, Jammu and Kashmir, India

Boreddy Jaya Chandra Reddy Division of Soil Science and Agricultural Chemistry, PJTAU, Hyderabad, Telangana, India

Jyotsana Kalsi Division of Plant Physiology, SKUAST Jammu, Jammu, Jammu and Kashmir, India

Murugesan Mohana Keerthi Department of Agronomy, School of Agriculture, SR University, Warangal, Telangana, India

Vijay Khajuria Farming System Research Centre, Chatha, Sher-e-Kashmir University of Agricultural Sciences and Technology, Jammu, Jammu and Kashmir, India

Stanzin Khenrab Division of Soil Science and Agricultural Chemistry, Faculty of Agriculture, Sher-e-Kashmir University of Agricultural Sciences and Technology of Jammu, Jammu, Jammu and Kashmir, India

Sarabdeep Kour Division of Soil Science and Agricultural Chemistry, Faculty of Agriculture, Sher-e-Kashmir University of Agricultural Sciences and Technology of Jammu, Jammu, Jammu and Kashmir, India

Shamal Shasang Kumar Crop Research Division, Ministry of Agriculture, Waterways & Sugar Industry (MOAW&SI), Sigatoka, Fiji

Vijay Kumar Rainfed Research Sub-station for Sub-tropical Fruits, Raya, Sher-e-Kashmir University of Agricultural Sciences and Technology, Jammu, Jammu and Kashmir, India

Mehraj-u-Din Lone Division of Soil Science and Agricultural Chemistry, FoA, Wadura, SKUAST-Kashmir, Srinagar, Jammu and Kashmir, India

Nahrir M. A. Lubis Soil Sciences Department, College of Food and Agricultural Sciences, King Saud University, Riyadh, Saudi Arabia

Tajamul Mansoor Division of Soil Science and Agricultural Chemistry, FoA, Wadura, SKUAST-Kashmir, Srinagar, Jammu and Kashmir, India

Sumyrah Mukhtar Division of Soil Science and Agricultural Chemistry, FoA, Wadura, SKUAST-Kashmir, Srinagar, Jammu and Kashmir, India

Shalendra Prasad Crop Research Division, Ministry of Agriculture, Waterways & Sugar Industry (MOAW&SI), Sigatoka, Fiji

Marina Qayyum Muhammad Nawaz Shareef University of Agriculture, Multan, Punjab, Pakistan

College of Life Sciences and Agriculture, Southwest University of Science and Technology, Mianyang, Sichuan, China

Muhammad Imran Rafique Soil Sciences Department, College of Food and Agricultural Sciences, King Saud University, Riyadh, Saudi Arabia

Lekkala Venakata Ravishankar Production and Plantations, Red Otter Frams Pvt Ltd, Nainital, Uttarakhand, India

Umair Riaz Muhammad Nawaz Shareef University of Agriculture, Multan, Punjab, Pakistan

Najmu Saqib Division of Soil Science and Agricultural Chemistry, FoA, Wadura, SKUAST-Kashmir, Srinagar, Jammu and Kashmir, India

Ami Chand Sharma Crop Research Division, Ministry of Agriculture, Waterways & Sugar Industry (MOAW&SI), Sigatoka, Fiji

Bhumika Sharma Division of Soil Science and Agricultural Chemistry, SKUAST Jammu, Jammu, Jammu and Kashmir, India

Peeyush Sharma Division of Soil Science and Agricultural Chemistry, SKUAST Jammu, Jammu, Jammu and Kashmir, India

Tamanna Sharma Division of Soil Science and Agricultural Chemistry, Sher-e-Kashmir University of Agricultural Sciences and Technology of Jammu, Jammu, Jammu and Kashmir, India

Uday Prasad Sharma Crop Research Division, Ministry of Agriculture, Waterways & Sugar Industry (MOAW&SI), Sigatoka, Fiji

Vikas Sharma Division of Soil Science and Agricultural Chemistry, Sher-e-Kashmir University of Agricultural Sciences and Technology of Jammu, Jammu, Jammu and Kashmir, India

Manjot Singh Sidhu P.G. Department of Agriculture, Khalsa College, Amritsar, Punjab, India

Mehraj Ud Din Sofi Division of Agronomy, Faculty of Agriculture, SKUAST Kashmir Wadura, Sopore, Jammu and Kashmir, India

Gourav Sudan Division of Soil Science and Agricultural Chemistry, Faculty of Agriculture, Sher-e-Kashmir University of Agricultural Sciences and Technology, Jammu, Jammu and Kashmir, India

Nikhil Thakur Division of Fruit Science, Sher-e-Kashmir University of Agricultural Sciences and Technology of Jammu, Jammu, Jammu and Kashmir, India

Muhammad Usama Department of Ecosystem and Conservation Sciences, University of Montana, Missoula, Montana, USA

Divyansh Verma Division of Soil Science and Agricultural Chemistry, Faculty of Agriculture, Sher-e-Kashmir University of Agricultural Sciences and Technology, Jammu, Jammu and Kashmir, India

Owais Ali Wani Division of Agronomy, Faculty of Agriculture, SKUAST-Kashmir, Wadura, Sopore, Jammu and Kashmir, India

Soil Health and Biochar Enhancement: Concept, Assessment, and Methods

Mohammad I. Al-Wabel, Jahangir Ahmad, Munir Ahmad, Mutair A. Akanji, Nahrir M. A. Lubis, Hamed A. Al-Swadi, Muhammad Imran Rafique, and Abdullah S. F. Al-Farraj

Abstract

The concept of soil health has unified several perspectives and concepts in soil science, that provides a clear definition of soil to function as a living system that sustains productivity, regulates water and nutrient cycles, support biodiversity and plays a major role in climate regulation. Early discussion in soil science involved scoping soil fertility in terms of nutrient supply alone that is closely related to agriculture production, without addressing the other functions of soil. This view later broadened, from "soil fertility" to "soil quality", which emphasized a wider range of soil functions such as environmental protection and water regulation. Today, the concept has progressed to emphasize soil health, which describes soils as a dynamic living system. The determination of soil health is based on a number of supporting parameters. These parameters involve soil physical traits such as bulk density and aggregate stability combined with chemical measures such as soil pH, electrical conductivity and cation exchange capacity, and biological properties such as soil organic matter, microbial activity and soil organic carbon, prevention of soil from degradation losses, increasing the nutrients content and soil carbon sequestration, attenuation of potentially toxic substances in soil, enhancing plant growth, biomass, and quality, and increasing the crop yields and profits of the agricultural sector. Among several approaches to improve these parameters, application of biochar has gained attention of its

M. I. Al-Wabel (✉) · M. Ahmad · M. A. Akanji · N. M. A. Lubis · H. A. Al-Swadi · M. I. Rafique · A. S. F. Al-Farraj
Soil Sciences Department, College of Food and Agricultural Sciences, King Saud University, Riyadh, Saudi Arabia
e-mail: malwabel@ksu.edu.sa

J. Ahmad
School of Earth, Atmosphere, and Environment, Monash University, Clayton, Victoria, Australia

O. A. Wani et al. (eds.), *Biochar for Sustainable Soil Health and Crop Production*, https://doi.org/10.1007/978-981-92-2230-8_1

beneficial traits that increase soil organic carbon, improve nutrient retention and also stimulate microbial growth alongside carbon sequestration. Thus, understanding the application of such indicators and amendments will be essential to contribute effectively to the awareness of soil health in achieving sustainable development goals.

1 Introduction

1.1 Importance of Soil in Agroecosystem and Natural Ecosystems

According to the soil science of America (as cited in Doran and Parkin 2015), soil is defined as "the layer(s) loose mineral and organic material that are affected by physical, chemical and biological processes at or near the planetary surface and usually holds liquids, gases and biota and support plant". From the agricultural point of view, soil health refers to the capacity of soil to sustain and promote plant development and production (Doran and Zeiss 2000). Fertile soil is considered to be able of providing enough water and nutrients needed to support the plant growth while (Voltr 2012). Soil develops through the weathering of rocks combined with living organism activities. These interactions form a dynamic mixture of minerals and organic matter, water, air and microorganisms. The complex interactions make soil evolve continually along with the influence of climate, vegetation topography and human activity. Since the process is too complex, soil is best understood as both natural body with physical chemical and biological properties and as a living ecosystem.

Early discussions of soil were focused and centered on agriculture specifically on improving crop yields. With time, scholars view this concept was narrow and overlooked the broader role of soil. Beyond supporting nutrient aspect, soil is integrated into a broader ecological network. Soil health concept placed the interconnection between lithosphere, biosphere and hydrosphere in a sustainable way that provides foundations to support all living things.

Soil health is closely related to biodiversity. The habitat that soil serves for microorganisms like bacteria, fungus, nematode and invertebrates that contribute hugely on nutrient cycle and decomposition of organic matter in the ecosystem. Together, these complex processes maintain ecological stability and contribute to the resilience of ecosystem. For this reason, soil cannot be regarded as a growth medium only, it has numerous roles that are fundamental to the sustainability of natural environments and managed landscape.

1.2 Role in Food Production, Carbon Cycling, and Biodiversity

Food production, biodiversity and carbon cycling are interconnected functions provided by soil and these functions are inseparable. FAO (2015) reported that over

90% of global food production relies on soil. Additionally, soil also is a habitat for microbial soil, which plays important roles in nutrient cycling, decomposition of organic matter to convert it into inorganic form that is usable by plant.

Human activities or anthropogenic input reduce soil health; these functions are being compromised. Polluted and compacted soil cause depletion in organic matter and microbial activity, thus directly reducing soil fertility, limiting water infiltration and increasing vulnerability to erosion (Kopittke et al. 2019). This led to low food production and increased external input such as synthetic fertilizer and water irrigation. Such inputs may further decrease soil health if used unsustainably. Hence, preserving soil health is crucial not only for productivity but also for long term viability of food system. As for soil carbon storage, more than 2500 gigatons of carbon are stored in soil which is one of the biggest carbon storages on earth (Lal 2004; Lehmann et al. 2020).

The most common of this carbon is in the form of soil organic matter, a dynamic influenced by land management and microbial activity. Healthy soil is a major influence on climate regulation that functions as carbon sink sequester as for the accumulation of soil organic matter (SOM) (FAO 2017; Chenu et al. 2019). However, when human activity such as tillage, deforestation or overgrazing, such activity can become net source or major contribution of greenhouse gas emissions (Smith et al. 2020). Soil quality loss is also known as soil degradation not only increases carbon release but also reduces the soil ability to store it, creating a feedback loop that exacerbates climate change.

1.3 Global Soil Degradation Trends and the Need for a Sustainable Soil Health Approach

Soil degradation by definition according to FAO (2020), "is a change in soil health status resulting in a diminished capacity of the ecosystem to provide goods and services for its beneficiaries". As one of the most pressing environmental issues in the twenty-first century, 33% of the world's soil is moderately to highly degraded, primarily due to soil compaction, erosion, salination, desertification and pollution (FAO 2015). Correspondingly, IPBES (2018) reported that soil degradation contributes to loss of biodiversity and ecosystems services, negatively impacting at least 3.2 billion people worldwide, more than 10% in economic losses of global GDP annually and declining of agricultural productivity, especially dry land (IPBES 2018).

The factors contributing to soil degradation differ across regions, for example in Africa and south Asia, water and wind erosion are the primary causes, in Central Asia and Middle East, soil salinization are the main causes, whereas Europe and North America, the decline of soil organic matter is the dominant issues (Bai et al. 2018).

The consequences of soil degradation are multifaceted. In agriculture, degraded soil reduces soil fertility that led to low production and diminished water holding capacity and increased reliance toward external input. Environmentally soil degradation, disrupts soil capacity as carbon sink. Soil degradation significantly contributed to greenhouse gas emissions releasing approximately 1.5 Gt of carbon per year thereby exacerbating climate change (Le Quéré et al. 2018).

Traditional approaches to soil management have often been insufficient, typically measured by the availability of macronutrients and crop yields. While this approach boosted short-term productivity, it neglected the broader ecological functions of soil, such as biodiversity support, hydrological regulation, and carbon sequestration. As a result, interventions that restored fertility temporarily often accelerated degradation in the long term. Like addition of soil nutrient such as fertilizer addresses chemical deficit, yet do not solve physical problems like compaction and nor restoring depleted soil microbial that required by a healthy soil (Bünemann et al. 2018).

One of the best strategies to address these issues is by the application of biochar, a carbon rich material that derives from a process called biomass pyrolysis. Scientific evidence has shown that applications of biochar as soil amendment improve several soil quality indicators (SQI) that are central to soil health. Biochar application improves soil aggregation, porosity, and water holding capacity, reducing soil compaction and the risk of erosion. As for chemical aspect, biochar improves cation exchange capacity, soil pH, and increase nutrient retention along with biological activity that significantly stimulates microbial activity and fosters stable soil organic carbon.

The growing acknowledgment of soil as more than a nutrient reservoir has reshaped the fundamental assumptions in soil and encouraged the adoption of soil health framework. Soil health concept is approached as a whole, it brings together physical, chemical and biological aspect to explain how soil function as a dynamic ecological system. Under these notions, soil is valued for its ability to sustain production while at the same time regulating water flow, moderating climate and supporting biodiversity.

In this chapter, soil health is discussed from conceptual and practical perspective, a distinction insight from soil physic, chemistry, and biology to provide integrated frameworks. The chapter will guide how the idea of soil health and the earlier notion of soil quality and fertility, along to review the main indicators and assessment approaches. It also views how soil health connects with wider global challenges.

2 Evolution of the Concept of Soil Health

Soil health is the capacity of soil to function as a living system that support plants, regulates water and nutrient cycles, carbon storage and sustain biodiversity. As for measurable indicators, scientist depends on Soil Quality Indicators (SQIs), which are properties that depict soil condition. These include physical properties such as texture, structure and bulk density, chemical measures like pH, EC and nutrient level including biological aspects, soil organic matter and microbial activity. These tools improve soil health concept into more practical and measurable and provide the foundation for understanding the development of this concept throughout history.

Over the past century the idea of soil health has evolved dramatically, reflecting changes in scientific understanding, environmental awareness, and agricultural practices. Back in the early to mid-twentieth century, the common terms were soil fertility which referred to the ability of soil to provide essential and non-essential nutrients for crop growth (Bünemann et al. 2018). In general, soil fertility primarily assessed through chemical parameters, for instance Nitrogen (N), Phosphorus (P), potassium (K), soil pH, Ec and stressed external input such as management practices, irrigation, fertilizer and liming to modified nutrients deficiency.

Between 1980s and early 1990s the raise of concern about environmental degradation and sustainable practices which led to changes of view into soil quality. Defined as 'the capacity of soil to function within ecosystem boundaries to sustain biological productivity, maintaining environmental quality, promote plant and animal health (Doran and Parkin 2015). Yet this concept is not fully understood or widely applied since interpreted as anthropocentric manner and often wrongly interpreted as soil's capacity to support crop production rather than its full ecological functions.

The perceptions of soil health started to change in the early of 2000s, it shifted to emphasized more on biological and ecological dimensions of soil. Scientists and experts turn to view soil as a living system that shaped by the interaction of soil physical, chemical, and biological aspect (Drenning 2021). This paradigm recognises soil health even more on its role in supporting ecosystems instead only on food production. The shift in discourse reflects a broader understanding on ecology, which value soil not only for fertility on productivity but also for biodiversity, resilience and climate regulation (Bünemann et al. 2018).

2.1 Influence of Ecological and System Thinking

The discourse shifted from soil fertility to soil health was strongly influenced by the rise of ecological and system thinking on the late twentieth century. The ecology of the soil ecosystem focuses on the interconnectedness with plants, microbes, climate, and human activities. Soil has been conceptualized as intricate adaptive system, wherein feedback mechanisms govern processes for instance, nutrient cycling, water infiltration, and the dynamics of organic matter.

The change of perspective from isolated properties to interdependent and interconnectedness process, emphasizing soil biodiversity, microbial ecology, and the role of organic substance in maintaining resilience of soil function. These ways of view, paved a way for more holistic and integrative concept of soil health connecting agriculture, environments, and climate science.

2.2 Key Global Initiatives

Recognition and awareness of soil health among global organizations and initiatives has growth exponentially. The USDA's Natural Resources Conversation Service

NRCS, considered as one of the leading organizations, define soil health as 'the continued capacity of soil to function as a vital living ecosystem that sustains plants, animals and human' with promoting farmer-based assessment tools, conservation tillage, and cover cropping., The FAO on the status of the world soil resources report (2015) emphasized soil health as the central element of global sustainability discussions. These was taken to emphasize the multifunctionality of soil and the need for monitoring framework.

Another significant step was the "4 per 1000" initiative, which launched at COP 21 climate negotiations in Paris 2015. This event proposed that increasing global soil carbon stocks by 0.4% per year could offset a huge portion of annual anthropogenic greenhouse gas emission (Chenu et al. 2019). Recently, the European union soil strategy (2021) and the integration of soil related goals in the UN sustainable development goals have reinforced the importance of soil health in achieving land degradation neutrality, food security and climate action (Montanarella and Panagos 2021).

2.3 Shift Toward Comprehensive, Integrative Definitions

The current scientist perspective on soil health adopts an integrative and systems-based approach. Such Contrast with earlier concept which focus on nutrient or productivity, modern concept emphasizes more integrated and multidimensional nature of soil that include soil physical, chemical composition and biological activity and its contribution toward ecosystem. Soil health now is more recognise as the central part for global sustainability challenges, food production in agriculture to climate mitigation, water security and environmental conservation (Keesstra et al. 2016; Lennartz and Liu 2019).

Several expert and scholars have emphasized the important value and to have a clear definition and measurable indicators. A well-defined principle will establish a strong correlation between advanced conceptual and real-world application (Karlen et al. 2019; Janzen et al. 2021). The comprehensive perspective that covers all ground and element of soil health will be positioned "soil role" in integrated framework that unites research and practices.

3 Soil Health Versus Soil Quality

3.1 Definitions and Conceptual Distinctions

Soil is not just an inert medium for plant roots; it is a dynamic and living system whose vitality determines ecosystem sustainability. Two central concepts of soil at its entirety revolves its health and soil quality are often used to describe its condition, though they differ in scope and emphasis.

Soil health refers to the vitality of the soil as a living ecosystem. It emphasizes the integration of physical, chemical, and especially biological properties that enable soil to function as a dynamic life-supporting system (Alshaal et al. 2018). A healthy soil is believed to foster plant productivity, sustains diverse microbial and faunal communities, regulates water and nutrient cycling, and contributes to broader ecosystem services such as climate regulation and resilience against disturbances.

Soil quality: Soil quality, on the other hand, is more function-oriented. It describes the capacity of a soil to carry out specific functions such as supporting crop yield, filtering pollutants, storing carbon, or providing a stable foundation for infrastructure (Seifu and Elias 2018). Moreover, the assessment of soil quality often relies on measurable indicators like pH, nutrient content, texture, or contamination levels. While important, this approach sometimes overlooks the biological and systemic dimensions of soil. Table 1 shows the key differences between soil health and soil quality.

3.2 Implications for Monitoring and Management

3.2.1 Soil Health Approach

Assessing soil health calls for a focus on biological activity and long-term sustainability. Methods such as reduced tillage, cover cropping, crop rotation, organic matter incorporation, and the application of microbial inoculants are used to strengthen soil biodiversity, improve structure, and build resilience (Angon et al. 2023). Such practices are closely aligned with agroecological and regenerative farming approaches.

3.2.2 Soil Quality Approach

Management of soil quality tends to be more short-term and utilitarian. Strategies may include correcting nutrient deficiencies with fertilizers, adjusting pH with lime, or reducing erosion with physical barriers (Sharma et al. 2025). While effective for achieving immediate goals (e.g., maximizing crop yield), such practices may not necessarily restore or maintain the soil's living components in the long run.

4 Literature Perspectives

Scholarly discourse on soil health and soil quality reveals varying and sometimes conflicting positions, reflecting both disciplinary orientations and methodological preferences.

4.1 Interchangeable Usage

Two discourses namely, soil health and soil quality have been understood and similarly used in soil, and environmental discussions over time. In such instances, both are broadly understood as descriptors of the soil's capacity to sustain productivity and ensure long-term ecological viability (Lal 2016; Keesstra et al. 2016; Therond

Table 1 Key differences between the health of a soil and its quality

Aspect	Soil quality	Soil health
Description/ definition	Its' the potent ability of a soil consistently performing specific functions (e.g., medium for plant growth and water flow)	Express soil as a living system with vitality and resilience to sustain ecosystem functions
Conceptual view	It emphasizes soil performance under given land use function-wise	It emphasizes biological integrity and long-term sustainability life-wise
Primary indicators	Include physical and chemical measures such as pH, nutrient levels, bulk density	Include biological measures such as microbial activity, earthworms, respiration, enzymes
Research perspective	Focuses on soil's ability to meet human needs	Recognizing soil as a dynamic, living component of nature
Application	Basically, used in evaluating productivity and environmental performance	Basically, used in guiding sustainable land management and resilience to stressors
Biological perspectives	It strongly focuses on soil organisms, biodiversity, and ecological processes	Often underemphasized; meanwhile, focus is more on measurable physical and chemical parameters
Resilience	Prioritizes soil's capacity to withstand stress (e.g., drought, erosion) and recover	May only measure short-term productivity
Function-wise	It is broader: Includes ecosystem interactions which include the cycling of soil nutrient, carbon storages, biodiversity support, and water regulation	It is narrower: Often tied to agricultural production, construction, or engineering suitability in the neighbourhood
Monitoring and evaluating indicators (M&E)	They include; organic matter, microbial activity, biodiversity indices, soil structure, infiltration rates	pH, nutrient levels, salinity, heavy metals, bulk density
Management approaches	Conservation tillage, cover crops, organic amendments, integrated pest management, agroecology	Fertilizer application, liming, soil amendments, erosion control

et al. 2017; Lehmann et al. 2020). This pragmatic conflation enables attention to be directed toward quantifiable soil attributes rather than theoretical differentiation. However, overlooking conceptual distinctions may obscure important nuances, thereby introducing inconsistencies in both scientific findings and as well practical implementation.

4.2 Distinctive Usage

With much research evidence, perspective has been brought towards the distinctive use of soil quality and soil health. For instance, Bhaduri et al. (2022) and Kumar et al. (2025) has been able to capture facts and ideological conveyed that when it comes to soil capacity to perform its agricultural and ecological (i.e. nutrient cycling and hydrological functions for crop production) soil quality best describes the phenomenon. *Soil health,* by contrast, is viewed as life-centered, portraying soil as a dynamic, living entity defined by biological activity, self-regulation, and resilience. These ideologies then show that a soil might not be healthy (i.e. deficient in "health" if its biodiversity or ecological integrity is compromised) and still be quality serving the above discussed functions.

4.3 Holistic Advocacy

There have been several research proposals on the issue of soil health, scientist have argued and tried to synthesize and convey their ideological findings on the subject of soil health, as it is important for crop and human sustainability (Friedrichsen et al. 2021; Larkin 2015). Researchers assures that soil health more accurately captures the ecological, biological, and socio-cultural dimensions of soil management, conceptualizing soil as an active, living system (captures diversity) rather than merely a substrate for production (Kingfisher 2022). Therefore, this orientation aligns closely with sustainable agriculture, climate adaptation initiatives, and ecosystem service frameworks in general, thereby positioning soil health as a robust foundation for long-term environmental governance and land-use policy.

4.4 Need for Clarity

Nevertheless, many scholars stress the importance of definitional precision. The interchangeable or ambiguous use of soil health and soil quality often leads to confusion in scientific research, extension practice, and policy development (Kingfisher 2022; Trimarco 2025). Adopting standardized definitions and frameworks would improve cross-study comparability, facilitate clearer communication among stakeholders, and ensure that soil management initiatives are guided by consistent and well-defined goals.

5 Key Takeaways

- Soil health and soil quality are related but not identical concepts.
- Soil health emphasizes vitality, biodiversity, and long-term sustainability of soil as a living system.
- Soil quality emphasizes the soil's capacity to perform specific functions, often evaluated through measurable indicators.
- Recognizing their distinctions is crucial for designing monitoring frameworks and management strategies that balance short-term productivity with long-term ecosystem resilience.
- Literature increasingly favours the soil health paradigm, as it fosters regenerative and sustainable land-use practices.

6 Objectives of Soil-Health Assessment

6.1 Evaluating Soil's Capacity to Perform Ecosystem Functions

The central objective of soil health assessment basically is to determine the soil's capacity across its properties, to sustain critical ecosystem functions, including cycling of nutrients, water balance, soil carbon sustainability, and biodiversity maintenance. As noted by Trivedi et al. (2016) and Gatica-Saavedra et al. (2023), this capacity can be evaluated through indicators such as soil organic matter, structural stability, pH, and microbial activity, as they collectively capture its biological and chemical properties. More recent contributions, such as those of Daba and Dejene (2018), extend this perspective by emphasizing soil as a dynamic living system whose condition expressed broader ecosystem services spanning from biodiversity conservation and water purification to climate regulation through carbon sequestration. Consequently, soil health assessment actually goes beyond the narrow scope of agricultural productivity, this positioning soil as a resilient and multifunctional component of the wider environment.

6.2 Detecting Degradation or Improvement

Another critical objective of soil health assessment is the early detection of degradation processes or signs of improvement. Long-term monitoring of soil indicators allows researchers and farmers to track changes in fertility of soils, structure, as well as biological activities. Rickson et al. (2015) underscore that soil degradation particularly erosion, nutrient depletion, and compaction remain a major constraint to global food security. However, studies have shown that the adoption of best management practices such as conservation tillage, organic amendments, and cover cropping can reverse degradation trends and restore soil functionality (Rahman et al. 2020; Francaviglia et al. 2023). By detecting these changes over time, soil assessments act as a diagnostic tool, providing evidence for either unsustainable practices that require intervention or successful strategies that warrant wider adoption.

6.3 Informing Sustainable Land-Use Practices

Soil health assessments also play a vital role in shaping sustainable land-use planning and agricultural management. Neemisha (2020) underscores the importance of soil organisms in sustaining ecological balance and delivering ecosystem services critical for crop production. This perspective makes clear that land-use choices must be closely linked to soil health considerations. Moreover, integrated soil approaches not only focus on productivity but also address wider issues such as climate change mitigation, ecosystem resilience, and farmer livelihoods. As such, soil health assessments provide the scientific foundation for promoting agricultural models like agroecology, regenerative farming, and mixed systems that seek to combine productivity with long-term sustainability.

6.4 Supporting Policies and Agricultural Innovation

The data for soil health is very useful at the farm level and contributes to policy formulation and agricultural innovation. García et al. (2022) stated that effective soil management requires enabling policies and governance frameworks that recognize soil as a natural ecological resource. At the same time, technological advances which include precision agriculture & digital soil-map approach provide new opportunities to translate soil health assessments into targeted, site-specific management solutions (Naorem et al. 2024). These innovations can help policymakers and practitioners design interventions that are both evidence-based and context-sensitive, ensuring that soil health improvement becomes central to agricultural development and environmental protection strategies.

6.5 Benchmarking and Comparative Analysis

Soil health assessment plays a very important role in benchmarking, allowing for systematic comparisons across fields, farms, and regions. Such benchmarks not only reveal best agricultural practices but also shows areas requiring urgent attention and remediation. As Zou et al. (2021) explain that, soil quality indices provide a valuable tool for evaluating the relative performance of different management systems. However, the inherent spatial and temporal variability of soils show their context-dependent character, indicating that uniform assessment approaches are inadequate. Meanwhile through comparative analysis, benchmarking can then support adaptive management, drives iterative improvement, and facilitates knowledge exchange across agricultural space and ecological contexts.

Collectively, these functions then show the multidimensional nature of soil health assessment. Rather than a singular or static exercise, it constitutes a continuous process that integrates ecological, agronomic, and socio-political operations. Therefore, focusing on ecosystem functionality, tracking degradation and recovery, guiding land-use strategies, informing policy and innovation, and enabling benchmarking, soil health assessment then offers a comprehensive framework for safeguarding the life-supporting potential of soils.

7 Soil Health Indicators

Towards practical assessing soil health, a multidimensional framework is required since no single soil parameter can adequately capture the complexity of soil functions. Accordingly, researchers employ an integrated set of biological, chemical, and physical indicators that together provide a comprehensive picture of soil status.

These indicators are essential for evaluating the soil's ability to sustain plant productivity, maintain ecological equilibrium, and withstand environmental stressors (Table 2). As illustrated in Fig. 1, the assessment process involves sequential steps of sampling, indicator selection, laboratory and field analysis, scoring, and the interpretation of results, consequentially leading to management recommendations.

7.1 Biological Indicators

Biological indicators are widely recognized as some of the most sensitive and responsive measures of soil health, since they directly reflect soil life and ecological functioning.

7.1.1 Microbial Biomass, Respiration, and Enzyme Activity

Decomposition of soil organic matter toward releasing soil nutrients, especially organic carbon, is mainly done by soil microorganism, which in-turn aids structural formation of the soil through the microbe movement within and outside the soil profile (Sahu et al. 2017). This automatically reveals relevance of organic matter in

Table 2 Common soil health indicators and their interpretations

Category	Indicator	Interpretation
Physical	Bulk density	Good: <1.4 g cm^{-3} → good porosity; Poor: >1.6 g cm^{-3} → compaction
	Aggregate stability	Good: >70% stable aggregates → erosion resistant; Poor: <40%
	Infiltration rate	Good: rapid infiltration, reduced runoff; Poor: slow infiltration, crusting
Chemical	Soil pH	Good: 6.0–7.5 → optimal nutrient availability; Poor: <5.5 or >8.5 → nutrient imbalance
	CEC	Good: High CEC (>15 cmol kg^{-1}) → nutrient retention; Poor: <5 cmol kg^{-1}
	EC (salinity)	Good: <2 dS m^{-1}; Poor: >4 dS m^{-1} → salinity stress
Biological	SOC	Good: >2% (depends on soil type); Poor: <1% → low fertility, poor structure
	Microbial biomass	Good: high microbial carbon pool; Poor: depleted biomass → low activity
	Earthworm abundance	Good: >200 m^{-2}; Poor: <50 m^{-2}

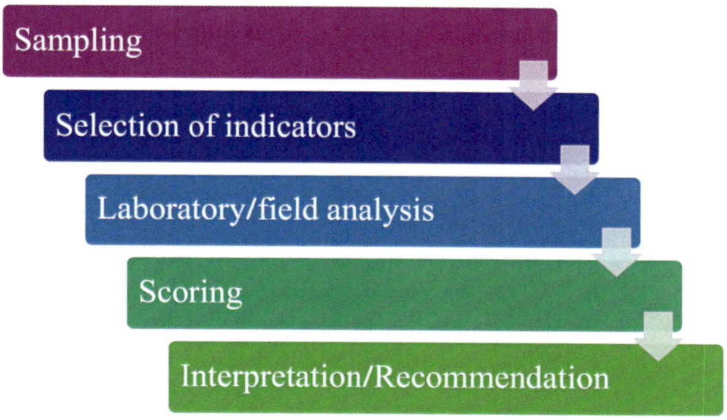

Fig. 1 A flowchart showing, sampling, selection of indicators, laboratory/field analysis, scoring, interpretation/management recommendations

soil function, serving as living fraction (microbial biomass) which enhances biological processes and pool. These processes of organic matter degradation and mineralization through enzymatic activities including phosphatate, dehydrogenase and urease (Adetunji et al. 2017), helps understand soil respiration including energy dynamics and carbon fluxes within soil. (Phillips et al. 2017), together shows the capacity of soil for bio-geochemical functions.

7.1.2 Earthworm Abundance and Nematode Populations

Earthworms are considered "ecosystem engineers" because of their ability to improve porosity, enhance aggregation, and redistribute nutrients (Devi and Khwairakpam 2022). Moreover, nematodes serve as useful bioindicators of soil food web structure. While some nematodes are plant-parasitic and detrimental, the diversity and balance of nematode groups can reveal the level of ecological stability or stress in soils (Lazarova et al. 2021). Their population dynamics therefore act as diagnostic tools for soil ecosystem health. While biological indicators are powerful early warning tools, their measurement can be technically demanding and costly, limiting their routine application in smallholder or resource-poor farming systems.

7.2 Chemical Indicators

Chemical properties represent another fundamental dimension of soil health, providing insight into fertility, nutrient balance, and environmental risks.

7.2.1 Soil pH, Organic Matter, and Nutrient Levels (N, P, K)

Soil pH is known to be an important soil factor that affects nutrient availability, microbial activity, and chemical balance. Organic matter is vital not only for nutrient storage and release but also for improving soil structure and water retention (Bashir et al. 2021). Macronutrients such as NPK remain central to fertility assessments, as deficiencies or imbalances directly constrain crop productivity.

7.2.2 Cation Exchange Capacity (CEC) and Salinity

Among soil indicators, cation exchangeable capacity (CEC) stands out, because it shows the efficient capacity of soil exchange and as well retains essential soil nutrients (the cations basically), which helps to determine soil fertility of any given soil. Its's noted that soil with high CEC values shows the presence of appreciable amount of nutrients, in turn shows better fertility (Ronson 2024), on the other hand a soil with low CEC is noted to have a decline soil nutrient, most time seen in sandy soils (known for their high sand content, and low clay). Moreover, soil degradation and reduced ability to support crop productivity can result from excessive salt content, a condition described as salinity, caused by the inherent or anthropogenic factors (Tessema et al. 2022). Its then proper to measure these soil indicators (CEC and Salinity) so as to determine the nutrient dynamics and efficiency for crop productivity per season.

7.3 Physical Indicators

Beyond the chemical and biological soil indicators, the soil physical indicators or properties e.g. bulk density, porosity, moisture content etc. helps to understand the hydrological and structural nature and dynamics of soils for crop productivity.

7.3.1 Bulk Density, Porosity, and Aggregate Stability

Foremost, the density (bulk density, BD), helps to know the level of compaction of a given soil, which can either enhance porosity, nutrient availability, plant root penetration and water infiltration within and down the profile of a soil (Shah et al. 2017). Specifically, porosity explains the percentage or level of soil pore spaces, which aids water, nutrient and microbial movement and activities in soils (Delgado and Gómez 2024). Lastly, one of the soil physical indicators known as aggregate stability has a major role in controlling soil erosion and water infiltration of any given soil, thus the need for its frequent evaluation (Gan et al. 2024).

7.3.2 Infiltration Rate and Water-Holding Capacity

Infiltration, being one of the soil water transmission properties explains the speed or rate through which water enters the soil profile, which could be affected by run-off and erosion and vice versa it determines erosion control measures and ground water replenishment (Zewide 2021) for that given soil. Then, the water-holding capacity (WHC) of soil expresses the tenacious ability of soils to hold water within itself for the benefit of plant uptake, which can be affected by climatic factor such as rainfall, temperature, and drought (Medrano et al. 2015).

Note, all the physical indicators evaluation, is believed to help regulate hydrological processes and address the impacts of climatic variability of soils.

These three indicators (biological, chemical, and physical indicators) collectively provide a holistic framework for soil health assessment. Biological measures capture the living dynamics of the soil, chemical indicators highlight fertility and nutrient balance, and physical measures reflect the structural and hydrological capacity of soils. Each category with a display of strengths and limitations. Integrating all three dimensions ensures a more comprehensive understanding of soil functionality. Such a multidisciplinary approach not only improves diagnostic accuracy but also strengthens the capacity of farmers, scientists, and policy-makers to design management practices that enhance soil resilience, productivity, and sustainability.

8 Popular Tools for Soil Health Assessment

In order to translate the health of the soil and its quality into adoptable practice, several assessment frameworks have been developed to integrate scientific rigor with farmer-oriented applications. Two notable examples are the Cornell Soil Health Test and India's Soil Health Card program.

8.1 Cornell Soil Health Test (CSHT)

Among all other soil health test measuring systems, Cornell Soil Health Test (CSHT) stands out, designed for a comprehensive soil evaluation that includes the biological, chemical and physical properties, just as afore mentioned and discussed above (Fine et al. 2017). It is relevant in all soil and crop aspects, including root, microbe, nutrient function, giving insight into strength and constraints of scored soil indicators, which then help to infer best methodology for addressing soil health limitation discovered.

8.2 India's Soil Health Card (SHC)

India's Soil Health Card program represents one of the world's largest soil evaluation campaigns. Every two years, farmers receive a card outlining 12 essential soil parameters including macro- and micronutrients, organic carbon, pH, and electrical conductivity along with tailored fertilizer and amendment recommendations. Unlike the CSHT, the SHC largely reflects the "soil quality" perspective, emphasizing nutrient status and functional indicators to enhance crop yields (Islam et al. 2017). Nonetheless, its mass implementation has successfully heightened soil awareness among millions of farmers, underscoring its value in large-scale agricultural policy.

9 Case Studies of Biochar and SQI

Several research works nation-wide expressed the synchronization of soil quality index (SQI) and biochar use. Foremost, a maize stalk biochar incorporated into soil has been found in China, improve soil acidity which is capable of reducing fertility in China soil (Yan et al. 2022; Hou et al. 2024). Moreover, the results not only prove that biochar improve soil pH and soil organic carbon but improve microbial activities, leading to higher SQI values and thus reveal measurable improvements in soil health. Specifically, there was a positive linear relationship between soil quality index (SQI) and biochar application amount in HO Compared with other treatments, the pH, EC, SOM, available phosphorus, sucrase and catalase activities were highest in HO3 (Yan et al. 2022). Report from India according to Munda et al. (2023) showed that rice husk made biochar tends to improve bulk density, electrical conductivity, as well improve soil quality of the India known for rainfed agriculture challenge.

Moreover, experiment reported from Brazil showed that biochar made from plants species *Qualea grandiglora*, *Qualea parviflora*, *Qualea multiflora* and *Tachigali vulgaris* found in Cerrado stricto sensu examined on a natural low nutrient poor tropical soil showed that biochar increased organic carbon & soil respiration, nutrient cycling & retention, and soil aggregation, thus increase SQI values (Petter et al. 2019). Lastly, considering the Sub-Saharan African, degraded soils has been improved as a result of biochar application. For instance, Kätterer et al. (2022) reported that application of biochar made from maize cob has aided the improvement of soil fertility in Kenya soil. Additionally, organic carbon, pH, exchangeable bases, and microbial activity, higher SQI, increased crop productivity at large has been reported by Oladele (2019) and Amadi et al. (2024) in Nigeria using rice husk and cassava peel biochar. All these findings reveal the impact of biochar on soil quality.

10 Challenges and Limitations

Though there has been a lot of progress and conceptualization in assessing the soil health, still there are a lot of challenges and limitations that are being faced:

There are many indicators to assess the soil, however, these indicators vary across regions, ecosystem, and management practices. There are no universal indicators that could be accepted worldwide. Soil responses vary according to the land use, seasonal changes, microbial activities, soil use monitoring system. The assessment on the lab level could also be limited by resources and could be expensive. Moreover, the laboratory results may be difficult to interpret at the field level by the farmers or implement at the field scale. Further, there are gaps between the policy framework and scientific metrics. High cost of soil analyses at the laboratory scale could be limitation to the widespread application of the soil health assessment.

11 Conclusion and Future Perspectives

The health of soil is the cornerstone in ensuring sustainable agriculture including the climate resilience and food security. In this chapter, the conceptual evolution of soil health, differentiation of soil health and quality, chemical, biological, and physical indicators for soil health were highlighted. However, several challenges including field-based, laboratory, lack of standardization, and adoption of these methods at the field levels were the obstacles. Development of harmonized, scalable, and cost-effective research approaches is the need of time to moving into the future. Integration of remote sensing and artificial intelligence for soil monitoring and interpretation should be considered. Economic activities aligned with the policies will strengthen the farmers. Soil health is not only maintained by agricultural activities, but it is also about the biodiversity, ecosystem, climate change, and future generations.

References

Adetunji AT, Lewu FB, Mulidzi R, Ncube B (2017) The biological activities of β-glucosidase, phosphatase and urease as soil quality indicators: a review. J Soil Sci Plant Nutr 17(3):794–807

Alshaal T, El-Ramady H, Elhawat N, El-Nahrawy S, Omara AED, Elsakhawy T et al (2018) Soil health and its biology. In: The soils of Egypt. Springer International Publishing, Cham, pp 175–185

Amadi N, Yabrade M, Okubor PC (2024) Impact of cassava peel on soil nutrient and microbial restoration of crude oil contaminated soil in Rivers State, Nigeria. Sci Afr 23(4):141–156

Angon PB, Anjum N, Akter MM, Shreejana KC, Suma RP, Jannat S (2023) An overview of the impact of tillage and cropping systems on soil health in agricultural practices. Adv Agric 2023(1):8861216

Bai Z et al (2018) Effects of agricultural management practices on soil quality: a review of long-term experiments for Europe and China. Agric Ecosyst Environ 265:1–7. Available at: https://doi.org/10.1016/j.agee.2018.05.028

Bashir O, Ali T, Baba ZA, Rather GH, Bangroo SA, Mukhtar SD et al (2021) Soil organic matter and its impact on soil properties and nutrient status. In: Microbiota and biofertilizers, vol 2: ecofriendly tools for reclamation of degraded soil environs. Springer International Publishing, Cham, pp 129–159

Bhaduri D, Sihi D, Bhowmik A, Verma BC, Munda S, Dari B (2022) A review on effective soil health bio-indicators for ecosystem restoration and sustainability. Front Microbiol 13:938481

Bünemann EK et al (2018) Soil quality – a critical review. Soil Biol Biochem 120:105–125. Available at: https://doi.org/10.1016/j.soilbio.2018.01.030

Chenu C et al (2019) Increasing organic stocks in agricultural soils: knowledge gaps and potential innovations. Soil Tillage Res 188:41–52. Available at: https://doi.org/10.1016/j.still.2018.04.011

Daba MH, Dejene SW (2018) The role of biodiversity and ecosystem services in carbon sequestration and its implication for climate change mitigation. Environ Sci Nat Resour 11(2):1–10

Delgado A, Gómez JA (2024) The soil: physical, chemical, and biological properties. In: Principles of agronomy for sustainable agriculture. Springer International Publishing, Cham, pp 15–30

Devi C, Khwairakpam M (2022) Earthworms as ecological engineers and its role in bioconversion of organic waste through vermicomposting. In: Earthworms and their ecological significance. Nova Science Publishers, p 169

Doran JW, Parkin TB (2015) Defining and assessing soil quality. In: Doran JW et al (eds) SSSA special publications. Soil Science Society of America and American Society of Agronomy, Madison, pp 1–21. Available at: https://doi.org/10.2136/sssaspecpub35.c1

Doran JW, Zeiss MR (2000) Soil health and sustainability: managing the biotic component of soil quality. Appl Soil Ecol 15:3–11

Drenning P (2021) Soil functions and ecosystem services: a literature review (Part 2/2). Available at: https://doi.org/10.13140/RG.2.2.23922.63685

FAO (2015) Status of the world's soil resources: main report. FAO: ITPS, Rome

FAO (2017) Unlocking the potential of soil organic carbon: outcome document of the Global Symposium on Soil Organic Carbon, 21–23 March 2017, FAO Headquarters, Rome, Italy. Food and Agriculture Organization of the United Nations, Rome

FAO (2020) Soil pollution: a hidden reality. Food and Agriculture Organization of the United Nations, Rome

Fine AK, Van Es HM, Schindelbeck RR (2017) Statistics, scoring functions, and regional analysis of a comprehensive soil health database. Soil Sci Soc Am J 81(3):589–601

Francaviglia R, Almagro M, Vicente-Vicente JL (2023) Conservation agriculture and soil organic carbon: principles, processes, practices and policy options. Soil Syst 7(1):17

Friedrichsen CN, Hagen-Zakarison S, Friesen ML, McFarland CR, Tao H, Wulfhorst JD (2021) Soil health and well-being: redefining soil health based upon a plurality of values. Soil Secur 2:100004

Gan F, Shi H, Gou J, Zhang L, Dai Q, Yan Y (2024) Responses of soil aggregate stability and soil erosion resistance to different bedrock strata dip and land use types in the karst trough valley of Southwest China. Int Soil Water Conserv Res 12(3):684–696

García SO, Santillan VS, Vivier VB, Anglés-Hernández M, Pérez ME, Prado B (2022) Soil governance and sustainable agriculture in Mexico. Soil Secur 7:100059

Gatica-Saavedra P, Aburto F, Rojas P, Echeverría C (2023) Soil health indicators for monitoring forest ecological restoration: a critical review. Restor Ecol 31(5):e13836

Hou X, He W, Zhang Y, Zhang N, Yan J, Chen Y (2024) Maize-straw biochar enhances soil properties and grain yield of foxtail millet in a newly reclaimed land. Agronomy 14(11):2465

IPBES (2018) The IPBES assessment report on land degradation and restoration. Zenodo. Available at: https://doi.org/10.5281/zenodo.3237393

Islam SDU, Bhuiyan MAH, Mohinuzzaman M, Ali MH, Moon SR (2017) A Soil Health Card (SHC) for soil quality monitoring of agricultural lands in south-eastern coastal region of Bangladesh. Environ Syst Res 6(1):15

Janzen HH, Janzen DW, Gregorich EG (2021) The 'soil health' metaphor: illuminating or illusory? Soil Biol Biochem 159:108167. https://doi.org/10.1016/j.soilbio.2021.108167

Karlen DL, Veum KS, Sudduth KA, Obrycki JF, Nunes MR (2019) Soil health assessment: past accomplishments, current activities, and future opportunities. Soil Tillage Res 195:104365. https://doi.org/10.1016/j.still.2019.104365

Kätterer T, Roobroeck D, Kimutai G, Karltun E, Nyberg G, Sundberg C, de Nowina KR (2022) Maize grain yield responses to realistic biochar application rates on smallholder farms in Kenya. Agron Sustain Dev 42(4):63

Keesstra SD, Bouma J, Wallinga J, Tittonell P, Smith P, Cerdà A et al (2016) Forum paper: the significance of soils and soil science towards realization of the UN sustainable development goals (SDGS). Soil Discuss 2016:1–28

Kingfisher L (2022) As the soil, so the human: narratives of ontological entanglement and soil management in regenerative agriculture. Wageningen University and Research eDepot, Netherlands

Kopittke PM et al (2019) Soil and the intensification of agriculture for global food security. Environ Int 132:105078. Available at: https://doi.org/10.1016/j.envint.2019.105078

Kumar KA, Jayanthi J, Singh RD, Sahu SK, Hasan A (2025) Exploring soil health and sustainability in the Northwestern Himalayas: assessing indicators amidst changing land use. Environ Earth Sci 84(8):1–17

Lal R (2004) Soil carbon sequestration to mitigate climate change. Geoderma 123(1):1–22. Available at: https://doi.org/10.1016/j.geoderma.2004.01.032

Lal R (2016) Soil health and carbon management. Food Energy Secur 5(4):212–222

Larkin RP (2015) Soil health paradigms and implications for disease management. Annu Rev Phytopathol 53(1):199–221

Lazarova S, Coyne D, Rodriguez MG, Peteira B, Ciancio A (2021) Functional diversity of soil nematodes in relation to the impact of agriculture—a review. Diversity 13(2):64

Le Quéré C et al (2018) Global carbon budget 2018. Earth Syst Sci Data 10(4):2141–2194. Available at: https://doi.org/10.5194/essd-10-2141-2018

Lehmann J et al (2020) The concept and future prospects of soil health. Nat Rev Earth Environ 1(10):544–553. Available at: https://doi.org/10.1038/s43017-020-0080-8
Lennartz B, Liu H (2019) Hydraulic functions of peat soils and ecosystem service. Front Environ Sci 7. Available at: https://doi.org/10.3389/fenvs.2019.00092
Medrano H, Tomás M, Martorell S, Escalona JM, Pou A, Fuentes S et al (2015) Improving water use efficiency of vineyards in semi-arid regions. A review. Agron Sustain Dev 35(2):499–517
Montanarella L, Panagos P (2021) The relevance of sustainable soil management within the European Green Deal. Land Use Policy 100:104950. Available at: https://doi.org/10.1016/j.landusepol.2020.104950
Munda S, Nayak AK, Shahid M, Bhaduri D, Chatterjee D, Mohanty S et al (2023) Soil quality assessment of lowland rice soil of eastern India: implications of rice husk biochar application. Heliyon 9(7):e17835
Naorem A, Patel A, Adak S, Singh P, Udayana SK (2024) Precision technologies and digital solutions: catalyzing agricultural transformation in soil health management. In: Digital agricultural ecosystem: revolutionary advancements in agriculture. Wiley, pp 175–190
Neemisha (2020) Role of soil organisms in maintaining soil health, ecosystem functioning, and sustaining agricultural production. In: Soil health. Springer International Publishing, Cham, pp 313–335
Oladele SO (2019) Changes in physicochemical properties and quality index of an Alfisol after three years of rice husk biochar amendment in rainfed rice–maize cropping sequence. Geoderma 353:359–371
Petter FA, Leite LFC, Machado DMD, Marimon BHD, Lima LBD, Freddi ODS, Araújo ASF (2019) Microbial biomass and organic matter in an oxisol under application of biochar. Bragantia 78(1):109–118
Phillips CL, Bond-Lamberty B, Desai AR, Lavoie M, Risk D, Tang J et al (2017) The value of soil respiration measurements for interpreting and modeling terrestrial carbon cycling. Plant Soil 413(1):1–25
Rahman MM, Alam MS, Kamal MZU, Rahman GM (2020) Organic sources and tillage practices for soil management. In: Resources use efficiency in agriculture. Springer Singapore, Singapore, pp 283–328
Rickson RJ, Deeks LK, Graves A, Harris JAH, Kibblewhite MG, Sakrabani R (2015) Input constraints to food production: the impact of soil degradation. Food Secur 7(2):351–364
Ronson PL (2024) Does soil fertility influence leaf venation? Doctoral dissertation. The Evergreen State College
Sahu N, Vasu D, Sahu A, Lal N, Singh SK (2017) Strength of microbes in nutrient cycling: a key to soil health. In: Agriculturally important microbes for sustainable agriculture: volume I: plant-soil-microbe nexus. Springer Singapore, Singapore, pp 69–86
Seifu W, Elias E (2018) Soil quality attributes and their role in sustainable agriculture: a review. Int J Plant Soil Sci 26(3):1–26
Shah AN, Tanveer M, Shahzad B, Yang G, Fahad S, Ali S et al (2017) Soil compaction effects on soil health and cropproductivity: an overview. Environ Sci Pollut Res 24(11):10056–10067
Sharma UC, Datta M, Sharma V (2025) Managing soil acidity. In: Soil acidity: management options for higher crop productivity. Springer Nature Switzerland, Cham, pp 427–522
Smith P et al (2020) How to measure, report and verify soil carbon change to realize the potential of soil carbon sequestration for atmospheric greenhouse gas removal. Glob Chang Biol 26(1):219–241. Available at: https://doi.org/10.1111/gcb.14815
Tessema N, Yadeta D, Kebede A, Ayele GT (2022) Soil and irrigation water salinity, and its consequences for agriculture in Ethiopia: a systematic review. Agriculture 13(1):109
Therond O, Duru M, Roger-Estrade J, Richard G (2017) A new analytical framework of farming system and agriculture model diversities. A review. Agron Sustain Dev 37(3):21
Trimarco T (2025) Soil health for the semi-arid west: a nexus of agricultural soil management and ecosystem services. Doctoral dissertation. Colorado State University

Trivedi P, Delgado-Baquerizo M, Anderson IC, Singh BK (2016) Response of soil properties and microbial communities to agriculture: implications for primary productivity and soil health indicators. Front Plant Sci 7:990

Voltr V (2012) Concept of soil fertility and soil productivity: evaluation of agricultural sites in the Czech Republic. Arch Agron Soil Sci 58:S243–S251

Yan S, Zhang S, Yan P, Aurangzeib M (2022) Effect of biochar application method and amount on the soil quality and maize yield in Mollisols of Northeast China. Biochar 4(1):56

Zewide I (2021) Review paper on effect of natural condition on soil infiltration. Chemistry 7(1):34–41

Zou X, Zhu X, Zhu P, Singh AK, Zakari S, Yang B et al (2021) Soil quality assessment of different Hevea brasiliensis plantations in tropical China. J Environ Manag 285:112147

Global Scenario of Food Production and Its Enhancement Via Biochar Application: Challenges, Gaps and Future Scenario

Abid Hussain, Adnan Fareed, Marina Qayyum, Umair Riaz, and Maher Iftekhar Ahmad

Abstract

Food production play a key role in the global food security and socio-economic stability. However, the challenges like climate change, land degradation and resource constraints continue to decline agricultural production. The current chapter critically evaluate the global scenario of food production, highlight the major trends and the significant challenges faced by the industry including technological gaps, policy in-efficiencies and equity problems. The chapter also explore the role of innovative solutions especially biochar application as a sustainable approach to address the aforementioned challenges. Biochar as a carbon (C) rich soil amendment show promising results in the improvement of soil fertility while enhancing the water retention and mitigating the climate change impact. The chapter discusses the efficacy of biochar to improve the agricultural productivity and soil health while reducing the reliance on the synthetic fertilizers. Moreover, it also explores the future scenarios while emphasizing on the sustainable intensification and agroecological transitions as a future pathway towards the resilient and climate smart food production system. By incorporating

A. Hussain (✉)
Muhammad Nawaz Shareef University of Agriculture, Multan, Punjab, Pakistan

Punjab Agricultural Research Board, Lahore, Punjab, Pakistan
e-mail: abid.hussain@mnsuam.edu.pk

A. Fareed · M. Qayyum
Muhammad Nawaz Shareef University of Agriculture, Multan, Punjab, Pakistan

College of Life Sciences and Agriculture, Southwest University of Science and Technology, Mianyang, Sichuan, China

U. Riaz · M. I. Ahmad
Muhammad Nawaz Shareef University of Agriculture, Multan, Punjab, Pakistan

O. A. Wani et al. (eds.), *Biochar for Sustainable Soil Health and Crop Production*, https://doi.org/10.1007/978-981-92-2230-8_2

biochar into modern agricultural practices, this chapter envisions a future where agriculture is more resilient, resource-efficient, and capable of supporting global food security in a changing climate.

1 Introduction

Food production ensure the global food security and the socioeconomic stability while agriculture serve as a primary driver of the global economy. The rapid growth of the global population in combination with the increasing urbanization and changing dietary pattern is putting an unbearable pressure on the agricultural productivity. To meet the demand for food while protecting the environment for an equitable access towards the resources give rise to an unprecedented challenge. Historically, the Green Revolution of the twentieth century transformed food production through the cultivation of high yield crops, modern irrigation techniques, chemical fertilizers, and mechanization (Sadaf et al. 2024). Despite these advances, the sustainability of such intensive farming methods has been under question due to their environmental and socioeconomic impacts particularly under the current climate change scenario (Średnicka-Tober et al. 2016). As the world grapples with the effects of climate change, land degradation, and resource depletion, it has become clear that the agricultural system must undergo sustainable reforms (Robinson 2018). This chapter aims to critically analyze the global scenario of food production and explore the potential for enhancing agricultural practices through innovative solutions, with a particular focus on biochar application (Lehmann and Joseph 2012). Biochar, a stable C rich material produced from the pyrolysis of organic matter has gained attention for its potential to improve soil fertility, increase water retention, and mitigate climate change (Hung et al. 2021). Recent studies have highlighted the ability of biochar to improve soil health, improve agricultural productivity and sequester C making it a promising agent of sustainable agriculture (Ding et al. 2016; Lehmann et al. 2021). This chapter explores the global trends in food production, highlighting the current status, challenges, and the critical gaps within the food production system (Gliessman 2015). The application of biochar can contribute to mitigating soil degradation, perhaps the most important issue in world agriculture, through improved nutrient retention, elevated microbiological activity and enhanced structure in soils (Kapoor et al. 2022). Biochar can also significantly benefit in reducing the C footprint in the agricultural subsector, offering a sustainable solution to meet the impacts of global change in food production (Lal 2010). There is proof that application in soils with biochar reduces the use of chemically formulated fertilizers as well as promotes environmentally friendly agricultural practices (Layek et al. 2022). The objective of this chapter is to discuss the integration of food production systems with biochar as part of the broader solutions to address the current as well as projected demands for global food production. Considering the important barriers in the form of climate change, resource scarcity, as well as land degradation overview in which biochar can bridge the important gaps in agricultural technologies, practices, as well as policies (Rajakumar and Sankar 2016). It will also include the future prospects of biochar application, among other sophisticated agricultural technologies in terms of developing more sustainable, resilient, as well as climate smart food production systems (Nair et al. 2017). Chapter organization involves detailed discussion with the description of the global food production status in relation to the current state with detailed assessment of significant issues in the agricultural sector. This chapter also aims to highlight the gaps in the food production chain, particularly evaluating the gaps that are technological, infrastructural, as well as policy gaps. This chapter also explores the scenarios in the future and agricultural sustainability through the incorporation of innovative products like biochar in reconfiguring food systems towards a more secure as well as sustainable future.

2 Current Global Status of Food Production

2.1 Global Food Output By Region and Crop Type

The worldwide food production is very diverse because of contrasting climate, soils, adoption technologies and differences in the local economies. The difference in production techniques have been observed among crop species that are cultivated in various regions. The top worldwide producer of rice is Asia, which shares over 90% of production, while China and India are the world's leading producers (Gliessman 2015). China and India are also leading producers of other crops such as wheat and maize. For instance, India produces rice, wheat, pulses, all kinds of fruits and vegetables. Asia has also diversified production systems which integrate the high output intensive systems, while North America is characterized by extensive, large-scale industrialized farm systems. Similarly United States of America is a top producer of maize and soybean where farms are extremely large in size, highly mechanized and use the latest technologies to gain maximum output. Canada is another world leader in food production where production of pulses, legumes especially barley is high and they also export to other countries (Sadaf et al. 2024). Farming in these places is usually capital-intensive, involving heavy investment, use of latest technology and inputs in the form of fertilizers and pesticides. Latin America specifically Brazil serve as an integral to global production in soybeans, coffee, sugarcane, and beef. Brazilian intensive farm production specifically through Amazon region deforestation, has elevated the country to one of the globe's leading producers to export their agricultural products. Argentina is also a significant exporter of soybeans along with maize and is significantly accountable for world food supplies. Despite concerns regarding sustainability as well as deforestation, Brazil remains integral to the world food supply market (Średnicka-Tober et al. 2016). Africa is a region where food production faces significant challenges, including soil degradation, water scarcity, and the impacts of climate change. However, countries like Nigeria, Ethiopia, and Kenya have managed to produce large quantities of crops such as maize, cassava, and sorghum. Despite these efforts, many African and Asian countries still face challenges with low productivity, limited

Table 1 Key global food production statistics (2020–2024)

Goods/commodities	2020 (Mt)	2021 (Mt)	2022 (Mt)	2023 (Mt)	2024 (Mt)
Maize	1134	1178	1160	1220	1260
Wheat	760	800	799	799	810
Rice	738	740	735	741	750
$\sum$ cereals	2775	2841	2828	2898	2951
Sugar crop(s)	19,785	1800	1853	1976	2000
Tubers and roots	1250	1271	1295	1315	1335
Vegetables	1000	1040	1055	1075	1095
Fruits	889	905	915	925	935
Oil seed crops	1356	1400	1454	1500	1550
Poultry	120	125	130	128	142
Pork	115	119	120	125	125
Beef	75	72	74	76	78
Mutton and goat	35	36	31	39	40
$\sum$ Meat	345	352	355	368	385
Milk	855	857	880	890	895
Eggs	75	81	82	85	86
Fish	81	84	90	95	98

Note: "Mt" denotes million tonnes

mechanization, and inconsistent access to modern agricultural practices. The key global food production statistics (2020–2024) offer a compiled database of various crops produce around the globe (Table 1).

2.2 Key Production Systems, Intensive Versus Subsistence

The agricultural systems can broadly be distinguished as intensive systems and subsistence systems each have a specific role to sustain world population through food. Intensive agriculture is characterized by high input and high output per unit of land. This system relies heavily on modern technologies such as mechanized farming equipment, chemical fertilizers, genetically modified organisms (GMOs) and automated irrigation setups. It is prevalent in developed countries such as the United States, Canada and Western Europe. Additionally, large scale farms uses machinery for planting, irrigation, and harvesting which significantly increase the productivity. The mechanization of agriculture has also reduced labor requirements and allowed for economies of scale, making food production more cost efficient (Lal 2010). However, the environmental impacts of intensive farming include soil degradation, water pollution and loss of biodiversity. The second is subsistence farming which is widespread agricultural system in developing nations especially in Africa and Asia. The farmers grow crops primarily for consumption in the household with a very little output to market. The efficiency in subsistence farming is low as the production is done through conventional means, limited use of modern technologies as well as farmer lack finances to buy new equipment or supplies. The farm systems are usually smallholder in size, relying primarily on rainfall, hence very vulnerable to climate change pressures such as drought as well as flooding (Rajakumar and Sankar 2016). There is added urgency to change subsistence farming in the majority of developing countries through adoption of sustainable production practices, increased access to markets, education of farmers to latest agricultural production technologies. However, this transition from subsistence to the more-intensity production systems is held back by restricted finance and limited infrastructure.

2.3 Major Producing Countries and Export Economies

The major food producing countries are also the principal world food traders who influence domestic and global markets. The leading food export countries in the world are the United States, China and Brazil who significantly influence the world food prices and trade flows. The United States is the largest exporter of maize, soybeans, and it plays a central role in feeding developing nations. Additionally, the United States exports include the processed food, dairy products and meat contributing to its agricultural dominance (Shiade et al. 2024). China is the global leader in rice production as well as pork production, furthermore it contributes towards the vegetable production. Whereas China produces huge amounts of food incountry, China has progressively become a net importer of principal commodities like corn and soybeans. The higher production of China is due to the increased food demand in country as well as the requirement to supply rapidly expanding urban centers (Lehmann et al. 2021). Brazil is also one of the world food giants, as it produces soybeans, coffee, sugar and beef. The country is particularly important in soybean production, as in recent years Brazil has become the world's leading nation in exporting soybeans. Brazilian agriculture is desirable due to extensive land endowments and growing agricultural export markets, although deforestation issues and sustainability problems in the environment remain controversial (Nair et al. 2017). The European Union is also the world's leading exporter in special farm products such as wine, dairy, olive oil, and cheese. Its agricultural sector is also highly subsidized(Sadaf et al. 2024).

2.4 Role of Technology and Mechanization

The use of technology and mechanization has revolutionized the food production in order to increase efficiency to meet the higher demands by the growing world population. The adoption of modern technologies in developed countries has facilitated improved production, reduced production costs and sustainable production methodologies (Farmonaut, 2024). Precision farminginvolve the implementation of technologies such as the use of satellites, sensors and drones which assists the farmers in monitoring crop health, apply pesticides and irrigate the crops with optimized water and fertilizer through drip irrigation system. The use of modern agriculture technology has reduced the waste production and minimized the use of available farm resources (Lal 2010). Mechanization is the key to large-scale, industrially diversified farming as the application of tractors and automatic systems for plantings, irrigation, harvesting has enhanced the efficiency in food production. The technologies are particularly significant in North America and Europe under large farms (Ding et al. 2016). A push for low-cost technologies to improve the production of smallholder farmers in developing nations is ongoing. However, smallholder farmers in Africa, South Asia and in Latin America are being exposed to mobile applications, drip irrigation systems and mini farm mechanical equipment (Madzivanzira et al. 2024). Additionally, low-cost technologies can be accepted very easily to local environments which are introduced to smallholder farmers to aid in the improvement of yields and reduce the application of the conventional time-consuming farming practices (Layek et al. 2022). Despite these developments, access to technology remains problematic in most places as majority of developing nations lack the affordability of modern equipment, technology, and access to technical services in addition to training is low (EOS, 2022). Therefore, significant room exist for smallholder farmers to adopt new technologies, which has hindered enhanced production and sustainability (Robinson 2018).

3 Major Challenges in Global Food Production

The major challenges in the global food production are closely influenced by the human based activities interventions and natural processes. Additionally, global map of food production hotspots and risk zones represent the areas which are affected by the influence of climate change (Fig 1). These issues endanger the safety of food supplies worldwide and question the sustainability of agriculture.

3.1 Weather Extremes and Climate Change

One of the most significant challenges to food production globally is climate change which promote the occurrence of extreme weather in tandem with changing trends in temperatures. These factors have significant lingering impacts for food production systems around the globe, while the weather changes also give birth to drought, floodingand temperature changes where the increased global temperatures have brought about increased severity and frequency of droughts in sub-Saharan Africa, South Asia, and the Americas. Contrastingly floods occurred in a few regions with increased precipitation and melting sea ice which severely impact crop yields and arable land (Robinson 2018). Additionally, impact on crop yield and regional sensitivities involve the shifts in temperature and precipitation patterns which directly affecting crop health, particularly for staples like maize, wheat, and rice. For instance, in South Asia the change in monsoon pattern have led to reduce rice yields while in the northern American Midwest, drought has impacted maize and

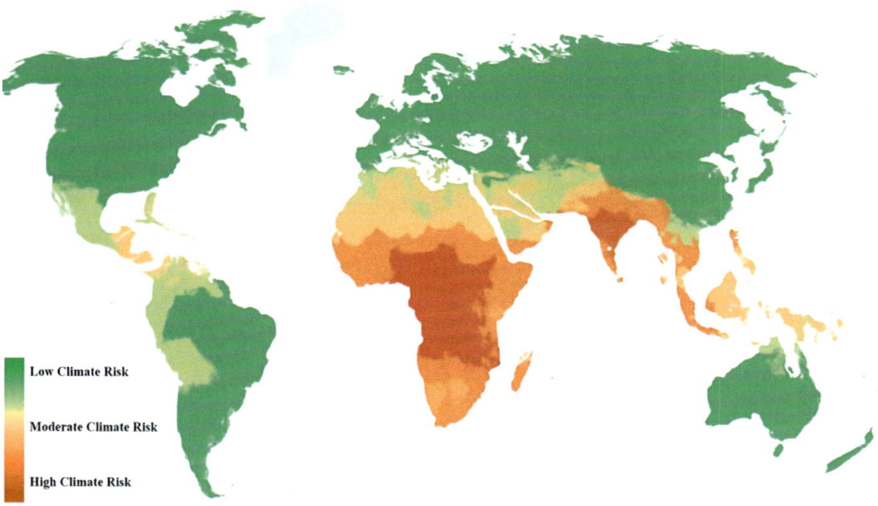

Fig. 1 Global map of food production hotspots and risk zones

soybean production (Średnicka-Tober et al. 2016). Global climate models forecasts that food production worldwide will drop by as much as 20% by 2050 due to climate stresses (Rajakumar and Sankar 2016).

3.2 Land and Soil Degradation

Land and soil degradation are a significant concern that affects the soil health, reduces crop production and enhances food insecurity particularly among developing nations. The degradation involves the soil erosion, salinization and nutrient depletion which is a result of unsustainable farming practices including overgrazing, deforestation, and improper irrigation. In regions such as the Horn of Africa and Central Asia, soil salinization occure due to excessive irrigation has reduced agricultural productivity (Kapoor et al. 2022). Furthermore, soil degradation is also a leading cause of declining agricultural yields in countries like India, China and Pakistan where intensive farming practices have led to the depletion of essential nutrients (Ding et al. 2016). The loss of arable land caused due to the rapid urban expansion into cities and infrastructure. The loss of arable land, particularly in fast-growing economies like China and Brazil, threatens future food production, however it is estimated that approximately 70% of the world's agricultural land is at risk of degradation, threatening food security for millions (Sadaf et al. 2024).

3.3 Scarcity and Overuse of Water

Water is extremely crucial to crop production while this resource is extremely scarce due to the climate change conditions. Globally overexploitation and inappropriate management of available water serve as an obstacles to sustainable crop production. Additionally, irrigation inefficiencies and groundwater depletion is caused due to the groundwater abstraction to feed irrigation systems in North India, the American Midwest and various parts of Middle East. The excessive use of irrigation to meet farm requirements has resulted in lowering groundwater tables, drying of rivers, reservoir and lakes (Layek et al. 2022). Therefore the ineffecient irrigation systems end up consuming huge volumes of water thus increasing the dryness in semi-arid and arid regions. Initiation of conflicts over water use and transboundary conflicts give rise to water scarcity due to water is aggravated by water-use conflicts. In addition, shared rivers between nations like India and Pakistan (Indus River) or Egypt and Sudan (Nile River) are exposed to conflicts over the use of waters which may affect agricultural production and food insecurity (Lal 2010). The human induced activities like pumping of ground water for irrigation purpose, industrial water consumption, population growth, urbanization, water pollution and poor water management also give rise to water scarcity.

3.4 Biodiversity Loss and Pest Pressure

The loss of biodiversity serve as a great challenges in achieving the soil health and the production of world food systems. Pollinators such as bees, butterfly species, and bats are crucial to ensure the fertilization of food crop production. Additionally, their populations are decreasing due to lost habitats, pesticide use and intensified climate change. Dwindling pollinators have great impacts on world food production, particularly on fruits, vegetable and nuts production. The literature document that 75% of the world's flowering plant species rely on pollinators to reproduce (Ding et al. 2016). Additionally, the excessive use of chemical pesticides has resulted in the evolution of resistant populations among pests, thus complicating crop-damaging pest management. The common pests like aphids, whiteflies and weevils have evolved to become resistant to more than one class of insecticides, hence farmers faces increased production cost with lower crop output (Rajakumar and Sankar 2016). This emerging challenge has motivated research into integrated management of pests as well as use of organic farming to decrease the application of chemical pesticides.

3.5 Socioeconomic and Political Instability

Political and socio-economic insecurity contributes to the food production in the world especially under conflict situations and unequal regions. Global food markets are very prone to disturbances caused by conflict as well as political tensions which give birth to the disturbances in trade, conflict zones and inequality in distribution of products. Food distribution and production is prevalent in the conflict zones such as Syria, Yemen, and parts in Sub-Saharan Africa which remain hampered by ruined infrastructures, displacement of farmland workers and economic volatilities (Robinson 2018). Trade disturbances in the form of trade wars or blockades also threaten global food security by reducing access to essential food supplies (Nair et al. 2017). Additionally, the influence from trade in global markets and market-based subsidies include world trade in food which is shaped by market forces and subsidies in many instances which is the disadvantage to agribusinesses in developing nations and against smallholder farmers. The volatility in prices of commodities in markets like wheat or maize can unduly impact poor farmers who lack the capacity to deal with this volatility. Agricultural subsidies from developed countries can also skew trade, as this makes trade competition in the world market impossible for developing countries' farmers (Sadaf et al. 2024).

4 Critical Gaps in the Food Production System

Modern Agriculture has benefited through the advancement in the food production technologies while a wide gap still exist that hinder world agriculture production. The gaps in technological access to gaps in infrastructure requir urgent treatment to ensure long-term food systems sustainability and improved global food security.

4.1 Technological Access and Knowledge Transfer

The gaps in technologies between developed regions and underdeveloped regions are among the primary challenges to improving food production in the world. The current technologies for farming have achieved huge success in developed countries which include the accessibility to these technologies remains minimal in most regions in the world. Additionally, differences between developed and developing areas report that agricultural production in the developed world such as United States, Canada, China and Europe is very intensive with extensive application of precision farming like artificial intellengance (AI) and satellite tracking systems. While the developing countries especially Africa and South Asia are still applying conventional systems of production which are not in a position to adopt latest technologies due to funds limitations, poor infrastructures, low levels of education and knowledge (Robinson 2018). Smallholder farmers in sub-Saharan Africa, are applying automatic farm equipment as well as high-tech irrigation systems, thus their production levels are low as a resilience to climate change (Lal 2010). The technological access also includes digital division which contributes to these inequalities since rural connectedness and digital illiteracy continues to hinder the adoption and utilization of new technologies in agriculture. A World Bank report in the year 2022 cited that approximately 40% of rural villages in developing nations lack access to mobile technology which greatly diminishing the likelihood that farmers will gain from innovations like mobile-based crop management or digital market access platforms (Nair et al. 2017).

4.2 Infrastructure and Supply Chains

The effective infrastructure is vital to guaranteeing that food is delivered to the consumer in time and in a safe environment. However, inappropriate storage, transport and cold chain facilities in most regions in the world contribute to huge losses, especially in underdeveloped nations. Poor storage, cold chain and transport capacity in most low-income countries means that much post-harvest loss is incurred, especially when dealing with perishable fruits, vegetables, as well as dairy products. The FAO has estimated that 1/3 of the produced food in the world is lost or wasted and most losses happen in developing nations due to storage and weak sources to transport food (FAO, 2023; Layek et al. 2022). Poor cold chain capacity in most places like South Asia as well as most places in Sub-Saharan Africa means that food is spoiled first even prior to entering the market. Additionally, fruit and vegetable losses after harvest are particularly significant in developing countries that have poor access to storage technologies. In the case of India, losses in wheat and rice may reach as high as 15–20% every year because storage is done improperly, storage facilities are few and temperature is not controlled (Rajakumar and Sankar 2016). Aside from impacting food security, losses also mean that huge amounts are wasted in terms of water, land, and labor that were initially applied to grow the produce.

4.3 Policy and Governance Gaps

The management and stable policy regimes are also important in attaining sustainable production patterns for food in order to end food insecurity. While large policy gaps coupled with gaps in planetary cooperation hamper food systems to function effectively in the world. The policy gaps involve the incoherent agricultural policy in most countries which may cause food production and distribution to become inefficient. Sustainable practices are not usually adequately integrated with overall economic or environmental policies, with the consequence that food systems are not improved as much as is desired (Shiade et al. 2024). Some countries still provide subsidies to unsustainable farming that erodes soil health as well as promotes increased use of chemical fertilizers, while others have adequately integrated their agricultural policies with sustainability. Moreover, the failure in global cooperation towards sustainability is the lack of joint global policies and cooperation towards sustainability in agriculture. Despite world stages like the UN's Food and Agriculture Organization (FAO) and Paris Climate Agreement having deliberated on agricultural issues, very little cooperation is seen over borders to address global food security challenges. There is increased need for global pledges towards sustainable production systems, global food market organization, and fair-trade regimes (Nair et al. 2017).

4.4 Deficits in Monitoring and Data

This lack of good data systems, combined with poor monitoring systems leads to food production process inefficiencies particularly in forecasting and in responding to imminent food security risks. The deficit in monitoring of data also involve inadequate agricultural data which is generally incomplete, disjointed, and obsolete, rendering it challenging to estimate the current state of food systems and plan. National agricultural databases are weak in most developing nations, whereby tracking crop production, monitoring pests and forecasting food gaps is problematic (Layek et al. 2022). Unreliable data also hinders the tracking of new or emerging risks like infestations by pests or impacts brought on by climate change. Additionally, adequate early warning systems for crop failures, climatic disasters and infestations can effectively prevent food crises. However, most countries especially in Africa and South Asia do not have viable monitoring systems to identify and respond to agricultural emergencies (Sadaf et al. 2024). The inadequacy in these systems is attributed to late responses that cause losses in food production and induced food insecurity. The improvement of early warning systems and collecting more inclusive data can increase food systems' resilience to climate change as well as other shocks in substantive ways.

5 Future Scenarios and Sustainable Pathways

The growing food system is confronted by diverse challenges, thus critically important to interpreting and setting future directions that will provide food security, sustainability, and resilience. Several measures and innovations are being pursued to meet climate change, shortages in available resources and achieving more equitable and sustainable production systems. The section lays out possible future prospects alongside the sustainable directions and application of biochar that can improve the food production to cope with the current food security issues.

5.1 Climate-Resilient Cropping Systems

Climate resilient crop produce is pivotal in the preservation of food security in the midst of fluctuating climatic conditions. Drought, flood, and temperature-tolerant produce shall form the cornerstone in preserving crop production in sensitive zones (Wolf et al. 2023). The approach includes drought resistant varieties to mitigate the issue of water scarcity. In addition, the improvements in breeding and gene modification have enabled scientists to produce varieties that can grow with minimal amounts of water. For instance, drought-resistant maize, rice, and wheat varieties are already being produced and trailed in sub-Saharan Africa and South Asia. These varieties can tolerate long periods without rain, enabling food supplies to be protected in places where water is becoming more limiting (Sadaf et al. 2024). Furthermore, regenerative practices and agroecology through the application of principles from ecology to guide agricultural production is another promising approach to climate resilience (Aguilar and Paulino 2025). Agroecological production approaches such as diversified cropping, agroforestry, and cover cropping help to improve soil health, decrease use of water, and increase biodiversity. Similarly, regenerative agricultural production is all focused-on rebuilding soil health, improving carbon storage, as well as reducing usage of synthetic fertilizer as well as pesticide application (Khangura et al. 2023). The aforementioned approaches are also supported by the application and incorporation of biochar within the soil which aim to improve the soil structure, soil organic matter content, water holding capacity and nutrient retention within the plants. Apart from helping in climate change mitigation, these approaches also contribute to attaining long-lasting sustainability in farming (Robinson 2018).

5.2 Technological Innovations

The technological progress is supporting the revolution in the farm industry to offer more productive, environmentally friendly and resilient farm systems. It involves the precision Farming, AI, internet of things (IoT) and CRISPR, while the approach uses sophisticated technologies like satellites, drones and IoT sensors to observe crop conditions as well as soil conditions in real time (Karunathilake et al. 2023). The above-mentioned techniques help farmers to use less water, fertilizer, and pesticide while boosting output through reduced wastage. Moreover, AI is also being deployed to decode huge volumes of agricultural data, helping farmers base their decision-making on sounder insights as well as predict the outcome more effectively (Lal 2010). The application of CRISPR gene editing technology is very promising towards the development of crop varieties with improved traits, like tolerance to pests, disease, as well as unfavorable weather. For instance, genetically improved wheat and rice with improved tolerance to drought and salt are already in trial in China and India (Shiade et al. 2024). Furthermore, crop production through vertical farming and hydroponics is utilized in the urban setup to increase the food besides the conventional farming techniques might not suffice to provide food (Mir et al. 2022).

5.3 Global Policy Reforms

A more sustainable food system for the world requires dramatic policy shifts in agriculture and more cooperation among nations. Future cooperation on trade and agricultural subsidies in developed nations have generated controversy for years. A significant number of subsidies tend to go towards unsustainable agricultural practices, promoting overuse of chemical inputs and monoculture (Bellmann 2019). Reforms that place their emphasis on paying subsidies to sustainable agricultural practices like agroecology, regenerative farming, as well as organic farming, will aid in encouraging environmentally friendly food production (Robinson 2018). Moreover, carbon markets can offer incentives to farmers to use low-carbon production practices, paying them for reductions in greenhouse gases through means like carbon storage in soils. In addition to this the global policy reforms also include the permission and subsidies to apply the biochar in the degraded soils where the organic matter content is less then 1% and soil are producing very low due to poor nutrient management. It is also important to have international trade cooperation to prevent food systems worldwide from being shaken by climate disruptions as well as by geopolitical conflicts. Free trade agreements that are fair can stabilize market food prices as well as increase food availability in food-scarce regions (Rajakumar and Sankar 2016).

Sustainable Development Goals (SDGs) and food sovereignty involve the achievement of particularly Goal 2—Zero Hunger, requires a radical shift in the management of food systems in the world (Charlton 2016). Food sovereignty, with local control over production and distribution of food, can help societies build more sustainable and resilient food systems. Strengthening local food systems, reduced dependence on world supply systems, and scaling up smallholder production are essential in ensuring food security as well as achieving the SDGs (Lal 2010).

5.4 Low-Carbon Food Systems and Circular Economy

A more environmentally friendly food production strategy is adopting a circular food system that minimizes waste while continually reusing resources to enhance environmental sustainability (Islam and Zheng 2025). The strategy includes recycling of waste, local food systems and production of bioenergy for a better resource management. Circular food systems can reduce food wastage through recycling into energy in the form of biogas, compost or animal feed (Sarangi et al. 2024). For instance, food wastage can be utilized in the production of biogas to produce energy, while organic waste can be composted to improve the health of soils. Bioenergy from organic residues such as crop residues and food wastage can offer energy to food production while reducing C intensity in food production (Ding et al. 2016). Scaling up low-input farming, compost, and biochar can be further applied to enhance the health of soils while minimizing the application of chemical fertilizers. These approaches can also sequester C which aims to combat climate change. Low-input farming systems that employed few synthetic inputs can also enhance resilience and decrease environment degradation (Rajakumar and Sankar 2016). However, the integration of such approaches as part of an overall sustainable food system will enable the sector to become more resilient and environmentally friendly.

6 Role of Biochar in Boosting Global Production Mechanism

Biochar has attained attention as a versatile material which is capable of improving global food production. It has unique properties which make it a valuable product in improving agricultural production, environmental sustainability and soil conditioner. Table 2 present comparison of current versus future agricultural practices which show that future agricultural practices could be improved by the application of biochar. Biochar can contribute to boost the agricultural production mechanism across various sectors such as

(i) Improvement of Soil Fertility and Productivity: Biochar serve as a soil conditioner where it is applied within the soil and improve fertility through the improvement of nutrient retention, water retention, water-nutrient holding capacity and reducing the acidity of soil. The improvements lead to a more effective nutrient cycling and a better plant growth while boosting the yield of crop. The capability of biochar to retain the nutrients like nitrogen (N), phosphorus (P) and potassium (K) which reduce the requirement of synthetic fertilizer resulting in cost cutting and increasing the economic benefit. As a result, biochar can assist to secure the food production in the areas that face challenges like poor soil fertility, soil degradation and water scarcity while making it an essential product for the sustainable global agricultural practices.

(ii) Carbon Sequestration and Climate Change Mitigation: Biochar serve a crucial role in the mitigation of climate change action by sequestering the C. Biochar remains stable for thousand years in the soil which sequester the C that could be released into the atmosphere as CO_2. The potential of biochar to sequester the C offer a significant means to reduce the greenhouse gas emissions (GHGs) which is essential in meeting the international climate targets. The integration of biochar into the global production system can contribute to the transition towards the low C economy while simultaneously improving the agricultural production.

(iii) Waste Management and Circular Economy: The production of biochar from the biomass residue serve as a unique opportunity for the waste management especially in the agriculture and forestry industry. Biomass waste like crop residue, wood chips and organic waste material is usually discarded which lead to the environmental pollution while the utilization such waste into biochar production serve as a recycle of these waste products while also transform into valuable product. The pyrolysis of biochar contributes to circular economy where waste is reduced and resources are continuously reused therefore improving the sustainability across the agricultural production.

(iv) Water and Wastewater treatment: Biochar has been proved to efficiently adsorb the organic pollutants, heavy metals and excessive nutrient from the wastewater which make it an effective maternal for the removal of contaminants from waste water. The removal of contaminants from the waste water improved the quality of water and reduce the pollution, therefore biochar contributes to a

Table 2 Comparison of current versus future agricultural practices

Aspect	Current agricultural practices	Future projections
Technology	Use of the basic machinery like tractors, combine harvesters and irrigation systems and pyrolysis machines	Widespread use of robotics, AI, drones and autonomous machinery for the precision farming and crop monitoring, porous biochar for more shelf life
Farming techniques	Conventional tillage, monoculture and irrigation	No-till or zero tillage, vertical farming and regenerative agriculture practices for the soil health and biodiversity
Crop management	Monoculture farming leading to the reduce soil fertility and increase pest risks	Diversified cropping, agroforestry and intercropping to improve biodiversity, soil health and pest resistance
Sustainability	Limited use of sustainable farming practices with more dependency on the synthetic fertilizer and pesticides	Precision agriculture, integrated pest management (IPM) and organic farming to reduce the use of agrochemicals and improve sustainability
Water consumption	High reliance on the traditional irrigation system with a significant water wastage	Smart irrigation system through the use of sensors and data analytics for the effective water use and drought resistant crop
Water management	About 2 billion people and million hectares are affected by the issue of water scarcity	Application of biochar will improve water retention in soils to mitigate the water stress
Nutrient management	The loss of essential nutrients due soil compaction and issue of structurer stability	Application of biochar will improve the soil structure, OMC and increase the retention of nutrients to support plant growth
Energy use	Fossil fuels dominate energy sources for farm operation and the transportation	Renewable energy sources like solar and wind with integrated into farming operations including agrivoltaics
Labor	Labor intensive, reliance on the manual labour for the planting, harvesting and maintenance	Automated labour through the robotics system and AI which reduce the need for human labour cost and increase efficiency
Soil health	Over reliance on the synthetic fertilizer leading towards the degradation of soil	Focus on the soil regeneration, use of cover crops and the reduced chemical inputs to maintain a healthy soil
Soil quality	Soil organic carbon (SOC) in degraded soils is low with organic matter content (OMC) less than 0.5%	Application of biochar can increase SOC up to 50% in degraded soils
Soil pH adjustment	Low or acidic pH affects approximately 40% of agricultural soils	Biochar can increase soil pH up to 1.5 units which neutralize the pH and improve the nutrient retention within soil
Climate	Limited adaptation to the changing climatic conditions with the emphasis on the drought resistant crops	Genetic engineering, CRISPR technology for the development of the climate resilient crops and the ecosystem-based adaptation practices
GHGs emissions	Agricultural activities under the conventional system contribute to 24% of global share	If biochar is applied widely it may can reduce the emission up to 15%

more resilient industrial process and support the efforts to ensure the clean water for agricultural production around the globe. Additionally, the ability of biochar to purify the water can also benefit the agricultural system by reducing the impact of industrial effluents on the farmland and soil ecosystem.

(v) Energy Production and Biofuel Development: Biochar is a byproduct of pyrolysis which is a process that generate biofuels like bio oil and syngas, however biofuels can be further utilized for the energy production, creation of sustainable energy sources which can also replace the utilization of fossil fuels. Additionally, the use of biochar in biofuel production in combination with the other bio energy sources can support the global energy production by reducing the reliance on the non-renewable energy sources and decreasing the GHGs from the energy sector as well. The process also aids to improve the energy security and support the shift towards renewable energy solutions.

(vi) Improving Livestock Productivity: In addition to the agricultural uses biochar is also applied for the improvement of livestock health and productivity. This product can be incorporated into the animal feed to improve the nutrient adsorption and digestion which leads to higher weight gain and reduced methane emissions from the livestock. The application of biochar in livestock could play a significant role in the improvement of the efficiency of livestock production at global scale, while it will reduce the environmental footprint of livestock industry and improve the food security.

(vii) Boosting Global Agricultural Resilience: Climate change presents the increasing challenges to the global food production like unpredictable weather patterns, floods and droughts. The application of biochar can improve the soil resilience by enhancing its structure and water retention which is a crucial factor for maintaining the productivity under the adverse conditions. The resilience allows the farmers to adapt the changing climatic conditions and to maintain the consistent crop yields therefore aiding to stabilize the food production system worldwide.

7 Conclusion and Recommendations

The world food production systems are at a crossroads due to unprecedented pressures from climate change, land degradation, scarcity of water and socio-political destabilization. In recent decades significant advances in the production have occurred through the Green revolution while most of these advances have accompanied unfavorable environment and socio-economic impacts. The world needs to move towards more sustainable, more equitable and more resilient food systems. The current chapter conclude that climatic change has thrown growing seasons off schedule, causing precipitation patterns to change, temperatures to increase, and extreme weather events to happen more often. All these are having significant effects on crop production, especially in sensitive areas such as sub-Saharan Africa, Southern Asia, and parts of the American Midwest (Rajakumar and Sankar 2016; Robinson 2018). In addition, the land is still degraded, with soils becoming unproductive through irresponsible farming and desertification. Urbanization and the construction of infrastructure contribute to this issue as most of the farmable land is taken up by industry and cities (Sadaf et al. 2024). Moreover, agricultural output is still subject to the threat posed by lack of water, primarily in the water-stressed zones. While improper irrigation technologies, in combination with over-extraction from groundwater levels, is augmenting this threat, with most zones suffering the potential to lose their waters (Lal 2010; Layek et al. 2022). The loss in biodiversity as well as growing pesticide resistance to modern farm systems among pests has posed newer challenges to crop health as well as food security (Rajakumar and Sankar 2016; Ding et al. 2016).

The risks offer the potential for innovation in technology like breeding drought-resistant crops, precise farming and vertical farming. In addition, the use of biochar, regenerative farming and agroecology hold promise to restore soils, reduce emissions and grow food more sustainably (Nair et al. 2017).

7.1 Key Strategic Areas for Research, Innovation and Resilience

The issues of global food production require some research innovations as there is no doubt that more research is required on climate-resilient crop varieties, sustainable agricultural practices and the use of biochar in farmland. There is a dire need to conduct research focused on breeding crop varieties that can grow in varying climate conditions, enhancing crop health, and minimizing water application (Robinson 2018; Sadaf et al. 2024). In addition, the technological innovation is vital to enhance agricultural efficiency while minimizing its damage to the environment. Precision agriculture, crop management through AI and the use of vertical farming will form the bedrock in achieving maximum output while achieving minimum footprints in the environment. Innovations in technologies are also important in enhancing the efficiency in food production systems like minimizing wastage and improving food storage (Lal 2010; Zaręba et al. 2021; Shiade et al. 2024). Moreover, the food security, food production and distribution are significant. Smallholder farmers in low-income countries ought to have access to technology, education, and capital in order to become more productive and resilient. Food sovereignty, also as means to empower local people to become proficient in their food systems, should themselves be promoted as means to offer more equitable access to food as well as to reduce reliance on worldwide systems for supplies (Lal 2010). The resilience in food systems will offer the foundation to combat climate change as well as geopolitical volatility. Investing in crop systems that are capable of withstanding climate change, in irrigation systems that are efficient, as well as in sustainable production systems in agriculture, will lower the impacts that climate change will have on world food production. Agricultural producers will need to have the capacity, the knowledge, as well as the finance to adapt (Nair et al. 2017; Robinson 2018). The intricacy of worldwide food production challenges requires multi-level, collaborative approach like societies and local authorities must play their part in embracing sustainable agricultural practices. Local farmers will be aided to obtain modern technologies in addition to knowledge that will improve their practices, while local food systems will aid the societies to build a resilience to external stimuli such as the influences of climate as well as market trends. Furthermore pervasive-level collaboration on the global level and countries are required to collaborate on food security concerns so that policies are in line with SDGs and climate-smart agricultural production. International trade collaboration, sustainable subsidies and mutual resources like water as well as land are required to create a more sustainable global food supply (Nair et al. 2017; Khan et al. 2021). The efficient food security policy needs to intersect environment, trade, as well as social policy. Cross-sector interaction among governments, non government organizations (NGOs), private sector and research institutions should develop innovative food production solution frameworks (Shafiq et al. 2023).

7.2 Routes to Future Food Production

The future of global food production requires sustainability and innovation from local farm to the global trade system. Furthermore, biochar should be applied to the soil as it will improve the soil health, reduce GHGs emissions and increase the sustainable food production. The future goal should focus on the development of climate resilient food production system under the umbrella of sustainable agriculture. The focus on sustainability, innovation and clustering of all sectors will built a more secure and sustainable food production (Fig. 2).

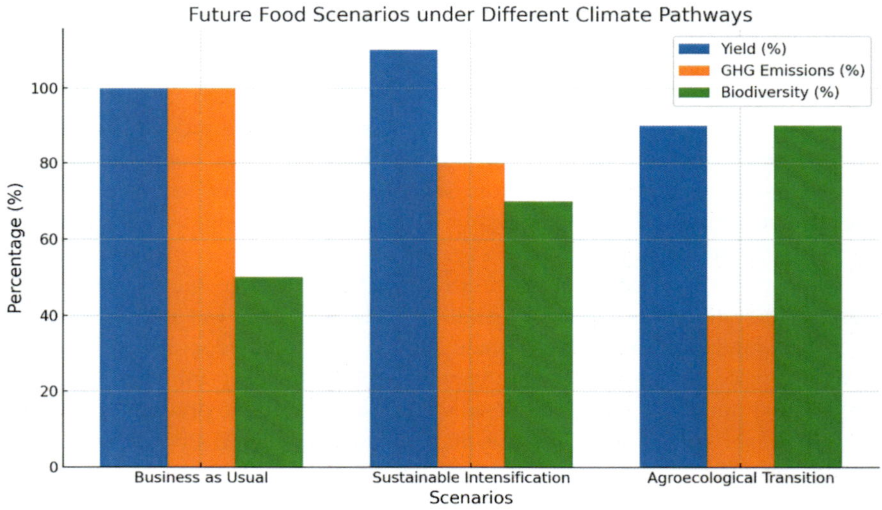

Fig. 2 Future food scenarios under different climate pathways

References

Aguilar G, Paulino S (2025) Different approaches for transformation of agri-food system in times of climate change: agroecology and regenerative agriculture. Agroecol Sustain Food Syst 49:1–28

Bellmann C (2019) Subsidies and sustainable agriculture: mapping the policy landscape. Chatham House, London

Charlton K (2016) Food security, food systems and food sovereignty in the 21st century: a new paradigm required to meet sustainable development goals. Nutr Diet 73:3–12

Ding Y, Liu Y, Liu S, Li Z, Tan X, Huang X et al (2016) Biochar to improve soil fertility. A review. Agronomy for sustainable development 36(2):36

EOS (2024) Agritech market overview and trends for 2025–2030. Retrieved from https://eos.com/blog/agritech-market-overview-and-trends-for-2025–2030

FAO (2023) World food and agriculture–statistical yearbook 2023. FAO. Retrieved from https://www.fao.org/statistics

Farmonaut (2024) Emerging technologies in agriculture 2025: top 7 new farming technologies. Retrieved from https://farmonaut.com/blogs/emerging-technologies-in-agriculture-2025-top-7-new-farming-technologies

Gliessman S (2015) A global vision for food system transformation. Agroecol Sustain Food Syst 39(7):725–726

Hung SH, Chou WY, Chang CY (2021) A study on practicing qigong and getting better health benefits in biophilic urban green spaces. Sustainability 13(4):1692

Islam MZ, Zheng L (2025) Why is it necessary to integrate circular economy practices for agri-food sustainability from a global perspective? Sustain Dev 33(1):600–620

Kapoor A, Sharma R, Kumar A, Sepehya S (2022) Biochar as a means to improve soil fertility and crop productivity: a review. J Plant Nutr 45(15):2380–2388

Karunathilake EMBM, Le AT, Heo S, Chung YS, Mansoor S (2023) The path to smart farming: innovations and opportunities in precision agriculture. Agriculture 13(8):1593

Khan N, Ray RL, Sargani GR, Ihtisham M, Khayyam M, Ismail S (2021) Current progress and future prospects of agriculture technology: gateway to sustainable agriculture. Sustainability 13(9):4883

Khangura R, Ferris D, Wagg C, Bowyer J (2023) Regenerative agriculture—a literature review on the practices and mechanisms used to improve soil health. Sustainability 15(3):2338

Lal R (2010) Enhancing eco-efficiency in agro-ecosystems through soil carbon sequestration. Crop Sci 50:S-120

Layek J, Narzari R, Hazarika S, Das A, Rangappa K, Devi S et al (2022) Prospects of biochar for sustainable agriculture and carbon sequestration: an overview for eastern Himalayas. Sustainability 14(11):6684

Lehmann J, Joseph S (eds) (2012) Biochar for environmental management: science and technology. Routledge

Lehmann J, Cowie A, Masiello CA, Kammann C, Woolf D, Amonette JE et al (2021) Biochar in climate change mitigation. Nat Geosci 14(12):883–892

Madzivanzira T, Mvumi BM, Nazare RM, Nyakudya E, Mtambanengwe F, Mapfumo P (2024) A review of appropriate mechanisation systems for sustainable traditional grain production by smallholder farmers in sub-Saharan Africa with particular reference to Zimbabwe. Heliyon 10(17):e36695

Mir MS, Naikoo NB, Kanth RH, Bahar FA, Bhat MA, Nazir A et al (2022) Vertical farming: the future of agriculture: a review. Pharm Innov J 11(2):1175–1195

Nair VD, Nair PR, Dari B, Freitas AM, Chatterjee N, Pinheiro FM (2017) Biochar in the agroecosystem–climate-change–sustainability nexus. Front Plant Sci 8:2051

Rajakumar R, Sankar J (2016) Biochar for sustainable agriculture–a review. Int J Appl Pure Sci Agric 2:173–184

Robinson GM (2018) New frontiers in agricultural geography: transformations, food security, land grabs and climate change. Bol Asoc Geógr Esp 78:1–48

Sadaf T, Sehar S, Hassan S (2024) Agricultural revolutions and food security. In: Food security in the developing world. Wiley, pp 95–122

Sarangi PK, Pal P, Singh AK, Sahoo UK, Prus P (2024) Food waste to food security: transition from bioresources to sustainability. Resources 13(12):164

Shafiq F, Anwar S, Zhang L, Ashraf M (2023) Nano-biochar: properties and prospects for sustainable agriculture. Land Degrad Dev 34(9):2445–2463

Shiade SRG, Fathi A, Minkina T, Wong MH, Rajput VD (2024) Biochar application in agroecosystems: a review of potential benefits and limitations. Environ Dev Sustain 26(8):19231–19255

Średnicka-Tober D, Obiedzińska A, Kazimierczak R, Rembiałkowska E (2016) Environmental impact of organic vs. conventional agriculture-a review. J Res Appl Agric Eng 61(4):204–211

Wolf MK, Wiesmeier M, Macholdt J (2023) Importance of soil fertility for climate-resilient cropping systems: the farmer's perspective. Soil security 13:100119

Zaręba A, Krzemińska A, Kozik R (2021) Urban vertical farming as an example of nature-based solutions supporting a healthy society living in the urban environment. Resources 10(11):109

Biochar and Soil Physical Properties: Enhancing Crop Yield Through Modulation of Soil Physical Properties with Biochar Application

Munir Ahmad, Jahangir Ahmad, Mohammad I. Al-Wabel, Muhammad Usama, and Abdullah S. F. Al-Farraj

Abstract

Biochar (BC) is a black solid product obtained through pyrolysis of organic feedstock under controlled environment. BC is characterized by high porosity, surface area, and abundance of functional groups at the surface. It has a great potential to alter the soil physiochemical characteristics and plays key role to produce better crop yield. BC could improve the soil porosity, aggregation, bulk density, water and nutrient holding capacity of soil, ultimately increasing soil fertility. In the sandy soil, biochar application reduces the leaching of nutrient with infiltration of water, while in the fine textured soils it promotes the drainage of water. Overall, BC applications enhance the physical soil properties with improving nutrients uptake and resilience of the crop under drought conditions or degraded soils. Various studies exhibited that due to the application of biochar in soil, the physical properties were improved, resulting in crop productivity enhancement in soils that are susceptible to stress conditions and low fertility. This chapter discusses the potential of biochar in improving soil physical properties and crop yield.

M. Ahmad (✉) · M. I. Al-Wabel · A. S. F. Al-Farraj
Soil Sciences Department, College of Food and Agricultural Sciences, King Saud University, Riyadh, Saudi Arabia
e-mail: amunir@ksu.edu.sa

J. Ahmad
School of Earth, Atmosphere, and Environment, Monash University, Clayton, Victoria, Australia

M. Usama
Department of Ecosystem and Conservation Sciences, University of Montana, Missoula, Montana, USA

O. A. Wani et al. (eds.), *Biochar for Sustainable Soil Health and Crop Production*, https://doi.org/10.1007/978-981-92-2230-8_3

1 Introduction

Biochar (BC), a black carbon-enriched organic product, porous in nature, is obtained through a process of pyrolysis of organic waste as a feed stock in controlled conditions (Sun et al. 2019) has high specific surface area, multiple surface functional groups, capability to tune (Smith 2016) resulted in bearing the high potential to adsorb heavy metals form the waste water as well as removal from soil and enhancement of carbon sequestration (Amonette et al. 2021). In the current era, BC is being used for various purposes including enhancement of soil health, climate change mitigation, environmental remediation, and maximum utilization of resources (Nepal et al. 2023; Wang et al. 2022; Yin et al. 2022). The optimistic effects after BC applications totally depend on which type of feedstocks has been used and conditions of the experimental terrain (Gu et al. 2022) and BC applications in soil with different ratios has been incorporated. The different studies conducted elaborated that BC can stabilize and regulate the soil water, nutrients, and salinity (Agegnehu et al. 2016; Madari et al. 2017). Various studies have demonstrated the positive impacts of biochar on soil physical properties, subsequently improving soil health and crop production. BC plays a significant role in reducing soil bulk density (SBD), with enhanced water and soil nutrients holding capacity leading to improved crop growth and productivity (Glab et al. 2016). Blanco-Canqui (2017) has reported that when BC is only implied the upper soil layer, only minor impact in the physical soil properties was recorded. BC can impact the soil matrix by changing size of soil pores. In the clay and loamy fine textured soil, BC plays a role in formation of large pores while caused the decrement of pore spaces in the sandy soils with coarse texture (Sun and Lu 2014; Liu et al. 2017c, d). When BC is combined with other fertilizers, it can improve the nutrient use efficiency acting as a slow released BC fertilizer (Mohammad et al. 2022). There are many studies which have elaborated that when BC was combined with nitrogen (N) fertilizers, it enhanced the soil fertility, soil aggregation, and plant nutrients uptake, ultimately leading to the better crop yield, N recovery efficiency (NRE), and water use efficiency (WUE). However, to optimize the crop yield, it is very critical to apply the N and BC in the appropriate amount in accordance with the crop's requirement and combinations ratio. For instance, a dose rate of 5 t ha^{-1} BC application combined with N 50 kg ha^{-1} is suggested to achieve the optimum efficiency from N fertilizer, while a dose of 0 t ha^{-1} of BC combined with a N rate of 100 kg ha^{-1} is recommended to achieve the optimum yield from maize crop (Omara et al. 2020). It has been proved through the recent studies that BC has a great capacity to enhance the compost performance, speeding up the humification of feedstock, boosting the diversity and activities of the microbes, and mitigating the greenhouse gas emission. Moreover, it is also a promising technique for the remediation of contaminated soil and water (Guo et al. 2020; Antonangelo et al. 2021). Therefore, this chapter highlights the potential of biochar application to promote soil health and crop productivity through improving soil physical characteristics.

2 Role of Physical Aspects of Soil in Plant Growth Induction

2.1 Bulk Density

Keeping SBD at a healthy level is important not only for growing crops but also for protecting the environment and supporting sustainable farming, especially as climate conditions continue to change (Eftene et al. 2020). SBD is a measure that reflects how compact the soil is and how much pore spaces it has. This is important because roots need space to grow, and both air and water need to move freely through the soil to support healthy plant growth. Ideally, soils should have about 50% pore space to allow these processes to happen efficiently. In contrast to clayey soils, sandy soils have high bulk density and are more compact than clay soil with dense and poor structure. Alteration in bulk densities depends on deep soil depth movement and the addition of management practices with the passage of time (Logsdon and Karlen 2004).

2.2 Soil Structure

Soil is an active porous medium that contains microbes and is heterogeneous in nature with dynamic specifications. Particles of soil, called soil solids, rarely exist individually. They exist together through forces and flocculating agents and form a dimensional framework called soil structure. Soil processes such as water retention, growth of roots, transport of liquids and gases in soil, decomposition and accumulation of soil organic carbon, as well as mineralization and immobilization as nutrient cycling are influenced by soil structure. A well-structured soil provides good aeration, water movement, and water and nutrient retention, which results in the provision of healthier crop productivity. Poorly structured soils result in waterlogging conditions, poor aeration, phytotoxins, and nutrient deficiencies in crops grown in those soils. Fluctuations in soil structure occurred as a result of forces, human activities, biological activities, as well as climatic events. Therefore, it is very important to consider that well-structured soil must be present in the root zone for proper root proliferation, growth, aeration, and movement of water (Sharma and Kumar 2023).

2.3 Soil Porosity

Porosity of soil is considered an important property because it is directly associated with many of the soil function such as exchange of gases into and out of the soil, retention of soil water and nutrients, and mobility of the soil water and nutrients. Soils with high bulk densities have low porosity. It has been reported that BC applications to the soil reduced the SBD and resulted in the increased porosity which favored the water and nutrient retention, root proliferation, aerobic biological activities, and microbial activities, as well as nutrient cycling, such as the nitrogen cycle. The factors such as permanent wilting point, field capacity, and the plant available water are also affected by soil porosity, Therefore, application of biochar in soil reduced bulk densities and improved porosity (Fernandes et al. 2021).

2.4 Water Holding Capacity

Water holding capacity (WHC) of any soil is defined as the capability of soil to hold the bioavailable contents of water in soil. Soil structure, porosity, SBD, organic matter contents are influenced by the soil WHC. Incorporation of organic soil amendments, for example organic matter, crop residues, compost, and straw influences properties of soil such as soil porosity, and finally enhancing WHC. Poorly structured soil has low WHCs with high bulk densities in comparison to the well-structured soils which has high porosity whereas low SBD was found. Soil WHC is of significant importance because plants can uptake more water as well as nutrients from the soil. Water is retained in soil in tiny pores with the help of the capillary action mechanism. It is important to maintain good WHC in soil for proper water management in agricultural systems on a sustainable basis (Meral et al. 2015).

2.5 Aggregate Stability

Soil aggregation is of key importance in physical soil health quality, which is associated to the soil use and management and its impacts on degradation and aggregation. Organic matter plays a significant role as a binding agent between the soil aggregates and soil aggregates protect the soil organic matter decomposition, consequently improving soil carbon storage and the provision of other essential nutrients. Stable aggregation in the soil results a decline in erosion, surface runoff, and crust genesis, along with soil degradation. Reduced disturbance of soil with aggregate stability improves water infiltration, soil organic carbon, and structure within the soil profile (Sithole et al. 2019).

2.6 Impact of Poor Physical Conditions on Root Development and Yield

Soils that have high bulk densities are compacted, have poor structures, and reduced spaces, which inhibits the movement of water and air between the particles, resulting in the inhibition of deep root proliferation as well as the prevalence of deprivation of water and nutrients in the development of roots. According to agrophysics, soils that are compacted in nature are dense, induces gradual growth of roots with stunted and less dense vegetation as compared to less compacted soils. Degraded and disturbed soils, along with these conditions, slow the production of crops and make them easily vulnerable to diseases and stress. Therefore, less and weak root growth and proliferations lead to low nutrient uptake by the plants and ultimately lead to stunted, weak plants. Therefore, conservative practices should be incorporated in soil, which helps to maintain good root health, ultimately resulting in healthier crops and sustainability in agroecosystems (Lipiec et al. 2003).

3 Application of Biochar as Soil Amendment

BC is generally produced through the process of pyrolysis while the temperature range for the production of this black, porous material ranges from 300 °C and goes up to 700 °C under controlled environment (Lehmann and Joseph 2009). BC is generally derived from the agricultural, municipal, forestry, and food waste. The production of BC from these feedstocks, puts the carbon in the recalcitrant form which is buried under the soil for up to hundreds of the years, suggesting the BC potential to combat the climate change and presenting a promising technology for negative greenhouse gas emission with other benefits (Wang et al. 2016; Tripathi et al. 2016; Smith et al. 2019). The properties of the produced BC for example the nutrients availability, mainly depends on the feedstock being used and the pyrolysis temperature and techniques (Nguyen et al. 2017). When the BC was produced with an increasing pyrolysis temperature, N, hydrogen (H), and oxygen (O), content of BC was decreased, however, carbon (C), potassium (K), phosphorus (P), calcium (Ca), pH, ash content, and surface area, of the BC was increased. These BC properties are derived from forcing carbon into a recalcitrant, condensed form and creation of carbonate/oxide mineral phases, deriving to more pH and ash contents. The selection of slow pyrolysis technique is supposed to be producing the BC with higher N, sulphur (S), P, magnesium (Mg), cation exchange capacity (CEC), Ca, and surface area as compared to the fast pyrolysis (Ippolito et al. 2020).

The BC application to soil influences chemical and physical characteristics of soil. These properties include the soil density and stability, air-water soil regime, and soil pH (Blanco-Canqui 2017). All these factors are responsible for effecting the root system of the plants including its rhizosphere and microbes present in this area (Igalavithana et al. 2017). It was reported that the application of BC has decreased the SBD and increased the soil moisture holding capacity. Meanwhile, the BC application also influences the nitrification and denitrification of bacteria. It has reported by the research that BC application has influenced the N_2O emission significantly. It was reported that in a soil with moisture content <70% the BC application has increased the emission of N_2O because of the stimulation of nitrifying bacteria. However, with the moisture content more than 70%, the reduction of N_2O to N_2 with anaerobic bacteria was reported (Liu et al. 2017a, b). Though, the application of BC is helpful in the reduction of SBD, however, it does not have significant influence in increasing the aeration in soil, and removal of the anaerobic zones (Blanco-Canqui 2017).

The BC application also influences the soil pore size, while the amount of moisture available to the plants has been the topic of conversation. It was reported that the direct impact of BC on the pore size was not observed however, it was studied with the perspective of holding of the moisture which depends on the soil granulometric composition. It was observed that in light-loamy soils, with the application of BC, the WHC of soil was increased, however, in the middle loamy soils no changes were observed whilst there was a decrease in the heavy soil (Aller et al. 2017). The capacity of BC to hold the moisture content favors the microbial communities even it is slightly available for the plants. Moreover, it's also beneficial for the microbial communities followed by a drought period. Various researches have been reported the impact of applications of the BC on aggregate stability of soil. The formation of aggregated around the BC grains are involved with the participation of microbes like mycelia and hyphae of fungi and actinobacteria. The hyphae may cause cohesion among the structural aggregates of different sizes and BC grains, leading to the aggregates formation (Zheng et al. 2018). It was also reported that soil pH is influenced to a great extent with the application of BC. The fluctuation in the soil pH could effectively fluctuate numerous agrochemical and biological characteristics for example; soil enzymes, minerals availability, and microbial communities (Gao et al. 2019; Krause et al. 2018; Palansooriya et al. 2019). There are several reasons to favor the increase in soil pH with the addition of BC such as pH of the BC is alkaline and when it is added to the soil it could raise the pH of the soil. Further, the pyrolysis temperature also influences the pH. It was reported that the pH of the BC derived from conocarpus waste increased with temperature from 9.67 (400 °C) to 12.38 (800 °C). However, the increment in the pH elevation was directly proportional to the ash content in the produced BC (Al-Wabel et al. 2013). Moreover, the BC has significant negative charges that also could link the hydrogen ions which may change the pH. Further, variation in the microbial communities were also reported due to the fluctuation on the soil pH (Krause et al. 2018; Zhang et al. 2018a, b, c; Palansooriya et al. 2019). The application of BC indicated the increased share of *Nitrospira* and *Nitrosomonas* in the nitrifying bacterial community and impacted the ammonium-oxidizing archaea and bacteria, which could be due to the availability of ammonia. Among different mechanism of BC, the one could be the adsorption of enzymes and their substrates on the active sites of BC (Bandara et al. 2017; Wang et al. 2018). The BC applications to the soil could change the solution pH, which may change the optimum value for a certain enzyme.

Different soil types respond differently to the BC applications for example a soil with sandy texture is likely to benefit from the BC application for a long period in comparison with fine textured soils (Hansen et al. 2016; Gao et al. 2016). Even within the sandy textured soil, the response of BC could be different based on the soil characteristics and produced BC (Clough et al. 2013). The availability of nutrients to the soil after the BC application depends on the properties and application rate of BC (Hossain et al. 2020). A research outcome elaborated that manure and biosolids derived BC likely to have more nutrients as compared to the BC produced from wood and straw feedstock, resulting in the enhanced nutrient availability, microbial biomass N, extractable P, and decrease nitrate leaching of the sandy soils (El-Naggar et al. 2019; Purakayastha et al. 2019). Moreover, it was stated in research that a bit higher rate of BC application could increase the NH_3 volatilization in soil and the magnitude of N_2O emission (Hossain et al. 2020). Contradictory, Shakoor et al. (2021) have reported no change with the BC application rate on N_2O emission. The variation in the soil nutrients with the addition of BC, could be due to the BC production temperature (Khadem et al. 2021).

4 Mechanisms of Biochar-Induced Changes in Soil Physical Properties

The properties of produced BC varied with pyrolysis temperature, feedstock, and application rate. These BCs have the capability to induce some changes in the soil under some favorable conditions.

4.1 Bulk Density Reduction and Root Penetrability

SBD is one of the major soil properties which is being influenced with the addition of BC. It was reported that BC application reduced SBD gradually indicating a negative correlation between BC application and SBD (Liu et al. 2016a, b). SBD is substantially associated to the soil properties and impact on plant growth of plant and crop production. Table 1 shows the type of BC, application rate and their impact on SBD.

4.2 Enhancement of Soil Porosity and Pore Connectivity

Generally, the porosity was increased with the application of BC however, it mainly hinges on the soil type and BC being added. Increasing the porosity with BC

Table 1 Impact of different biochars on soil bulk density

Type of biochar	Pyrolysis temperature (°C)	Soil type	Application rate	Type of study	Impact of biochar	References
Peanut shell	550	Ultisol	25%, 50%, 75%, 100%	Laboratory	Soil bulk density was reduced	Githinji (2014)
Green waste	450	Sand (bauxite)	40 and 80 Mg/ha	Laboratory	Soil bulk density was reduced	Jones et al. (2010)
Corn stover	350	Alfisol	10 and 11.3 Mg/ha	Laboratory	Soil bulk density was reduced	Herath et al. (2013)

Table 2 Impact of various types of biochar on soil porosity

Type of biochar	Pyrolysis temperature (°C)	Type of soil	Application rate of biochar	Study type	Porosity (% cm^3 cm^{-3})	References
Pine	650	Alliphanic	0	Laboratory	69.64	Abd El-Rahman et al. (2011)
			12		71.28	
			25		72.05	
Peanut shell	500	Loamy sand	0	Laboratory	0.50	Githinji (2014)
			50		0.61	
			100		0.78	
Corn fodder	350	Silt loam	1.13	Column	10	Herath et al. (2013)
			1.7		13	
			1		19	

addition has positive impact on heat, gas, and exchange of water in the soil (Omondi et al. 2016). BC application to sandy soil resulted in increased water retention and decreased the saturated hydraulic conductivity in the soils with sandy texture though saturate hydraulic conductivity was increased and overflow was reduced in the soil with clayey texture (Edeh et al. 2020). Moreover, the addition of BC to the surface soil resulted the clogged pores and ultimately, soil porosity was reduced. Table 2 show the application of BCs and their effect on the porosity.

4.3 Improved Water Retention and Infiltration

Water infiltration is the one of the important factors controlling the lands runoff. Therefore, higher application rate of BC lowers the infiltration rate (Villagra-Mendoza and Horn 2018; Zhao et al. 2021). Hussain et al. (2020) reported that application of BC to the coarse textured soil reduced the infiltration rate of water while application of BCs derived from different sources enhanced the water infiltration rate in the compacted soils. Soil pores were reduced with the application of BC ultimately preventing the incremental infiltration rate (Chen et al. 2022). The BC incorporation in the subsoils further reduces the rate of infiltration. Moreover, the BC particles, application rate, and pattern of application are the factors that are controlling the infiltration rate (Zhan et al. 2021). The cracks present in the surface layer caused the reduction of infiltration rate while the addition of BC to the soil gives a finer structure to the soil resulting in the increased water retention capacity (Gopal et al. 2019). Similarly, the formation of aggregates with the addition of BC helped the water entrance and movement by maintaining the certain pathway, moreover, soil cover became more efficient (Chen et al. 2022).

4.4 Improved Soil Structure and Aggregate Stability

The BC has the huge potential to improve the soil aggregates because of its strong interaction with other soil minerals particles, especially the BC which is produced at low pyrolysis temperature. The carbon fraction in the BC interacts with the inorganic and organic particles of the soil and thus contributing in modifying and amending the soils aggregates. BC helps to increase the wet aggregates which are more stable in sandy soil as compared to the clayey soils. Organic particles in BC tend to bind with larger particles and thus can contribute to enhance the soil aggregation in the soils with coarse texture, in comparison to fine textured soils (Burrell et al. 2016).

5 Biochar in Crop Yield

The research outcomes have elaborated the BC potential in improving the crop performance and crop productivity when applied to farmlands (Table 3). Figure 1 shows a schematic flow of BC in betterment of crop growth and production. The application of BC in the subtropical Taihu Lake region in the China to the wheat crop has significantly improved the NUE and grain yield (GY) (He et al. 2018). Correspondingly, the BC application also improved the wheat yield, soil fertility, and soil physical characteristics in the semi-arid Mediterranean conditions of soil (Olmo et al. 2014). Foster et al. (2016) reported the BC application into an alkaline soil in northern Colorado improved significantly the nutrients availability, microbial activities along with the improvement of maize yield. Further, it was reported in many studies that BC supports the root growth and is also beneficial for the soil microbial communities along with promotion of carbon sequestration in soil (Abiven et al. 2015; Obia et al. 2016; Pandian et al. 2016; Brassard et al. 2016; Smith 2016; Madari et al. 2017; Sheng and Zhu 2018). Liu et al. (2016a, b) reported that the BC application in the subtropical province Zhejiang, China increased the bioavailability of P and K ultimately increased the rice yield. It was reported by Kätterer et al. (2019) in Kenya, a 10-year study proved the improved soil fertility and physical conditions (i.e., porosity, pH, plant available N and P, WHC of soil, and GY of soybean and maize). Moreover, the application of BC to the legumes based intercropping system also improved microbial activities and soil fertility (Liu et al. 2017c, d). BC has shown positive impact on N by reducing the leach down of nitrate and carbon sequestration in the soil layers (Shafqat et al. 2021). Moreover, higher N use efficiency was reported in the cotton crop, which was attributed to enhance the microbial activities and N cycling in the soils (Manzoor et al. 2021). BC based slow released fertilizers minimize the N losses as they release the N slowly and make it bioavailable for longer period of time with more efficiency (Cantarella et al. 2018; Manzoor et al. 2021). Therefore, the N losses can be minimized by the use of BC based slow released fertilizers (Garcia et al. 2018; Perveen et al. 2021).

Table 3 Biochar application rates and crop yield response

Biochar feedstock	Tested crop	Biochar application rate	Impact of biochar	References
Acacia bark	Maize	10 L m^{-2}	Crop growth and yield was improved	Yamato et al. (2016)
Peanut hull		0 to 22.4 t ha^{-1}		
Oak tree	Soybean	10 t ha^{-1}	Dry matter yield was improved	Lee et al. (2015)
Eucalyptus	Common bean	10–50 t ha^{-1}	Bean yield was enhanced	Raboin et al. (2016)
Maize straw	Mung bean	0–100 t ha^{-1}	A max. Yield was noted with 25 t ha^{-1}	Rab et al. (2016)
Coppiced woodlands	Wheat	30 and 60 t ha^{-1}	Plant growth and yield was improved	Amer (2016)
Soft and hard wood		0–5% w/w		
Rice straw		2 t ha^{-1}		
Litchi branch	Oilseed rape	10–30 t ha^{-1}	Yield was improved	Jiang et al. (2020)
Wood biomass	Rice	10% and 20%	Improved yield and other traits	Medynska-Juraszek et al. (2021)
Forest wood		5.5 and 11 t ha^{-1}		
Wood residues		0–16 t ha^{-1}		
Rice husk		4.13 g m^{-2}		
Wood	Carrot	0–20 t $acre^{-1}$	Crop weight was improved	Carpenter and Nair (2014)
Tobacco	Sweet potato	5 t ha^{-1}	Tuber weight, vine length, and yield was improved	Indawan et al. (2018)
Wheat straw		0–40 t ha^{-1}		
Grass	Onion	0.5 kg m^{-2}	Yield was enhanced	Aneseyee and Wolde (2021)
Biochar and N, P, K	Potato	7.5 t ha^{-1}	Improved tuber yield	Mollick et al. (2020)

Various mechanisms have been reported in the literature proving the improved plant growth with the addition of BC including increased availability of nutrients, enhanced soil properties, microbial population or gene expression pattern which could impact the plant growth (Zhao et al. 2019; Cui et al. 2021; You et al. 2021).

6 Case Studies and Experimental Evidence

Various developed countries around the globe have adopted the practice of applying BC in agricultural soils, and the outcomes are magnificent. BC has the potential to enhance soil properties, which supports productivity and sustainability in agriculture (Table 4). Regarding soil carbon sequestration, the performance of BC is outstanding. A field study was conducted in an Australian subtropical grassland. The

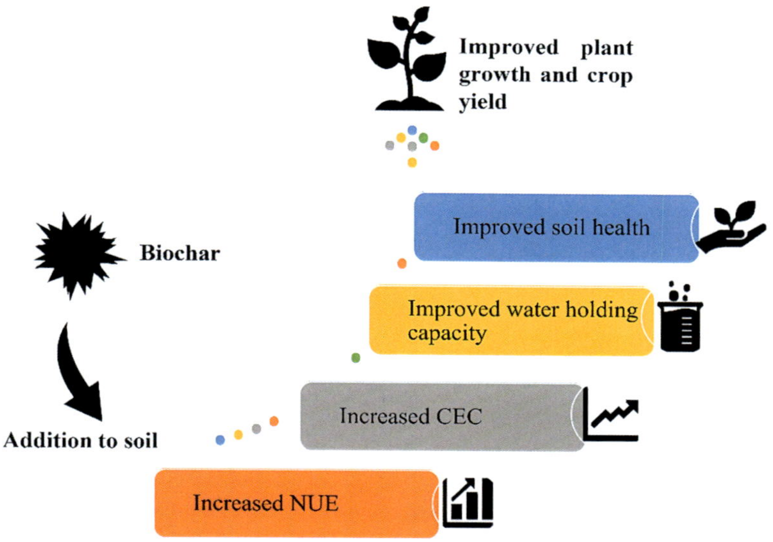

Fig. 1 Schematic flow of biochar in improving plant growth and yield

BC was applied to soil, and it has been interpreted that at the tenth year of BC application, there was 25% more young root-borne carbon remaining in the soil than in the control (Weng et al. 2017). The field trial was conducted in Zambian soil. The study outcomes revealed that as a consequence of corn cob derived BC incorporation, the availability of plant available P was elevated, and parallel to this, the direct input of K was noticed in the tropical soils (Martinsen et al. 2014). In another experiment conducted in the pumpkin field in Nepal, the results revealed that the application of a combination of cow urine (6.3 m^3 ha^{-1}) and BC (0.75 t ha^{-1}) performed 300% better than only urine and 85% better than only BC. In this way, due to the genesis of organic coating, the BC potential for nutrients capture and exchanges was increased (Schmidt et al. 2015). Outcomes of a field trial where wheat bran BC was applied on tomatoes, conducted in Italy, resulted in the elevated levels of soil organic carbon, soil CEC, and the availability of plant essential nutrients such as P, K, NH_3 and, Mg, and due to this, the demand for chemical fertilizer and water was declined (Vaccari et al. 2015). A study from Portugal explored the use of various BCs in a field trial to remediate soils contaminated with heavy metals under field conditions. Rice husk BC (RHB) improved soil WHC, but most soil physical properties remained largely unchanged after 2 years. Crop wood BC (CWB) and RHB showed strong heavy metal accumulation when buried in polluted soils, demonstrating effective metal adsorption. However, applying 8 t ha^{-1} of BC did not reduce the bioavailable heavy metals in the highly contaminated soils from the Guadiamar River. Over time, BC's chemical composition changed, with decreased aryl carbon and increased oxygen-containing groups, potentially enhancing metal binding. These results emphasize BC's stability and evolving properties in

Table 4 Effects of biochar application on key soil physical properties

Soil type	Biochar applications	Impact of biochar	References
Sandy soil	20% biochar by volume	Marked reduction in bulk density (~40%) Total porosity increased by half Noticeable enhancement in water retention	Chang et al. (2021)
Loamy sand	Miscanthus biochar at 6.25–25 Mg ha^{-1} combined with 25 Mg ha^{-1} of sugarcane processing residue	Slight bulk density reduction Porosity improved within 9 months Long-term gains in micropore development, soil stability, and water availability Increased water flow through soil	de Jesus et al. (2020)
Not specified	Biochar alone and mixed with crop residue	Rise in medium and large-sized soil aggregates, leading to improved structure	Xue et al. (2022)
Sand, silt loam, clay	Biochar blended into soils of differing textures	Enhanced availability of soil water, particularly in sandy conditions	Zhang et al. (2021)
Silty loam (Haplic Luvisol)	Applied biochar at 0, 10, and 20 t·ha^{-1}	Bulk density dropped by 12% Total porosity increased by 12%	Toková et al. (2020)
Red soil	Biochar sourced from bamboo, rice, and tobacco biomass	Clear reduction in soil compactness Improved capillary and overall porosity	Qian et al. (2020)

polluted soils. The study suggests higher BC doses or combining BC with phytoremediation for improved soil reclamation in trace element-polluted areas (Campos et al. 2021). This study examined the impact of BC and zeolite soil conditioners on P use efficiency in maize and oats under Mediterranean conditions in Spain and Portugal. A field research was conducted with two types of crops including maize in summer and oats in winter, where a pot experiment with maize were conducted using different P fertilizer rates (0–200 kg P_2O_5 ha^{-1}). Six treatments, including BC, zeolites, and controls with and without P, were tested. It was noticed that there is no significant impact of soil conditioners on crop dry matter (DM) yield or tissue nutrient composition. Field maize DM yields ranged from 14.3 to 15.6 tha^{-1} (2018) and 11.8–13.7 tha^{-1} (2019); oats yielded 2.1–2.4 and 3.0–3.2 tha^{-1}. BC improved soil organic carbon, while zeolites increased K due to their inherent properties. P application enhanced crop yield and nutrient uptake in both field and pot trials. In pots, P also boosted labile soil P and plant tissue nutrient levels. The findings suggest BC and zeolites may have limited short-term benefits in high-quality, intensively managed Mediterranean soils (Arrobas et al. 2022). A research conducted in Egypt by Al-Sayed et al. (2022) investigated the effects of BC produced from pear waste on the growth of roselle (*Hibiscus sabdariffa*). The BC was produced through pyrolysis at 350 °C and applied to sandy loam soil at a rate of 20 tha^{-1}. Results showed a significant improvement in plant performance, with fresh and dry weights of roselle sepals increasing by 27.98% and 35.78%, respectively, compared to the untreated control. Although the study highlighted the positive impact on crop yield, it did not report data on water and nutrient retention. A study by Alghashm et al. (2018) in China examined the effects of BC derived from digested food residue from an anaerobic digester on cabbage (*Brassica chinensis* L.) grown in alkaline soil. The BC was produced at temperatures ranging from 400 °C to 900 °C and applied at a rate of 1% to the soil. BC produced at 600 °C significantly enhanced plant growth, increasing plant length by 57.78%, number of leaves by 110.26%, fresh weight by 163.81%, and leaf area by 147.94% than control. Additionally, P adsorption by the BC from 23.04 mg kg^{-1} at 400 °C was increased to 70.90 mg kg^{-1} at 900 °C. In another study conducted in Pakistan, where rhizobacteria was co-applied with BC-compost mixture and compost mixed biochar under the stress of Pb in the mint plants resulted in increased availability of nutrients such as K (63%), N (46%), P (39%), chlorophyll content (37%) along with the uplift of dry root weight (58%) and dry leaves weight (32%). Overall, an increment was recorded in the soil quality and fertility, whereas, mint plant productivity was also noticed more (Zafar-ul-Hye et al. 2021). In research conducted in Pakistan, it was reported that the BC produced at low temperature and high cation exchange capacity proved to be better soil amendment as compared to the one produced at high temperature, because of its high cation exchange capacity fitted for soils of Pakistan which are alkaline in nature using bagasse as a feed stock along with the biochar derived from municipals waste and animal manure which are capable to provide nutrition to plants (Rasul et al. 2017). Additionally, nutrient uptake improved with higher BC doses; P absorption reached up to 12 mg kg^{-1} and K up to 210 mg kg^{-1} in the 1% treatments. These results suggested that BC from organic waste sources can significantly enhance wheat growth and nutrient uptake in sandy soils. In the United States, applying food waste BC at 5% to sandy soils showed no noticeable effect on tomato growth (Mazac 2016), and maize cob BC had little impact on maize yield, despite improving N retention (Nair et al. 2017). Conversely, in the Czech Republic, food waste BC increased lettuce root biomass by 25% and enhanced photosynthetic efficiency (Mustafa et al. 2022). Research from Asia revealed encouraging outcomes. In China, BC from cooked rice, soybean, and steamed bread boosted wheat growth by nearly 30% (Jia et al. 2021), while palm shell BC significantly enhanced the growth of Suaeda salsa in saline soils by up to 115% (Jeffery et al. 2011). In Bangladesh, banana peel BC increased water spinach yield and leaf number, though higher doses reduced plant height (Islam et al. 2019); rice hull BC also improved groundnut nodulation by 32% and biomass by 14% (Shikha et al. 2023). In Pakistan, maize grown with corn cob BC under drought conditions exhibited increased shoot and root growth, particularly at 3% BC and 40% field capacity (Ali et al. 2021). In the Middle East, studies from Qatar showed that orange peel and mixed vegetable waste

BCs enhanced soil water retention by up to 20% and improved chickpea growth by increasing soil porosity and nutrient availability (Abdelaal et al. 2021; Pradhan et al. 2020). Cucumber green waste BC also reduced water loss by 71% and promoted cucumber growth (Ghiat et al. 2022). European trials demonstrated benefits as well: walnut shell BC in Turkey improved germination and growth of triticale, peas, and mung beans at higher application rates (Uslu et al. 2020). In Germany, peanut hull BC raised quinoa yields, nutrient uptake, and decreased greenhouse gas emissions (Cayuela et al. 2013), while rice hull BC boosted rice straw yield by 14% in Japan (Godlewska et al. 2017). In Australia, rice hull BC dramatically increased biomass production in lettuce and Chinese cabbage, up to 903% (Nair et al. 2017), and in India, it enhanced oyster mushroom yields by more than 100% (Haefele et al. 2011). Lastly, in South America, maize stalk BC improved bean yields by up to 140% on chromium-contaminated soils in Colombia, while modifying N uptake and increasing C:N ratios (Nair et al. 2017).

Multiple global meta-analyses have assessed the agronomic and environmental impacts of BC application. On average, BC increased plant productivity and crop yields by 10–16%, with particularly pronounced yield gains in tropical regions compared to temperate zones (Dai et al. 2020; Jeffery et al. 2017). Synergistic effects were observed when BC was applied alongside fertilizers, resulting in further yield enhancements (Ye et al. 2020). Improvements in plant physiological traits were also evident, including increased photosynthetic rate (up to 27%), stomatal conductance, transpiration, and chlorophyll content (He et al. 2020). Root development showed marked improvements, with increases in root biomass (32%), root length (52%), and nodulation (25%), although a slight reduction in root fungal colonization was noted (Xiang et al. 2017). In soil microbial communities, BC generally promoted microbial biomass carbon and N, with increases ranging from 20% to 26% (Zhou et al. 2017; Zhang et al. 2018d), although some carbon-cycling enzymes exhibited reduced activity. BC also contributed to elevated soil organic carbon levels, with reported increases of up to 52%, alongside modest enhancements in microbial activity (Liu et al. 2016a, b; Bai et al. 2019). Soil physical properties were positively affected as well such as porosity and reductions in SBD (−8%) were observed (Omondi et al. 2016), while plant water use efficiency improved by approximately 20% (Gao et al. 2020). Nutrient dynamics also shifted under BC application. While soil ammonium and nitrate levels occasionally declined, likely due to sorption effects, plant-available P increased substantially (+45%), and N uptake by crops improved by 11% (Gao et al. 2019; Liu et al. 2018). In a research study, BC was derived from walnut shells and then applied in an incubation trial containing soil contaminated with the heavy metals, and results revealed that it has a strong potential to inhibit bioavailability of lead and copper along with the increment of soil properties such as soil organic carbon, soil cation exchange capacity and pH (Peng et al. 2020). BC combined with composting materials could be used to minimize greenhouse gas emissions in contrast to control. Zhang et al. (2020) in a study revealed that the co-pyrolysis of hen manure and saw dust with straw at 400 °C–500 °C has reduced 12.4% ammonia in contrast to the control. Similarly, in an experiment, Awasthi et al. (2020) reported that co-pyrolysis of poultry manure and wheat straw with bamboo proved to have the potential for the reduction of carbon dioxide, methane, ammonia, and nitrous oxide as compared to the control. Zahra et al. (2021) revealed that BC derived from cow manure when applied to the maize crop with an application rate of 5 t ha^{-1} in calcareous soil proved the improved soil fertility, soil organic matter with the provision of essential nutrients to the plants, and reduced pH of calcareous soil. It not only affects the soil properties, as well as it increases plant growth and yield, such as grain yield, biological yield, harvest index, nutrient contents of shoots and roots. In a research experiment, it was reported that a combination of rice husk BC with poultry litter applied with a rate of 10 Mg ha^{-1} in the maize crop resulted in increased soil organic matter contents along with the plant growth (Alarefee et al. 2021). In study, it was reported that co-application of BC with fertilizer and compost resulted in increased soil organic matter, N, P, and K have been noticed; therefore, it has been concluded that this type of combination as a soil amendment in calcareous soil is of great value because of the decreased soil pH (Naeem et al. 2018). Likewise, co-application of BC with vermicompost and fertilizer was studied. The BC was prepared from the vineyard and wood chips, when applied to the soil, it increased the soil properties such as cations, anions, and other essential nutrients. The co-application of BC and inorganic fertilizer has an optimistic effect on the properties of soil and leaf nutrient content. However, due to vermicompost application, a 22% increment in plant growth and yield by 116% in comparison to BC occurred, along with the accumulation of nitrate in leaf by 81% in contrast to ammonium nitrate. Hence, from this, it has been evaluated that the co-application of BC and vermicompost has performed better and increased swiss chard yield (Rivelli and Libutti 2022). In a study, two-port experiments were conducted. The two pot experiments' results demonstrated that, after being incorporated into paddy soil, rice straw BC resulted in enhanced mass decomposition and mineralization of N. Integrating rice straw BC and leguminous residue resulted enhanced rice yield components and raised soil fertility, which in turn raised grain and straw yields as well as the efficiency of N absorption and usage in rice shoots. Nevertheless, the reddish paddy soil's NH_4^+ and NO^{3-}contents dropped when rice straw BC was added, either with or without leguminous cover crop residues (Xie et al. 2022). Further field studies are recommended to confirm these findings (Xie et al. 2022). This meta-analysis examined the effects of co-applying BC and fertilizers on crop yields, based on 627 paired observations from 57 studies. Results showed that BC alone improved crop yield by 25.3%, and inorganic fertilizer alone by 21.9%, when combined, yields increased further by 37.4%, and up to 179.6% when both inorganic and organic fertilizers were used with BC, though this latter figure is based on limited data and should be interpreted with caution. The greatest yield benefits were observed in very acidic soils ($pH \leq 5$), especially when BC was produced at 401°C–500 °C and had a C: N ratio between 31 and 100. The timing and rate of BC application had minimal effect on outcomes. These findings support the growing practice of integrating BC with fertilizers to improve crop productivity, enhance fertilizer efficiency, and promote sustainable farming with reduced environmental impact (Bai et al. 2022).

7 Challenges and Limitations

The benefits of BC are often soil-specific, meaning not all soils respond positively. For example, Ghorbani and Amirahmadi (2024) observed that in some well-structured soils, BC did not significantly enhance yields and, in some instances, and led to nutrient imbalances. In clay-rich soils, Brtnicky et al. (2021) the application of BC for a longer period reduced water retention due to pore clogging and limited capillary movement, which negatively affected plant growth. The application of biochar changes the rate of soil formation (Verheijen et al. 2010). Volatile compounds, poly aromatic hydrocarbons, and dioxins are released as a result of BC applications, these compounds could cause the production of toxic materials which could be responsible for air and soil pollution. (Brtnicky et al. 2021). It was published in a research that the BC produced at high temperature has more ash contents as compared to the BC produced at low temperature therefore the negative impacts has been triggered on the crops cultivated in soil containing BC produced with more ash contents (Butnan et al. 2015). BC can adsorb N while restricting the iron uptake, which otherwise would create a diverse effect on plant growth. (Younas et al. 2024). BC production at low temperature favors the immobilization of N caused by microbes, while at high temperature more cation exchange capacity and surface areas are resulted causing the adsorption of ammonia. Therefore, co-application of BC with compost and fertilizers is a good strategy to overcome challenges and limitations (Haider et al. 2024; Rajkovich et al. 2012).

8 Conclusion

The health of soil is a combination of soil physiochemical and biological properties due to which soil becomes capable of producing and playing a role in the ecosystem, and ensures the provision of environmental quality. Biochar is an effective and cost-efficient soil amendment to improve soil health and crop production. Biochar can potentially improve soil structure, water holding capacity, and nutrients retention, consequently boosting crop growth and productivity.

References

Abd El-Rahman HH, Abedo AA, Salman FM, Mohamed MI, Shoukry MM (2011) Partial substitution of cumin seed meal by Jatropha meal as a potential protein source for feed. Afr J Biotechnol 10(68):15456–15461. https://doi.org/10.5897/AJB11.1595

Abdelaal A, Pradhan S, AlNouss A, Tong Y, Al-Ansari T, McKay G, Mackey HR (2021) The impact of pyrolysis conditions on orange peel biochar physicochemical properties for sandy soil. Waste Manag Res 39:995–1004

Abiven S, Hund A, Martinsen V, Cornelissen G (2015) Biochar amendment increases maize root surface areas and branching: a shovelomics study in Zambia. Plant Soil 395(1):45–55. https://doi.org/10.1007/s11104-015-2533-2

Agegnehu G, Bass AM, Nelson PN, Bird MI (2016) Benefits of biochar, compost and biochar–compost for soil quality, maize yield and greenhouse gas emissions in a tropical agricultural soil. Sci Total Environ 543:295–306

Alarefee AH, Ishak CF, Karam DS, Othman R (2021) Efficiency of rice husk biochar with poultry litter co-composts in oxisols for improving soil physico-chemical properties and enhancing maize performance. Agronomy, 11(12):2409

Alghashm S, Qian S, Hua Y, Wu J, Zhang H, Chen W, Shen G (2018) Properties of biochar from anaerobically digested food waste and its potential use in phosphorus recovery and soil amendment. Sustainability 10:4692

Ali L, Manzoor N, Li X, Naveed M, Nadeem SM, Waqas MR, Khalid M, Abbas A, Ahmed T, Li B, Yan J (2021) Impact of corn cob-derived biochar in altering soil quality, biochemical status and improving maize growth under drought stress. Agronomy 11:2300

Aller D, Rathke S, Laird D, Cruse R, Hatfield J (2017) Impacts of fresh and aged biochars on plant available water and water use efficiency. Geoderma 307:114–121

Al-Sayed HM, Ali AM, Mohamed MA, Ibrahim MF (2022) Combined effect of prickly pear waste biochar and azolla on soil fertility, growth, and yield of roselle (*Hibiscus sabdariffa* L.) plants. J Soil Sci Plant Nutr 22:3541–3552

Al-Wabel MI, Al-Omran A, El-Naggar AH, Nadeem M, Usman AR (2013) Pyrolysis temperature induced changes in characteristics and chemical composition of biochar produced from conocarpus wastes. Bioresour Technol 131:374–379

Amer MM (2016) Effect of biochar, compost tea and magnetic iron ore application on some soil properties and productivity of some field crops under saline soils conditions at North Nile Delta. Egypt J Soil Sci 56:169–186

Amonette JE, Blanco-Canqui H, Hassebrook C, Laird DA, Lal R, Lehmann J, Page-Dumroese D (2021) Integrated biochar research: a roadmap. J Soil Water Conserv 76(1):24A–29A

Aneseyee AB, Wolde T (2021) Effect of biochar and inorganic fertilizer on the soil properties and growth and yield of onion (*Allium cepa*) in tropical Ethiopia. Sci World J 2021:5582697

Antonangelo JA, Sun X, Zhang H (2021) The roles of co-composted biochar (COMBI) in improving soil quality, crop productivity, and toxic metal amelioration. J Environ Manag 277:111443. https://doi.org/10.1016/j.jenvman.2020.111443

Arrobas M, Decker JV, Feix BL, Godoy WI, Casali CA, Correia CM, Ângelo Rodrigues M (2022) Biochar and zeolites did not improve phosphorus uptake or crop productivity in a field trial performed in an irrigated intensive farming system. Soil Use Manag 38(1):564–575

Awasthi MK, Duan Y, Awasthi SK, Liu T, Zhang Z (2020) Influence of bamboo biochar on mitigating greenhouse gas emissions and nitrogen loss during poultry manure composting. Bioresource technology, 303, 122952.3. Blanco-Canqui H (2017) Biochar and soil physical properties. Soil Science Society of America Journal, 81(4):687–711

Bai X, Huang Y, Ren W, Coyne M, Jacinthe P, Tao B, Hui D, Yang J, Matocha C (2019) Responses of soil carbon sequestration to climate-smart agriculture practices: a meta-analysis. Glob Chang Biol 25(8):2591–2606

Bai SH, Omidvar N, Gallart M, Kämper W, Tahmasbian I, Farrar MB, Singh K, Zhou G, Muqadass B, Xu CY, Koech R (2022) Combined effects of biochar and fertilizer applications on yield: a review and meta-analysis. Sci Total Environ 808:152073

Bandara T, Herath I, Kumarathilaka P, Seneviratne M, Seneviratne G, Rajakaruna N et al (2017) Role of woody biochar and fungal-bacterial co-inoculation on enzyme activity and metal immobilization in serpentine soil. J Soils Sediments 17(3):665–673

Blanco-Canqui H (2017) Biochar and soil physical properties. Soil Sci Soc Am J 81:687

Brassard P, Godbout S, Raghavan V (2016) Soil biochar amendment as a climate change mitigation tool: key parameters and mechanisms involved. J Environ Manag 181:484–497. https://doi.org/10.1016/j.jenvman.2016.06.063

Brtnicky M, Datta R, Holatko J, Bielska L, Gusiatin ZM, Kucerik J, Hammerschmiedt T, Danish S, Radziemska M, Mravcova L, Fahad S (2021) A critical review of the possible adverse effects of biochar in the soil environment. Sci Total Environ 796:148756

Burrell LD, Zehetner F, Rampazzo N, Wimmer B, Soja G (2016) Long-term effects of biochar on soil physical properties. Geoderma 282:96–102. https://doi.org/10.1016/j.geoderma.2016.07.019

Butnan S, Deenik JL, Toomsan B, Antal MJ, Vityakon P (2015) Biochar characteristics and application rates affecting corn growth and properties of soils contrasting in texture and mineralogy. Geoderma 237:105–116. https://doi.org/10.1016/j.geoderma.2014.08.010

Campos P, Knicker H, Miller AZ, Velasco-Molina M, De la Rosa JM (2021) Biochar ageing in polluted soils and trace elements immobilisation in a 2-year field experiment. Environ Pollut 290:118025

Cantarella H, Otto R, Soares JR, De Brito Silva AG (2018) Agronomic efficiency of NBPT as a urease inhibitor: a review. J Adv Res 13:19–27. https://doi.org/10.1016/j.jare.2018.05.008

Carpenter BH, Nair A (2014) Effect of biochar on carrot production. Iowa State Univ Res Demonstr Farms Prog Rep 13:1

Cayuela ML, Sánchez-Monedero MA, Roig A, Hanley K, Enders A, Lehmann J (2013) Biochar and denitrification in soils: when, how much and why does biochar reduce N_2O emissions? Sci Rep 3:1–7

Chang Y, Rossi L, Zotarelli L et al (2021) Biochar improves soil physical characteristics and strengthens root architecture in muscadine grape (*Vitis rotundifolia* L.). Chem Biol Technol Agric 8:1–11

Chen Z, Kamchoom V, Apriyono A, Chen R, Chen C (2022) Laboratory study of water infiltration and evaporation in biochar-amended landfill covers under extreme climate. Waste Manag 153:323–334

Clough TJ, Condron LM, Kammann C, Müller C (2013) A review of biochar and soil nitrogen dynamics. Agronomy 3(2):275–293

Cui L. Ippolito JA, Noerpel M, Scheckel KG, Yan J (2021) Nutrient alterations following biochar application to a Cd-contaminated solution and soil. Biochar, 3(4):457–468

Dai Y, Zheng H, Jiang Z, Xing B (2020) Combined effects of biochar properties and soil conditions on plant growth: a meta-analysis. Sci Total Environ 713:136635

de Jesus DS, Glaser B, Pano BLP, Cerri CEP (2020) Biochar and sugar cane filter cake interaction on physical and hydrological soil properties under tropical field conditions. Biochar 2:195–210

Edeh IG, Mašek O, Buss W (2020) A meta-analysis on biochar's effects on soil water properties-new insights and future research challenges. Sci Total Environ 714:136857. https://doi.org/10.1016/j.scitotenv.2020.136857

Eftene A, Ignat P, Chiurciu IA, Manea A, Raducu D, Dumitru S (2020) Soil bulk density as important management factor and ecosystem services well function

El-Naggar A, Lee SS, Rinklebe J, Farooq M, Song H, Sarmah AK et al (2019) Biochar application to low fertility soils: a review of current status, and future prospects. Geoderma 337:536–554

Fernandes JD, Garófalo CLH, Carvallo GHO, Alves TG (2021) Chemical and physical changes of soil amended with biochar. Water Air Soil Pollut 232(8):338

Foster EJ, Hansen N, Wallenstein M, Cotrufo MF (2016) Biochar and manure amendments impact soil nutrients and microbial enzymatic activities in a semi-arid irrigated maize cropping system. Agric Ecosyst Environ 233:404–414

Gao S, Hoffman-Krull K, Bidwell A, DeLuca T (2016) Locally produced wood biochar increases nutrient retention and availability in agricultural soils of the San Juan Islands, USA. Agric Ecosyst Environ 233:43–54

Gao S, DeLuca TH, Cleveland CC (2019) Biochar additions alter phosphorus and nitrogen availability in agricultural ecosystems: a meta-analysis. Sci Total Environ 654:463–472

Gao Y, Shao G, Lu J, Zhang K, Wu S, Wang Z (2020) Effects of biochar application on crop water use efficiency depend on experimental conditions: a meta-analysis. Field Crop Res 249:107763

Garcia PL, Gonzalez-Villalba HA, Sermarini RA, Trivelin PCO (2018) Nitrogen use efficiency and nutrient partitioning in maize as affected by blends of controlled-release and conventional urea. Arch Agron Soil Sci 64:1944–1962. https://doi.org/10.1080/03650340.2018.1469746

Ghiat I, Pradhan S, Govindan R, Mackey HR, Mckay G, Al-Ansari T (2022) Biochar: a sustainable approach of green waste management in agricultural practices under controlled microclimate. Chem Eng Trans 92:331–336

Ghorbani M, Amirahmadi E (2024) Biochar and soil contributions to crop lodging and yield performance-a meta-analysis. Plant Physiol Biochem 215:109053
Githinji L (2014) Effect of biochar application rate on soil physical and hydraulic properties of a sandy loam. Arch Agron Soil Sci 60(4):457–470
Glab T, Palmowska J, Zaleski T, Gondek K (2016) Effect of biochar application on soil hydrological properties and physical quality of sandy soil. Geoderma 281:11–20. https://doi.org/10.1016/j.geoderma.2016.06.028
Godlewska P, Schmidt HP, Ok YS, Oleszczuk P (2017) Biochar for composting improvement and contaminants reduction. A review. Bioresour Technol 246:193–202
Gopal P, Bordoloi S, Ratnam R, Lin P, Cai W, Buragohain P, Garg A, Sreedeep S (2019) Investigation of infiltration rate for soil-biochar composites of water hyacinth. Acta Geophys 67:231–246
Gu YY, Zhang HY, Liang XY, Fu R, Li M, Chen CJ (2022) Impact of biochar and bioorganic fertilizer on rhizosphere bacteria in saline–alkali soil. Microorganisms, 10(12):2310
Guo X, Liu H, Zhang J (2020) The role of biochar in organic waste composting and soil improvement: a review. Waste Manag 102:884–899. https://doi.org/10.1016/j.wasman.2019.12.003
Haefele SM, Konboon Y, Wongboon W, Amarante S, Maarifat AA, Pfeiffer EM, Knoblauch C (2011) Effects and fate of biochar from rice residues in rice-based systems. Field Crop Res 121:430–440
Haider FU, Khan I, Farooq M, Cai L, Li Y (2024) Co-application of biochar and plant growth regulators improves maize growth and decreases cd accumulation in cadmium-contaminated soil. J Clean Prod 440:140515. https://doi.org/10.1016/j.jclepro.2023.140515
Hansen V, Müller-Stöver D, Munkholm LJ, Peltre C, Hauggaard-Nielsen H, Jensen LS (2016) The effect of straw and wood gasification biochar on carbon sequestration, selected soil fertility indicators and functional groups in soil: an incubation study. Geoderma 269:99–107
He T, Liu D, Yuan J, Luo J, Lindsey S, Bolan N, Ding W (2018) Effects of application of inhibitors and biochar to fertilizer on gaseous nitrogen emissions from an intensively managed wheat field. Sci Total Environ 628:121–130
He Y, Yao Y, Ji Y, Deng J, Zhou G, Liu R, Shao J, Zhou L, Li N, Zhou X, Bai SH (2020) Biochar amendment boosts photosynthesis and biomass in C3 but not C4 plants: a global synthesis. GCB Bioenergy 12(8):605–617
Herath HMSK, Arbestain MC, Hedley M (2013) Effect of biochar on soil physical properties in two contrasting soils: an Alfisol and an Andisol. Geoderma 209–210(210):188–197. https://doi.org/10.1016/j.geoderma.2013.06.016
Hossain MZ, Bahar MM, Sarkar B, Donne SW, Ok YS, Palansooriya KN et al (2020) Biochar and its importance on nutrient dynamics in soil and plant. Biochar 2(4):379–420
Hussain R, Bordoloi S, Gupta P, Garg A, Ravi K, Sreedeep S, Sahoo L (2020) Effect of biochar type on infiltration, water retention and desiccation crack potential of a silty sand. Biochar 2:465–478
Igalavithana AD, Lee SE, Lee YH, Tsang DC, Rinklebe J, Kwon EE et al (2017) Heavy metal immobilization and microbial community abundance by vegetable waste and pine cone biochar of agricultural soils. Chemosphere 174:593–603
Indawan E, Lestari SU, Thiasari N (2018) Sweet potato response to biochar application on suboptimal dry land. J Degraded Min Lands Manage 5:1133
Ippolito JA, Cui L, Kammann C, Wrage-Mönnig N, Estavillo JM, Fuertes-Mendizabal T et al (2020) Feedstock choice, pyrolysis temperature and type influence biochar characteristics: a comprehensive meta-data analysis review. Biochar 2(4):421–438
Islam M, Halder M, Siddique MAB, Razir SAA, Sikder S, Joardar JC (2019) Banana peel biochar as alternative source of K for plant productivity and sustainable agriculture. Int J Recycl Org Waste Agric 8:407–413
Jeffery S, Verheijen FG, van der Velde M, Bastos AC (2011) A quantitative review of the effects of biochar application to soils on crop productivity using meta-analysis. Agriculture, ecosystems & environment, 144(1):175–187

Jeffery S, Abalos D, Prodana M, Bastos AC, Van Groenigen JW, Hungate BA, Verheijen F (2017) Biochar boosts tropical but not temperate crop yields. Environ Res Lett 12(5):053001
Jia H, Ben H, Wu F (2021) Effect of biochar prepared from food waste through different thermal treatment processes on crop growth. PRO 9:276
Jiang C, Bo J, Xiao X, Zhang S, Wang Z, Yan G, ... & He H (2020) Converting waste lignin into nano-biochar as a renewable substitute of carbon black for reinforcing styrene-butadiene rubber. Waste management, 102:732–742
Jones BEH, Haynes RJ, Phillips IR (2010) Effect of amendment of bauxite processing sand with organic materials on its chemical, physical and microbial properties. J Environ Manag 91(11):2281–2288. https://doi.org/10.1016/j.jenvman.2010.06.013
Kätterer T, Roobroeck D, Andrén O, Kimutai G, Karltun E, Kirchmann H, Nyberg G, Vanlauwe B, de Nowina KR (2019) Biochar addition persistently increased soil fertility and yields in maize-soybean rotations over 10 years in sub-humid regions of Kenya. Field Crop Res 235:18–26
Khadem A, Raiesi F, Besharati H, Khalaj MA (2021) The effects of biochar on soil nutrients status, microbial activity and carbon sequestration potential in two calcareous soils. Biochar 3(1):105–116
Krause HM, Hüppi R, Leifeld J, El-Hadidi M, Harter J, Kappler A et al (2018) Biochar affects community composition of nitrous oxide reducers in a field experiment. Soil Biol Biochem 119:143–151
Lee SS, Shah HS, Awad YM, Kumar S. and Ok YS (2015). Synergy effects of biochar and polyacrylamide on plants growth and soil erosion control. Environmental Earth Sciences, 74(3): pp.2463–2473
Lehmann J, Joseph S (2009) Biochar for environmental management: an introduction. In: Lehmann J, Joseph S (eds) Biochar for environmental management: science, technology and implementation, 1st edn. Routledge, New York, pp 1–12
Lipiec J, Medvedev VV, Birkas M, Dumitru E, Lyndina TE, Rousseva S, Fulajtar E (2003) Effect of soil compaction on root growth and crop yield in Central and Eastern Europe. Int Agrophys 17(2):61–69
Liu Y, Lu H, Yang S, Wang Y (2016a) Impacts of biochar addition on rice yield and soil properties in a cold waterlogged paddy for two crop seasons. Field Crop Res 191:161–167
Liu Z, Dugan B, Masiello CA, Barnes RT, Gallagher ME, Gonnermann H (2016b) Impacts of biochar concentration and particle size on hydraulic conductivity and DOC leaching of biochar-sand mixtures. J Hydrol 533:461–472
Liu L, Wang Y, Yan X, Li J, Jiao N, Hu S (2017a) Biochar amendments increase the yield advantage of legume-based intercropping systems over monoculture. Agric Ecosyst Environ 237:16–23
Liu Z, Dugan B, Masiello CA, Gonnermann HM (2017b) Biochar particle size, shape, and porosity act together to influence soil water properties. PLoS One 12(6):1–19. https://doi.org/10.1371/journal.pone.0179079
Liu C, Liu F, Ravnskov S, Rubæk GH, Sun Z, Andersen MN (2017c) Impact of wood biochar and its interactions with mycorrhizal fungi, phosphorus fertilization and irrigation strategies on potato growth. J Agron Crop Sci 203(2):131–145
Liu Q, Liu B, Zhang Y, Lin Z, Zhu T, Sun R et al (2017d) Can biochar alleviate soil compaction stress on wheat growth and mitigate soil N_2O emissions? Soil Biol Biochem 104:8–17
Liu Q, Zhang Y, Liu B, Amonette JE, Lin Z, Liu G, Ambus P, Xie Z (2018) How does biochar influence soil N cycle? A meta-analysis. Plant Soil 426(1–2):211–225
Logsdon SD, Karlen DL (2004) Bulk density as a soil quality indicator during conversion to no-tillage. Soil Tillage Res 78(2):143–149
Madari BE, Silva MAS, Carvalho MTM, Maia AHN, Petter FA, Santos JLS, Tsai SM, Leal WGO, Zeviani WM (2017) Properties of a sandy clay loam Haplic Ferralsol and soybean grain yield in a five-year field trial as affected by biochar amendment. Geoderma 305:100–112. https://doi.org/10.1016/j.geoderma.2017.05.029
Martinsen V, Mulder J, Shitumbanuma V, Sparrevik M, Børresen T, Cornelissen G (2014). Farmer-led maize biochar trials: Effect on crop yield and soil nutrients under conservation farming. Journal of Plant Nutrition and Soil Science, 177(5):681–695

Manzoor S, Rahman MHU, Haider G, Ghafoor I, Ahmad S, Afzal M, Nawaz F, Iqbal R, Yasin M, Haq T, Danish S, Ghaffar A (2021) Biochar and slow release nitrogen fertilizer improve growth, yield, NUE, fiber quality of cotton and reduce nitrogen losses under arid climatic conditions. Environ Sci Pollut Res 11:1–14. https://doi.org/10.1007/s11356-021-16576-6

Mazac R (2016) Assessing the use of food waste biochar as a biodynamic plant fertilizer. In: Proceedings of the National Conference on Undergraduate Research (NCUR). University of North Carolina Asheville, Asheville, North Carolina April 7–9, 2016, p 43

Medynska-Juraszek A, Latawiec A, Królczyk J, Bogacz A, Kawałko D, Bednik M, Dudek M (2021) Biochar improves maize growth but has a limited effect on soil properties: evidence from a three-year field experiment. Sustainability 13:3617

Meral R, Demir AD, Demir Y, Malasli MZ, Turan V (2015) The improvement of soil water holding capacity, infiltration rate, and aggregate stability with different soil conditioners. Fresensius Environ Bull 24(11):3550–3555

Mohammad G, Petr K, Reinhard WN, Marek K, Elnaz A, Daniel B et al (2022) Interaction of biochar with chemical, green and biological nitrogen fertilizers on nitrogen use efficiency indices. Agronomy 12:9. https://doi.org/10.3390/agronomy12092106

Mollick OA, Paul AK, Alam I, Sumon MM (2020) Effect of biochar on yield and quality of potato (*Solanum tuberosum*) tuber. Int J Bio-Resour Stress Manag 11:445–450

Mustafa A, Brtnicky M, Hammerschmiedt T, Kucerik J, Kintl A, Chorazy T, Naveed M, Skarpa P, Baltazar T, Malicek O, Holatko J (2022) Food and agricultural wastes-derived biochars in combination with mineral fertilizer as sustainable soil amendments to enhance soil microbiological activity, nutrient cycling and crop production. Front Plant Sci 13:1028101

Naeem MA, Khalid M, Aon M, Abbas G, Amjad M, Murtaza B, Khan WUD, Ahmad N (2018) Combined application of biochar with compost and fertilizer improves soil properties and grain yield of maize. J Plant Nutr 41(1):112–122

Nair VD, Nair PKR, Dari B, Freitas AM, Chatterjee N, Pinheiro FM (2017) Biochar in the agroecosystem-climate-change-sustainability nexus. Front Plant Sci 8:2051

Nepal J, Ahmad W, Munsif F, Khan A, Zou Z (2023) Advances and prospects of biochar in improving soil fertility, biochemical quality, and environmental applications. Front Environ Sci 11:1114752

Nguyen TTN, Xu CY, Tahmasbian I, Che RX, Xu ZH, Zhou XH, Wallace HM, Bai SH (2017) Effects of biochar on soil available inorganic nitrogen: a review and meta-analysis. Geoderma 288:79–96

Obia A, Mulder J, Martinsen V, Cornelissen G, Børresen T (2016) In situ effects of biochar on aggregation, water retention and porosity in light-textured tropical soils. Soil Tillage Res 155:35–44. https://doi.org/10.1016/j.still.2015.08.002

Olmo M, Alburquerque JA, Barrón V, Del Campillo MC, Gallardo A, Fuentes M, Villar R (2014) Wheat growth and yield responses to biochar addition under Mediterranean climate conditions. Biol Fertil Soils 50:1177–1187

Omara P, Aula L, Oyebiyi FB, Eickhoff EM, Carpenter J, Raun WR (2020) Biochar application in combination with inorganic nitrogen improves maize grain yield, nitrogen uptake, and use efficiency in temperate soils. Agronomy 10(9):1241. https://doi.org/10.3390/agronomy10091241

Omondi MO, Xia X, Nahayo A, Liu X, Korai PK, Pan G (2016) Quantification of biochar effects on soil hydrological properties using meta-analysis of literature data. Geoderma 274:28–34

Palansooriya KN, Wong JTF, Hashimoto Y, Huang L, Rinklebe J, Chang SX et al (2019) Response of microbial communities to biochar-amended soils: a critical review. Biochar 1:3–22

Pandian K, Subramaniayan P, Gnasekaran P, Chitraputhirapillai S (2016) Effect of biochar amendment on soil physical, chemical and biological properties and groundnut yield in rainfed Alfisol of semi-arid tropics. Arch Agron Soil Sci 62(9):1293–1310. https://doi.org/10.1080/03650340.2016.1139086

Peng X, Deng Y, Liu L, Tian X, Gang S, Wei Z, Yue K (2020) The addition of biochar as a fertilizer supplement for the attenuation of potentially toxic elements in phosphogypsum-amended soil. Journal of Cleaner Production, 277:124052

Perveen S, Ahmad S, Skalicky M, Hussain I, Rahman MHU, Ghaffar A, Bashir MMS, Batool M, Brestic M, Fahad S, Hassan M, Sabagh AE (2021) Assessing the potential of polymer coated urea and sulphur fertilization on growth, physiology, yield, oil contents and nitrogen use efficiency of sunflower crop under arid environment. Agronomy 11(269):1–12. https://doi.org/10.3390/agronomy11020269

Pradhan S, Abdelaal AH, Mroue K, Al-Ansari T, Mackey HR, McKay G (2020) Biochar from vegetable wastes: agro-environmental characterization. Biochar 2:439–453

Purakayastha T, Bera T, Bhaduri D, Sarkar B, Mandal S, Wade P et al (2019) A review on biochar modulated soil condition improvements and nutrient dynamics concerning crop yields: pathways to climate change mitigation and global food security. Chemosphere 227:345–365

Qian Z, Tang L, Zhuang S et al (2020) Effects of biochar amendments on soil water retention characteristics of red soil at South China. Biochar 2:479–488

Rab A, Khan MR, Haq SU, Zahid S, Asim M, Afridi MZ, Arif M, Munsif F (2016) Impact of biochar on mungbean yield and yield components. Pure Appl Biol 5:632–640

Raboin LM, Razafimahafaly AHD, Rabenjarisoa MB, Rabary B, Dusserre J, Becquer T (2016) Improving the fertility of tropical acid soils: liming versus biochar application? A long term comparison in the highlands of Madagascar. Field Crop Res 199:99–108

Rajkovich S, Enders A, Hanley K, Hyland C, Zimmerman AR, Lehmann J (2012) Corn growth and nitrogen nutrition after additions of biochars with varying properties to a temperate soil. Biol Fertil Soils 48:271–284. https://doi.org/10.1007/s00374-011-0624-7

Rasul F, Ahmad A, Arif M, Mian IA, Ali K, Qayyum MF, Shackley S (2017) Biochar for agriculture in Pakistan. In Sustainable agriculture reviews (pp. 57–114). Cham: Springer International Publishing

Rivelli AR, Libutti A (2022) Effect of biochar and inorganic or organic fertilizer co-application on soil properties, plant growth and nutrient content in Swiss chard. Agronomy 12(9):2089

Schmidt HP, Pandit BH, Martinsen V, Cornelissen G, Conte P, Kammann CI (2015) Fourfold increase in pumpkin yield in response to low-dosage root zone application of urine-enhanced biochar to a fertile tropical soil. Agriculture, 5(3):723–741

Shafqat W, Jaskani MJ, Maqbool R, Chattha WS, Ali Z, Naqvi SM, Haider MS, Khan IA, Vincent CI (2021) Heat shock protein and aquaporin expression enhance water conserving behavior of citrus under water deficits and high temperature conditions. Environ Exp Bot 181:104270. https://doi.org/10.1016/j.envexpbot.2020.104270

Shakoor A, Shahzad SM, Chatterjee N, Arif MS, Farooq TH, Altaf MM et al (2021) Nitrous oxide emission from agricultural soils: application of animal manure or biochar? A global meta-analysis. J Environ Manag 285:112170

Sharma PK, Kumar S (2023) Soil structure and plant growth. In: Soil physical environment and plant growth. Springer, Cham. https://doi.org/10.1007/978-3-031-28057-3_5

Sheng Y, Zhu L (2018) Biochar alters microbial community and carbon sequestration potential across different soil pH. Sci Total Environ 622–623:1391–1399. https://doi.org/10.1016/j.scitotenv.2017.11.337

Shikha FS, Rahman MM, Sultana N, Mottalib MA, Yasmin M (2023) Effects of biochar and biofertilizer on groundnut production: a perspective for environmental sustainability in Bangladesh. Carbon Res 2:10

Sithole NJ, Magwaza LS, Thibaud GR (2019) Long-term impact of no-till conservation agriculture and N-fertilizer on soil aggregate stability, infiltration and distribution of C in different size fractions. Soil Tillage Res 190:147–156

Smith P (2016) Soil carbon sequestration and biochar as negative emission technologies. Glob Chang Biol 22(3):1315–1324

Smith P, Adams J, Beerling DJ, Beringer T, Calvin KV, Fuss S, Griscom B, Hagemann N, Kammann C, Kraxner F, Minx JC, Popp A, Renforth P, Vicente JLV, Keesstra S (2019) Impacts of land-based greenhouse gas removal options on ecosystem services and the United Nations sustainable development goals. Annu Rev Environ Resour 44:255–286

Sun F, Lu S (2014) Biochars improve aggregate stability, water retention, and pore-space properties of clayey soil. J Plant Nutr Soil Sci 177(1):26–33. https://doi.org/10.1002/jpln.201200639

Sun X, Zhong T, Zhang L, Zhang K, Wu W (2019) Reducing ammonia volatilization from paddy field with rice straw derived biochar. Sci Total Environ 660:512–518

Toková L, Igaz D, Horák J, Aydin E (2020) Effect of biochar application and re-application on soil bulk density, porosity, saturated hydraulic conductivity, water content and soil water availability in a silty loam Haplic Luvisol. Agronomy 10:1005

Tripathi M, Sahu JN, Ganesan P (2016) Effect of process parameters on production of biochar from biomass waste through pyrolysis: a review. Renew Sust Energ Rev 55:467–481

Uslu OS, Babur E, Alma MH, Solaiman ZM (2020) Walnut shell biochar increases seed germination and early growth of seedlings of fodder crops. Agriculture 10:427

Vaccari FP, Maienza A, Miglietta F, Baronti S, Di Lonardo S, Giagnoni L, Genesio L (2015) Biochar stimulates plant growth but not fruit yield of processing tomato in a fertile soil. Agriculture, Ecosystems & Environment, 207:163–170

Verheijen F, Jeffery S, Bastos AC (2010) Biochar application to soils. Institute for Environment and Sustainability, Luxembourg

Villagra-Mendoza K, Horn R (2018) Effect of biochar addition on hydraulic functions of two textural soils. Geoderma 326:88–95

Wang J, Xiong Z, Kuzyakov Y (2016) Biochar stability in soil: meta-analysis of decomposition and priming effects. Glob Change Biol Bioenergy 8:512–523

Wang C, Chen S, Wu L, Zhang F, Cui J (2018) Wheat straw-derived biochar enhanced nitrification in a calcareous clay soil. Pol J Environ Stud 27:1297–1305. https://doi.org/10.15244/pjoes/76504

Wang H, Shao D, Ji B, Gu W, Yao M (2022) Biochar effects on soil properties, water movement and irrigation water use efficiency of cultivated land in Qinghai-Tibet plateau. Sci Total Environ 829:154520

Weng Z, Van Zwieten L, Singh BP, Tavakkoli E, Joseph S, Macdonald LM, Rose TJ, Rose MT, Kimber SW, Morris S, Cozzolino D (2017) Biochar built soil carbon over a decade by stabilizing rhizodeposits. Nat Clim Chang 7(5):371–376

Xi J, Li H, Xi J, Tan S, Zheng J, Tan Z (2020) Effect of returning biochar from different pyrolysis temperatures and atmospheres on the growth of leaf-used lettuce. Environmental Science and Pollution Research, 27(28):35802–35813

Xiang Y, Deng Q, Duan H, Guo Y (2017) Effects of biochar application on root traits: a meta-analysis. GCB Bioenergy 9(10):1563–1572

Xie Z, Shah F, Zhou C (2022) Combining rice straw biochar with leguminous cover crop as green manure and mineral fertilizer enhances soil microbial biomass and rice yield in South China. Front Plant Sci 13:778738

Xue P, Fu Q, Li T et al (2022) Effects of biochar and straw application on the soil structure and water-holding and gas transport capacities in seasonally frozen soil areas. J Environ Manag 301:113943

Yamato M, Okimori Y, Wibowo IF, Anshori S, Ogawa M (2016) Effects of the application of charred bark of *Acacia mangium* on the yield of maize, cowpea and peanut, and soil chemical properties in South Sumatra, Indonesia. Soil Sci Plant Nutr 52:489–495

Ye L, Camps-Arbestain M, Shen Q, Lehmann J, Singh B, Sabir M (2020) Biochar effects on crop yields with and without fertilizer: a meta-analysis of field studies using separate controls. Soil Use Manag 36(1):2–18

Yin J, Zhao L, Xu X, Li D, Qiu H, Cao X (2022) Evaluation of long-term carbon sequestration of biochar in soil with biogeochemical field model. Science of the Total Environment, 822:153576

You X, Suo F, Yin S, Wang X, Zheng H, Fang S, Li, Y (2021) Biochar decreased enantioselective uptake of chiral pesticide metalaxyl by lettuce and shifted bacterial community in agricultural soil. Journal of hazardous materials, 417:126047

Younas Z, Rahman U, Ikram M, Raja NI (2024) Biochar and micronutrients availability: problem and future prospects. In: Biochar-assisted remediation of contaminated soils under changing climate. Elsevier, pp 101–127

Zafar-ul-Hye M, Tahzeeb-ul-Hassan M, Wahid A, Danish S, Khan MJ, Fahad S, Brtnicky M, Hussain GS, Battaglia ML, Datta R (2021) Compost mixed fruits and vegetable waste biochar with ACC deaminase rhizobacteria can minimize lead stress in mint plants. Sci Rep 11:6606

Zhan S, Song M, Li Z, Ma H (2021) Effects of different straw biochars on soil water infiltration and evaporation. J Soil Water Conserv 35:294–300

Zhang G, Guo X, Zhu Y, Liu X, Han Z, Sun K et al (2018a) The effects of different biochars on microbial quantity, microbial community shift, enzyme activity, and biodegradation of polycyclic aromatic hydrocarbons in soil. Geoderma 328:100–108

Zhang K, Sun P, Faye MCA, Zhang Y (2018b) Characterization of biochar derived from rice husks and its potential in chlorobenzene degradation. Carbon 130:730–740

Zhang Y, Xu X, Cao L, Ok YS, Cao X (2018c) Characterization and quantification of electron donating capacity and its structure dependence in biochar derived from three waste biomasses. Chemosphere 211:1073–1081

Zhang L, Jing Y, Xiang Y, Zhang R, Lu H (2018d) Responses of soil microbial community structure changes and activities to biochar addition: a meta-analysis. Sci Total Environ 643:926–935

Zhang H, Marchant-Forde JN, Zhang X, Wang Y (2020) Effect of cornstalk biochar immobilized bacteria on ammonia reduction in laying hen manure composting. Molecules, 25(7), 1560.

Zhang J, Amonette JE, Flury M (2021) Effect of biochar and biochar particle size on plant-available water of sand, silt loam, and clay soil. Soil Tillage Res 212:104992

Zhang Y, Shengzhe E, Wang Y, Su S, Bai L, Wu C, Zeng X (2021) Long-term manure application enhances the stability of aggregates and aggregate-associated carbon by regulating soil physicochemical characteristics. Catena, 203, 105342

Zhao Z, Nie T, Zhou W (2019) Enhanced biochar stabilities and adsorption properties for tetracycline by synthesizing silica-composited biochar. Environmental Pollution, 254:113015

Zhao W, Liu Y, Li Z, Bao Z (2021) Effects of biochar on infiltration and evaporation of soil water with sand mulching. J Biobaased Mater Bioenergy 15:369–373

Zheng H, Wang X, Luo X, Wang Z, Xing B (2018) Biochar-induced negative carbon mineralization priming effects in a coastal wetland soil: roles of soil aggregation and microbial modulation. Sci Total Environ 610:951–960

Zahra MB, Fayyaz B, Aftab ZEH, Haider MS (2021) Mitigation of degraded soils by using biochar and compost: a systematic review. Journal of Soil Science and Plant Nutrition, 21(4):2718–2738

Zhou H, Zhang D, Wang P, Liu X, Cheng K, Li L, Zheng J, Zhang X, Zheng J, Crowley D, van Zwieten L, Pan G (2017) Changes in microbial biomass and the metabolic quotient with biochar addition to agricultural soils: a meta-analysis. Agric Ecosyst Environ 239:80–89

Biochar's Role in Soil Chemical Properties and Its Impact on Yield Enhancement

Shamal Shasang Kumar, Shalendra Prasad, Ami Chand Sharma, Uday Prasad Sharma, and Owais Ali Wani

Abstract

Biochar is an integrative approach to soil management that improves the soils chemical properties, nutrient dynamics and aims at achieving sustainable agricultural productivity with the use of carbon (C)-rich organic amendments across diverse agricultural systems. This paper explores the key roles of biochar that modifies important soil chemical properties including soil pH, cation exchange capacity (CEC), nutrient dynamics, organic carbon (OC) content and examines its cascading effects on crop yield enhancement. The alkaline nature of biochar helps stabilize soil acidity, mainly in degraded and highly weathered tropical soils, thus increasing the bioavailability of essential macronutrients viz., phosphorus (P), potassium (K), calcium (Ca) and magnesium (Mg), respectively. The porous structure and the surface functional groups of biochar contribute to increase in CEC, thereby enhancing the soil's ability to retain nutrients and reduce leaching. Biochar serves as a store house for the slow releasing nutrients and facilitates conducive environment for beneficial microbiological activity, fostering nutrient cycling and root development. These soil improvements often translate into higher crop yields, especially when biochar is used synergistically with composts or fertilizers. Therefore, biochar represents a sustainable soil fertility management strategy which enhances agricultural productivity while also contributing to long-term C sequestration and improved soil health.

S. S. Kumar (✉) · S. Prasad · A. C. Sharma · U. P. Sharma
Crop Research Division, Ministry of Agriculture, Waterways and Sugar Industry (MOAW&SI), Sigatoka, Fiji

O. A. Wani
Division of Agronomy, Faculty of Agriculture, SKUAST-Kashmir, Wadura, Sopore, Jammu and Kashmir, India

O. A. Wani et al. (eds.), *Biochar for Sustainable Soil Health and Crop Production*, https://doi.org/10.1007/978-981-92-2230-8_4

Keywords

Biochar · Soil chemical properties · Nutrient availability · Crop yield · Sustainable agriculture · Soil fertility · C sequestration

1 Introduction

The increasing pressure to meet global food demand, combined with ongoing soil degradation, climate change (CC) and declining resource availability, has made the search for sustainable agricultural inputs more urgent than ever. One such promising solution is biochar, a stable, C-rich product obtained through the pyrolysis of organic biomass in limited oxygen (O_2) (Wani et al. 2023). Biochar improves soil chemical properties and increases crop yields, making it a valuable soil amendment (Murtaza et al. 2021; Singh et al. 2022). Soil pH, nutrient availability, and CEC are crucial for crop productivity. In tropical and subtropical regions, degraded soils often display low fertility, acidity, and poor nutrient-holding capacity (Asio and Asio 2021). Biochar offers an effective strategy to ameliorate these conditions through its influence on soil chemical processes. With its porous structure, high surface area, and diverse functional groups, biochar effectively retains nutrients and regulates soil pH (Leng et al. 2021).

One of biochar's most significant contributions to soil chemistry is its ability to increase soil pH, particularly in acidic soils (Tusar et al. 2023). This effect is attributed to the carbonates, oxides, and hydroxides present in biochar, which neutralize soil acidity (Arwenyo et al. 2023). The resulting pH increase reduces the solubility and toxicity of harmful metals such as aluminum (Al^{3+}) and manganese (Mn^{2+}), while enhancing the availability of essential nutrients, including P, Ca and Mg (Huang et al. 2025). Consequently, crops grown in acidic soils amended with biochar often exhibit improved growth and yield. Beyond pH correction, biochar also enhances the soil's CEC, a critical property reflecting the soil's ability to retain and exchange positively charged nutrient ions. Biochar's high surface area and abundance of O_2-containing functional groups contribute to this increase in CEC, facilitating the retention of key nutrients such as K, ammonium (NH_4^+), and calcium (Ca) (Garg et al. 2025). This retention reduces nutrient leaching, particularly in sandy or low-organic matter (OM) soils, and ensures greater nutrient availability for plant uptake over time. In addition, biochar can directly supply nutrients to plants, depending on the type of feedstock and production conditions. For example, biochars produced from manure or crop residues typically contain higher levels of N, P and K than those derived from wood (Whalen et al. 2024). Moreover, biochar serves as a slow-release nutrient source, helping to match nutrient availability with plant demand and thereby enhancing nutrient use efficiency (NUE) (Waqar et al. 2022).

Another important effect of biochar on soil chemical properties is its contribution to soil organic carbon (SOC). Owing to its high C stability, biochar promotes long-term C storage in the soil, which supports microbial activity, enhances nutrient cycling, and helps mitigate greenhouse gas (GHG) emissions (Kumar et al. 2025).

This stable C pool also improves the soil's buffering capacity and resilience to environmental stresses, thereby fostering sustainable crop production. The positive impact of biochar on crop yield is closely linked to these improvements in soil chemical properties. Several studies have reported higher yields in crops such as maize, rice, wheat, tomato, and root vegetables following biochar application (Naeem et al. 2018; Farhangi-Abriz et al. 2021; Liu et al. 2022). Biochar also influences soil microbial activity, which plays a key role in plant productivity. Its porous structure provides microhabitats for soil microorganisms, offering protected surfaces for colonization. These microbes are essential for nutrient transformations, including N-fixation, P solubilization, and OM decomposition (Mahala et al. 2020). However, the effectiveness of biochar is not universal and depends on multiple factors. The feedstock type and pyrolysis temperature strongly affect the chemical composition and functional properties of biochar (Wani et al. 2023). Biochars produced at low temperatures (300–400 °C) generally contain more volatile matter and labile C but lower ash content, whereas high-temperature biochars (500–700 °C) have larger surface areas, more stable C, and higher pH (Tomczyk et al. 2020). In addition, soil characteristics such as texture, fertility status, and the type of crop also influence how biochar interacts within the soil-plant system.

In sandy soils, biochar enhances nutrient and water retention by increasing CEC and soil porosity (Li et al. 2021). In clayey soils, it can reduce bulk density (BD) and improve aeration, which is beneficial for root penetration (He et al. 2022). In acid affected tropical soils, biochar's alkalizing effect is particularly advantageous, creating a more balanced nutrient environment. Thus, biochar's role in enhancing yield is often context-specific, requiring careful consideration of site conditions and objectives (Sharma et al. 2025). Biochar also supports the objectives of climate-smart and regenerative farming. In addition to enhancing crop productivity, it promotes C storage, valorization of organic waste, and the mitigation of agricultural greenhouse gas emissions (Kamyab et al. 2024). Producing biochar from agri/forestry residues offers an opportunity to create value from waste streams while closing nutrient loops in farming systems. Despite its potential, biochar adoption faces hurdles such as high production costs, inconsistent quality, limited access to technology, and low farmer awareness. Inconsistent results also highlight the need for site-specific research and clear guidelines on application rates and integration with other amendments. Addressing these challenges requires strong policy support and capacity building. Governments and research bodies should promote biochar through subsidies, farmer training, and inclusion in sustainable land policies (Fridahl et al. 2020). Public-private partnerships can drive innovation, making biochar more affordable, scalable, and environmentally sound.

In conclusion, biochar represents a transformative soil conditioner having the capacity to enhance the chemical properties and support productive agriculture, particularly in degraded and nutrient-deficient soils. By improving pH, nutrient availability, CEC, and SOC content, biochar creates a more favorable chemical environment for plant growth. Its multifaceted benefits spanning agronomic, environmental, and climate domains position biochar as a cornerstone of sustainable soil management practices. As global agricultural systems seek resilience and efficiency, the strategic use of biochar presents an opportunity for long-term food security and ecological sustainability.

2 Effects on Soil Chemical Properties

The application of biochar modifies soil chemical composition in multiple ways, all of which contribute to enhanced soil fertility and crop performance. Both short- and long-term nutrient dynamics are affected by biochar through mechanisms illustrated in Fig. 1. These effects are influenced by site-specific soil characteristics as well as the physicochemical properties of the biochar, which depend on the type of feedstock and pyrolysis conditions. Beyond immediate improvements in fertility, biochar also enhances soil structure, stimulates microbial activity, and promotes long-term C storage. Details on how biochar impacts specific soil chemical parameters and the mechanisms underlying these changes are covered in detail in the ensuing subsections.

2.1 Soil pH and Acidification

One of the most consistent and well-documented effects of biochar on soil chemistry is its ability to increase soil pH, particularly in acidic soils (Paz-Ferreiro et al. 2020). This improves nutrient bioavailability and delays acidification. The main cause of this pH variation is the alkaline nature of many biochars, which include significant concentrations of basic cations including Ca, Mg, K and sodium (Na) in addition to carbonate and oxide compounds produced during pyrolysis (Fidel et al. 2017). Exchangeable acidity and aluminum (Al^{3+}) toxicity, which are major constraints to root development and nutrient uptake in acidic soils, can be alleviated by biochar through the neutralization of hydrogen (H^+) and Al^{3+} ions (Rahman et al. 2018). For example, applying biochar at 10 t ha^{-1} has been shown to significantly raise soil pH, likely because biochar consumes protons upon incorporation (Shi et al. 2019). Rice husk biochar (RHB), which is alkaline (pH = 8.15) and contains a high proportion of ash (~32%), releases hydroxide ions (OH^-) from its base cations

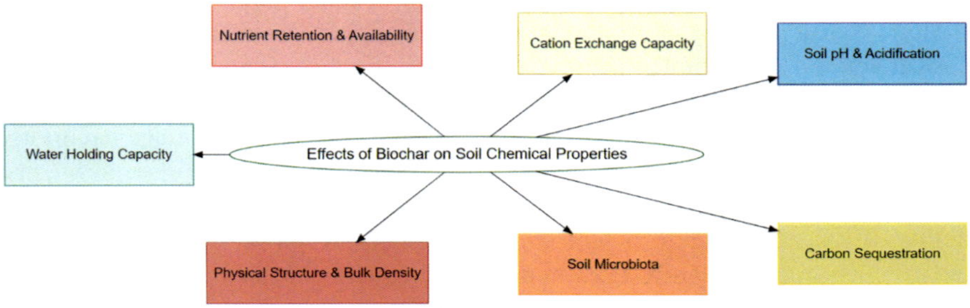

Fig. 1 Biochar affects soil chemical properties

during hydrolysis, further contributing to soil pH enhancement (Eduah et al. 2020). In a similar study, Panhwar et al. (2020) reported that combining RHB with biofertilizer at 4 t ha^{-1} each increased soil pH by 1.39 units. Al^{3+} toxicity in acidic soils can be mitigated by biochar through several mechanisms, including electrostatic attraction, complex formation, cation exchange, adsorption, and absorption (Shetty et al. 2021). Mosharrof et al. (2022) demonstrated that using biochar together with lime and varying P fertilizer rates significantly reduced exchangeable Al, indicating that raising soil pH with biochar and lime can lessen Al toxicity (Ch'ng et al. 2019). Additionally, biochar-facilitated Al complexation may further reduce aluminum stress on plants (Sasmita et al. 2017). The degree of pH increase and Al reduction depends on factors such as feedstock type, pyrolysis temperature, soil buffering capacity, and application rate (Ippolito et al. 2020). In neutral to alkaline soils, caution is necessary since excessive pH elevation may result in micronutrient deficits, especially in iron (Fe), zinc (Zn), and Mn^{2+}, even though this impact is most advantageous in acidic soils (Riaz et al. 2020).

2.2 Cation Exchange Capacity

Application of biochar often leads to measurable improvements in soil CEC, enhancing the soil's ability to retain and supply essential nutrients such as K^+, NH_4^+, Ca^{2+} and Mg^{2+} (Adhikari et al. 2024). This increase in CEC is primarily driven by biochar's large surface area, high porosity, and abundance of negatively charged functional groups, including carboxyl and hydroxyl groups, which develop during oxidation and soil aging (Murtaza et al. 2023). Xu et al. (2012) reported that soils amended with biochar produced from various feedstocks showed CEC increases ranging from 19% to 83%. Chintala et al. (2014) found that maize stover biochar applied at 52, 104, and 156 t ha^{-1} increased CEC by 87%, 120%, and 142%, respectively, while switchgrass biochar at the same rates raised CEC by 58%, 89%, and 122%. Similarly, Cornelissen et al. (2013) observed a 60% increase in CEC in soils treated with maize cob and wood-derived biochars. Feedstock's mainly woody materials commonly produce higher CEC than crop wastes, pyrolysis conditions (lower temperatures generally yield biochars with more oxygen-containing functional groups) and soil type all have a significant impact on the amount of CEC improvement (López-Cano et al. 2018; Bakshi et al. 2020; Kumar et al. 2022). Higher CEC in nutrient-poor tropical soils indicates increased resistance to nutrient loss, especially under conditions of intense precipitation (Mesele et al. 2024). The CEC impact can be enhanced over time by biochar's interaction with microbes and soil OM (SOM), which can further increase its surface reactivity (Palansooriya et al. 2019).

2.3 Nutrient Retention and Availability

By physically adsorbing nutrients, lowering leaching losses, and improving nutrient cycling, biochar can significantly increase nutrient accessibility and retention capacity (Hossain et al. 2020). Its high CEC guarantees that positively charged nutrients stay exchangeable for plant uptake, while its internal pore network and surface functional groups enable the collection of dissolved nutrients (Singh et al. 2023). Concerning N, biochar can adsorb Al ions and encourage nitrification while lowering gaseous nitrous oxide (N_2O) losses through microbial interactions (Clough et al. 2013). Several studies have shown that incorporating biochar into soils can substantially reduce N_2O emissions. For example, Rondon et al. (2005) reported a 50% reduction in N_2O emissions under soybean systems and an 80% decrease under grass systems. Similarly, biochar treatments have been observed to lower N_2O emissions from 1768 to 45–699 µg N_2O-N m^{-2} h^{-1} (Wang et al. 2013) and suppress emissions by 21.3–91.6% (Stewart et al. 2013). Biochar also influences P availability, particularly in acidic soils, where its alkalizing effect reduces P fixation by iron and Al^{3+} oxides (Peng et al. 2023). Nelson et al. (2011) demonstrated that, in the absence of added P, biochar reduced soil P concentrations, whereas combining biochar with a P source increased available P. Furthermore, when applied together with N fertilizer, biochar enhanced extractable P, suggesting that additional N may be required in biochar-amended soils to optimize crop productivity. Biochar application decreased K leaching by 25% and preserved higher levels of accessible P throughout the growth season in a controlled experiment with vegetable crops (Hardie et al. 2015). Responses differ, too, as biochar can occasionally immobilize specific micronutrients in high-pH soils, decreasing their availability for plant uptake (Ullah et al. 2022). Combining biochar with either organic or inorganic fertilizers frequently results in synergistic effects, where fertilizers meet immediate nutrient needs while charcoal reduces nutrient losses (Partey et al. 2014).

2.4 Water Holding Capacity

Moisture is an important an integral part of the soil environment, the amount of which depends on soil texture and rainfall intensity. Biochar's sponge-like and porous structure helps soils hold onto more water, which is particularly beneficial in drought-prone regions and sandy soils (Khan et al. 2024). Both the direct water storage in the pores of biochar and its enhancement of soil aggregation and porosity lead to improvements in WHC (Sun and Lu 2014). Soil WHC is influenced by the microporous structure of biochar. Studies have shown that biochar can improve WHC under field conditions. Verheijen et al. (2019) observed that incorporating biochar lowered soil bulk density (SBD) and improved WHC, with the effects being more pronounced in sandy soils compared to loamy soils. Similarly, Brockhoff et al. (2010) revealed that incorporation of 5% and 25% biochar raised water retention in sandy soils by 260% and 370%, respectively, relative to untreated controls. The resulting improvement in crop productivity was likely linked to higher soil water content (SWC) (Colantoni et al. 2016). However, field applications of biochar at 20 t ha^{-1} did not lead to a substantial rise in crop yield. Seemingly, Novak et al. (2009) demonstrated incorporating 2% biochar derived from different switchgrass feedstocks into loamy sand soil improved its WHC. They explored the impact of feedstock type and pyrolysis conditions, though all assessments were conducted at the same 2% application rate. When the retort temperature was varied between 250 and 750 °C, WHC increased by 7–16%. An additional observation indicated an 11% improvement in water retention, although this was not confirmed through appropriate control methods (Karhu et al. 2011). Such improvements in WHC emphasize the effectiveness of biochar in water-limited environments (Karhu et al. 2011). Also by keeping nutrient-rich soil solution in the root zone, improved WHC not only promotes plant development under water stress but also lowers watering requirements and improves nutrient uptake (Cheraghi et al. 2023). However, in heavy clay soils, where water retention is already high and excessive moisture could lead to anaerobic conditions, the benefits of WHC may be less noticeable (Abdallah et al. 2021).

2.5 Physical Structure and Bulk Density

Biochar often improves soil physical structure by reducing SBD and increasing porosity, facilitating root penetration and enhancing soil aeration (Blanco-Canqui 2017). Biochar particles can improve aggregation and moisture retention in sandy soils by filling in the spaces between sand grains, whereas they can break up compacted layers and lessen compaction in clay soils (Li et al. 2021). Particle size and application rate have been found to affect the degree of improvement, with studies demonstrating 3–31% decreases in BD and 14–64% increases in total porosity following the incorporation of biochar (Blanco-Canqui 2017). More workability and erosion resistance can be achieved by improving soil consistency, which includes raising plastic and liquid limitations. Additionally, the low-density, light nature of biochar makes tillage simpler and lowers the energy required for soil preparation (Blackwell et al. 2012). Applying high rates of finely powdered biochar may increase the risk of wind erosion in very loose soils unless it is combined with other organic amendments (Shackley et al. 2010). Quin et al. (2014) observed that biochar derived from woody residues had a stronger effect in reducing SBD in coarse-textured soils compared to soils with higher clay content. Similarly, Laird et al. (2010), who observed substantial reductions in BD in biochar-amended soils after 500 days, even when applied at relatively low rates. Figure 2 illustrates the influence of biochar type and application rate on SBD across soil types.

2.6 Soil Microbiota

Biochar's influence on soil microbiota is complex, serving as a habitat and a chemical condition modifier (Lehmann et al. 2011). Microorganisms can find sanctuary in its extensive surface area and interconnected pores, which shields them from desiccation and predation (Arshad 2024). Biochar indirectly promotes the development and diversity of microorganisms by enhancing pH, nutrient retention, and WHC (Dai et al. 2021). Following biochar application, multiple studies have documented enhanced bacterial and fungal diversity, along with greater populations of beneficial

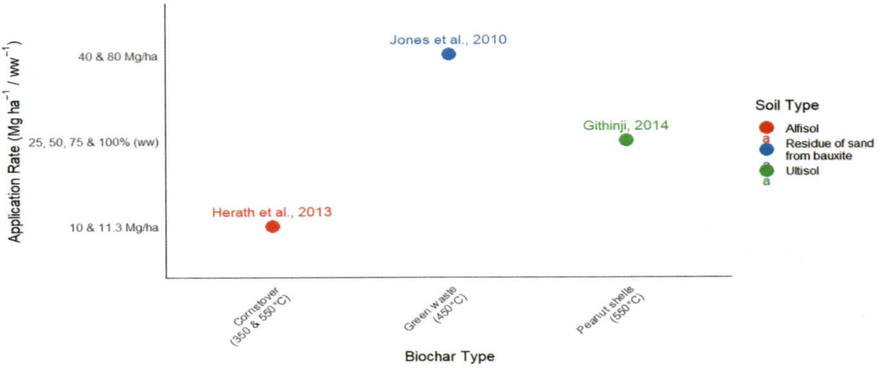

Fig. 2 Biochar type and application rate influences SBD across different soil types

microorganisms such as mycorrhizal fungi and N-fixing bacteria (Dzvene and Chiduza 2024). Additionally, biochar can decrease soil toxicity and absorb allelopathic chemicals, which increases microbial activity (Sujeeun and Thomas 2023). Not all effects, though, are favorable; some studies have revealed short-term decreases in microbial biomass or a shift toward slower nutrient mineralization, particularly when biochar is applied at very high rates or is rich in aromatic C that decomposes slowly (Asirifi et al. 2021). In a study, Xu et al. (2023) indicated that biochar exerts only limited effects on most microbial phyla, with notable impacts observed primarily on *Acidobacteria* and *Gemmatimonadetes*. These changes were strongly associated with the intrinsic properties of the biochar and the consequent alterations in soil conditions, indicating that biochar's effect on soil microbial diversity can be modulated by its characteristics. Additionally, several studies have reported that biochar application increases bacterial community diversity. This effect may be attributed to the additional C-sources and addition of ash provided by biochar, which can serve as substrates supporting the proliferation of diverse bacterial phyla and species (Thies et al. 2015; Ibrahim et al. 2021).

2.7 Carbon Sequestration

Soil C sequestration is the process of capturing and storing atmospheric carbon dioxide (CO_2) in the soil in the form of SOC and soil inorganic carbon (SIC) (Kumar et al. 2020). This process not only helps mitigate CC by reducing GHGs in the atmosphere but also improves soil health, enhancing fertility, water retention, and structure. Biochar amendment to the soil not only enhances the fertility and vitality of the soil, but also sequesters the C. Recycling biomass into biochar is regarded as one of the simplest and most effective approaches for C biosequestration (Mathews 2008). Applying biochar has a special and long-term advantage in that it helps sequester C in the soil. Biochar can remain in soils for decades to centuries because it is primarily made up of stable, aromatic C compounds that are impervious to microbial breakdown (Chen et al. 2019). This stability helps balance GHGs emissions from agriculture and lowers CO_2 emissions from decomposing organic materials. Biochar has the ability to promote the development of organo-mineral complexes, which will further stabilize the organic C in both native and biochar-derived soil (Kelly et al. 2017). Following the application of biochar, long-term experiments conducted in temperate and tropical locations have demonstrated consistent increases in SOC stocks as well as decreased emissions of CO_2, methane (CH_4), and N_2O (Van Zwieten et al. 2015; Guenet et al. 2021). As shown in Table 1, biochar enhanced soil C or dissolved organic carbon (DOC) storage, thereby sequestering C in the soil and contributing to reduced GHGs emissions. Nevertheless, the permanence of C storage depends on environmental conditions, biochar type, and soil management practices (Lal 2013). Improperly produced biochar, or application to erosive soils, may limit sequestration benefits.

3 Impact on Crop Yield & Nutrient Use Efficiency

3.1 Yield and Nutrient Use Efficiency

Biochar has been widely reported to enhance soil fertility and promote crop performance, although its effects vary depending on crop type and soil characteristics. For instance, application of biochar at 68 t ha^{-1} increased biomass production in rice (*Oryza sativa* L.) and cowpea (*Vigna unguiculata* (L.) Walp.) by 20% and 50%, respectively, while a higher dose of 136.75 t ha^{-1} doubled cowpea biomass (Glaser et al. 2002). Similarly, biochar incorporation improved both biomass accumulation and grain yield in durum wheat (*Triticum durum* L.) by up to 30%, though no significant change was observed in grain N content (Vaccari et al. 2011). In Ghana, maize (*Zea mays* L.) cultivated on charcoal-amended soils recorded a 91% increase in grain yield and a 44% rise in biomass compared to neighboring non-amended fields (Oguntunde et al. 2004). Likewise, in degraded soils in Kenya, Eucalyptus-derived biochar application led to a twofold increase in maize yield (Kimetu et al. 2008). In Laos, biochar application produced variable yet promising outcomes. For example, biochar significantly improved grain yield of upland rice in soils with low P availability, illustrating its potential to optimize P-use efficiency. Conversely, in N-deficient soils, biochar addition reduced leaf chlorophyll content, suggesting that without supplementary N, biochar may adversely affect crop growth (Nelson et al. 2011). These results indicate that the benefits of biochar are context-specific, relying on existing soil nutrient status and fertilizer management practices. Globally, numerous studies have reinforced biochar's capacity to increase crop yields in nutrient-depleted soils. Van Zwieten et al. (2010) reported improved wheat productivity on highly weathered Australian soils, whereas Zhang et al. (2012) observed similar benefits in maize production in China. Collectively, these findings highlight biochar as a valuable soil amendment, particularly in degraded or nutrient-limited systems, provided its application aligns with the soil's specific nutrient constraints. The positive effects of biochar on crop performance are also reflected in its role in

Table 1 Biochar-mediated C-sequestration

Feedstock	Soil type	Experiment type	Biochar rate (t/ha)	N fertilization (kg/ha)	Time (days)	Soil carbon/DOC	CO_2	CH_4	N_2O	References
Corn stalk	Paddy inceptisol	Pot	24	250	116	Soil DOC increased	–	Decreased by 63%	–	Feng et al. (2012)
Corn stalk	Paddy ultisol	Pot	24	250	116	Soil DOC increased	–	Decreased by 61%	–	
Wheat straw	Rice paddy	Field	40	300	150	–	–	Increased by 34%	Decreased by 40–51%	Zhang et al. (2010)
Wheat straw	Rice paddy	Field	40	0	150	–	–	Increased by 41%	Decreased by 21–28%	
Wood	Wheat soil	Field	30	222	420	–	No difference	No difference	Decreased by 26–76%	Castaldi et al. (2011)
Wood	Wheat soil	Field	60	222	420	–	No difference	No difference	Decreased by 59–88%	
Sawdust	Corn field	Pot	2–60%	–	100	–	Decreased	Decreased ≥20%	Decreased	Spokas and Reicosky (2009)
Oak	Sand	Pot	10% (4.6 g/kg)	–	365	Carbon lost in 100 years: 18%	–	–	–	Zimmerman (2010)
Pine	Sand	Pot	10% (4.6 g/kg)	–	365	Carbon lost in 100 years: 14%	–	–	–	
Cedar	Sand	Pot	10% (4.6 g/kg)	–	365	Carbon lost in 100 years: 7%	–	–	–	
Bubinga	Sand	Pot	10% (4.6 g/kg)	–	365	Carbon lost in 100 years: 8%	–	–	–	
Gamma	Sand	Pot	10% (4.6 g/kg)	–	365	Carbon lost in 100 years: 27%	–	–	–	
Sugar	Sand	Pot	10% (4.6 g/kg)	–	365	Carbon lost in 100 years: 11%	–	–	–	
Crop waste	Paddy	LCA	–	–	–	–	Decreased by 50–80%	–	–	Gaunt and Lehmann (2008)

improving NUE. Its highly porous structure and extensive surface area allow biochar to retain nutrients such as NH_4^+, NO_3^- and PO_4^{3-}, minimizing leaching and prolonging their availability for plant uptake (Wang et al. 2015). As a potent organic amendment, biochar can significantly enhance NUE in crops like rice (Oladele et al. 2019). For instance, Haque et al. (2022) reported that the agronomic efficiency (AE) of N, P, and K improved with the application of RHB combined with 25% oil palm empty fruit bunch biochar (EFBB). However, these benefits depend on factors such as soil nutrient status, fertilizer input, plant nutrient absorption, and yield response (Nagabovanalli Basavarajappa et al. 2021). Similarly, Lee et al. (2021) found that combining biochar with NPK fertilizer enhanced N, P, and K utilization, while Arif et al. (2017) demonstrated that biochar improved P-use efficiency (Fig. 3).

3.2 Combined Amendments

Combining biochar with other soil amendments can further optimize soil fertility and boost crop performance. In acidic Ethiopian soils, the co-application of biochar with lime or inorganic fertilizers enhanced soil chemical properties and maize grain yield, sometimes exhibiting similar effectiveness across different amendment combinations (Lulu et al. 2022; Takala et al. 2025). These observations highlight biochar's capacity to function synergistically with other amendments, particularly in soils constrained by low pH or limited nutrient availability. In a similar vein, Sasmita et al. (2017) reported that biochar, applied either alone or alongside inorganic

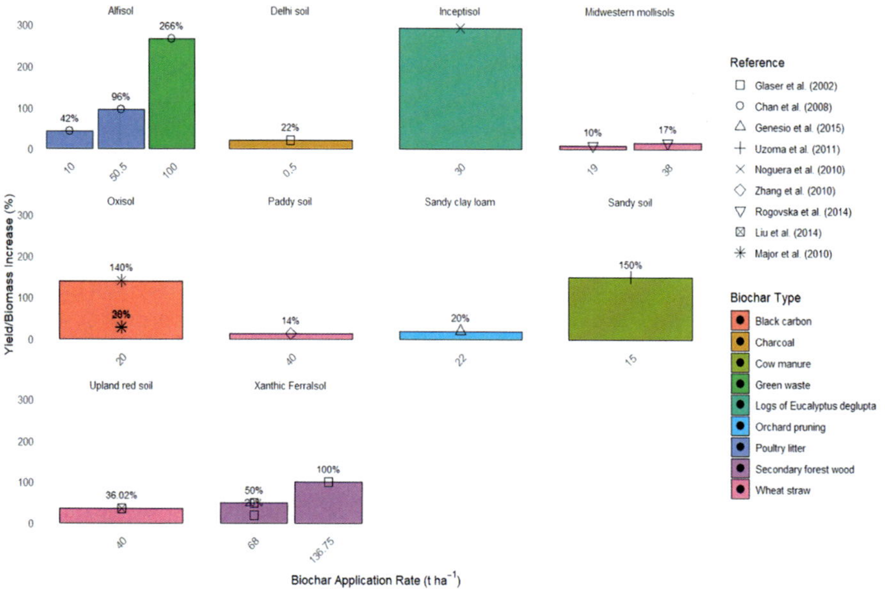

Fig. 3 Biochar-mediated changes in crop performance

fertilizers, elevated soil-available P in acidic soils, demonstrating a synergistic interaction that promotes P availability. Syuhada et al. (2016) observed that soil levels of K, Ca and Mg responded significantly to biochar dosage, regardless of whether it was applied individually or in combination with inorganic fertilizers. Likewise, Hussien Ibrahim et al. (2020) found that sorghum plant height increased by 20.1% and 13.7% under 5% and 10% biochar treatments, respectively, compared to untreated soils. Moreover, Bai et al. (2022) showed that the combined use of biochar with either inorganic or organic fertilizers in acidic soils substantially enhanced crop productivity compared to the use of biochar or fertilizers alone. Evidence from meta-analyses supports this synergistic effect: biochar application alone has been shown to increase crop yield by approximately 25.3%, while co-application with inorganic fertilizers contributed an additional 10% improvement. Other reviews have reported yield enhancements of around 10% following biochar addition (Biederman and Harpole 2013; Jeffery et al. 2011). More recent findings suggest even greater benefits, with Rivelli and Libutti (2022) reporting notable yield gains when biochar was co-applied with inorganic fertilizers during the first year of application.

3.3 Environmental Benefits

In recent decades, increasing attention has been directed toward the substantial GHG emissions associated with agriculture. Biochar has emerged as a promising approach for mitigating these emissions, particularly CO_2, CH_4 and N_2O (Kammann et al. 2017). Its efficacy is largely attributed to its extensive surface area, which facilitates gas adsorption and reduces their release into the atmosphere (Sajjadi et al. 2019). For instance, Wang et al. (2011) reported that biochar application to agricultural soils significantly decreased N_2O emissions, although no substantial impact on CO_2 emissions was observed. Additionally, CH_4 emissions from livestock manure were reduced by up to 70% following biochar incorporation (Mukherjee et al. 2014). A meta-analysis of 14 studies by Cayuela et al. (2014) further indicated an average 49% reduction in N_2O emissions after biochar amendment, while Clough et al. (2013) observed a 70–80% decrease in N_2O emissions in maize cultivation systems. In paddy soils, Qi et al. (2021) found that biochar suppressed methanogenic archaea populations, with CH_4 mitigation primarily resulting from inhibited methanogenesis, enhanced methanotrophic activity, and improved soil porosity (Nan et al. 2020). Organic contaminants from industrial, agricultural, and urban sources present serious threats to ecosystems and human health. Biochar has demonstrated considerable potential for reducing these pollutants through adsorption and catalytic degradation mechanisms. For example, Chen et al. (2018) showed that rice straw-derived biochar effectively decreased concentrations of polycyclic aromatic hydrocarbons (PAHs) in water. Beyond adsorption, biochar improves water retention, nutrient availability, and microbial activity, creating favorable conditions for microbe-mediated organic matter degradation (Mukherjee et al. 2014). Huong et al. (2020) reported that biochar-activated peroxymonosulfate markedly enhanced the breakdown of bisphenol A (BPA) and tetracycline in wastewater, with efficiency increasing at higher biochar dosages. Similarly, the degradation of the insecticide imidacloprid, which is toxic to human and animal cardiorespiratory and nervous systems, was attributed to biochar-mediated hydroxyl radical generation (Huang et al. 2023). A comprehensive review by Gasim et al. (2022) highlighted that doping biochar with heteroatoms such as N, S, and P increases its active sites, thereby improving its adsorption capacity. This observation is consistent with findings by Miao et al. (2023), who attributed the enhanced catalytic activity of doped biochars to their improved electron-donating properties. Such improvements were particularly evident in tetracycline degradation (Gasim et al. 2022). Additionally, the aromatic structure of pyrolyzed C was identified as a key factor contributing to pollutant degradation, by both strengthening catalytic activity through π–π^* interactions and providing resilience against unwanted oxidative processes (Gasim et al. 2022).

4 Mechanisms Underpinning Biochar's Effects

Biochar influences soil properties and crop productivity through a series of interconnected physical, chemical, and biological mechanisms.

4.1 pH Buffering Via Alkaline Biochars

One of the main ways in which biochar improves soil quality is through its capacity to buffer soil acidity (Dai et al. 2014). Many biochars, especially those produced from crop residues, livestock manure, or woody biomass, exhibit alkaline characteristics due to their high ash content, which is rich in carbonates, oxides, and hydroxides of base cations such as Ca, Mg, K, and Na (Kuryntseva et al. 2023). When incorporated into acidic soils, biochar neutralizes surplus hydrogen ions, thereby elevating soil pH toward a range more conducive to nutrient availability and plant growth (Bedassa 2020). This pH adjustment not only enhances the accessibility of key macronutrients, including phosphorus, but also reduces the mobility and toxicity of heavy metals and aluminum (Karim et al. 2022). Consequently, biochar acts as a natural soil amendment, mitigating acidity issues and fostering a more balanced chemical environment for both plant roots and soil microbial communities (Tusar et al. 2023).

4.2 Cation Exchange Capacity Enhancement

Biochar additionally enhances soil CEC, which reflects the soil's ability to retain and exchange positively charged nutrient ions such as K^+, Ca^{2+} and Mg^{2+} (Adekiya et al. 2025). This improvement primarily results from biochar's surface functional groups, including carboxyl and hydroxyl moieties, which form during pyrolysis and are further modified through oxidation upon soil interaction. These negatively charged sites on biochar surfaces increase nutrient retention, thereby limiting nutrient leaching and improving NUE (Xiao et al. 2018). Elevated CEC ensures that essential nutrients remain available within the root zone for extended periods, allowing crops to access them as needed. Over time, biochar's influence on CEC strengthens as it undergoes aging and oxidation processes, reinforcing its role in maintaining long-term soil fertility (Wang et al. 2020).

4.3 Porosity and WHC

Another significant advantage of biochar is its highly porous architecture, which enhances soil physical characteristics such as porosity and WHC (Hardie et al. 2014). Biochar particles possess a complex network of micro- and mesopores formed during pyrolysis, providing additional space for retaining water and dissolved nutrients (Wong and Ogbonnaya 2021). When incorporated into coarse-textured or sandy soils, biochar promotes soil aggregation, decreases SBD, and improves water infiltration, thereby boosting soil moisture retention (Obia et al. 2016). Figure 4 illustrates studies examining the influence of different types and application rates of biochar on soil porosity across various soil textures. These physical enhancements are especially advantageous in drought-prone regions, where maintaining soil moisture is critical for crop survival and productivity. By serving as a water reservoir, biochar buffers plants against fluctuations in water availability, alleviating stress during dry periods and supporting consistent growth (Gharred et al. 2022).

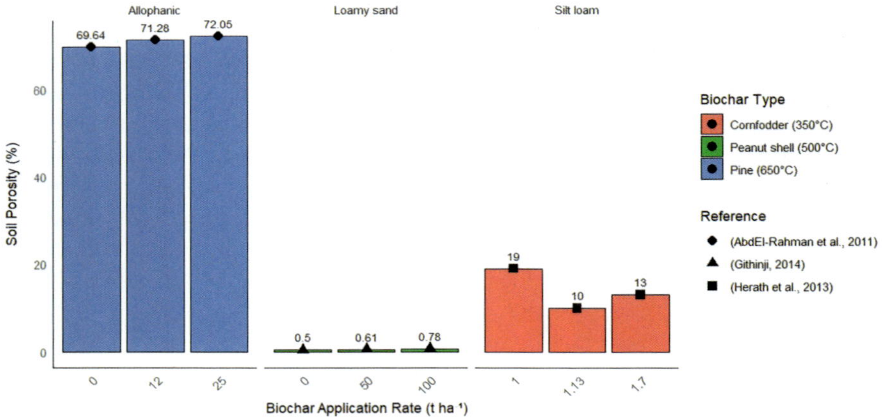

Fig. 4 Effects of different types and application rates of biochar on soil porosity across various soil types

4.4 Nutrient Adsorption and Microbial Facilitation

Biochar also exerts a dual influence on soil nutrient dynamics through nutrient adsorption and the promotion of microbial activity. Its extensive surface area and diverse functional groups enable biochar to retain nutrients such as NH_4^+, NO_3^-, and PO_4^{3-}, thereby minimizing losses via leaching and volatilization (Hossain et al. 2020). Additionally, biochar provides a physically and chemically favorable habitat for soil microorganisms. Nutrients retained on biochar surfaces become more accessible to microbes, which in turn drive essential processes including nutrient cycling, SOM decomposition, and overall soil fertility improvement (Pan et al. 2021). For instance, microbial activity on biochar can accelerate OM mineralization, releasing nutrients in forms readily available to plants (Das et al. 2022). This synergistic interaction between biochar's physicochemical properties and microbial functions enhances soil nutrient dynamics and plant nutrition. The porous architecture of biochar creates protected microenvironments, allowing microbial communities to establish and thrive while being shielded from predators and desiccation (Jaafar et al. 2015). These microhabitats promote microbial diversity, increase colonization by beneficial organisms, and enhance microbial-mediated processes such as N-fixation, P solubilization, and enzyme activity (Kobua et al. 2025). By supporting the proliferation of advantageous microbes, biochar indirectly stimulates plant growth through improved nutrient availability and suppression of soil-borne pathogens (Abid et al. 2023). Furthermore, biochar's pH buffering capacity and its role in stabilizing SOM create conditions favorable for microbial activity, making it a key amendment for sustaining healthy soil biological function (Fig. 5).

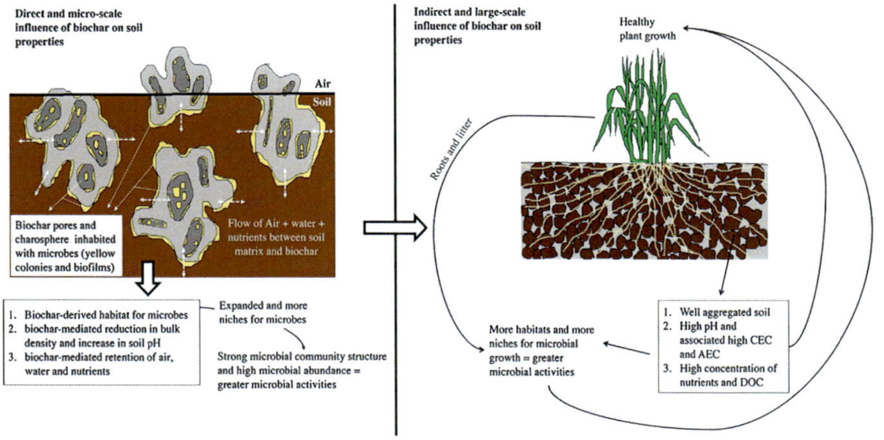

Fig. 5 A conceptual model depicting how biochar directly affects microbial activity at the micro-scale and indirectly influences larger-scale processes by modifying soil properties, enhancing habitat availability, and expanding ecological niches

4.5 Carbon Stability and Long-Term Sequestration

In contrast to many organic amendments that decompose rapidly, biochar exhibits exceptional stability in soils due to its aromatic C structure formed during high-temperature pyrolysis (Singh et al. 2012). This recalcitrant form of C resists microbial breakdown, enabling biochar to remain in the soil for decades to centuries. Consequently, biochar not only provides immediate improvements in soil fertility but also functions as a long-term C reservoir, helping mitigate CC by sequestering atmospheric CO_2 in stable forms (Lorenz and Lal 2014). Incorporation of biochar further enhances SOC stocks and can reduce GHG emissions, such as N_2O and CH_4, by influencing soil microbial activity (Abagandura et al. 2019). This dual function of improving soil quality while contributing to CC mitigation underscores biochar's potential as a sustainable soil management strategy.

5 Conclusion

In conclusion, biochar has emerged as a highly effective soil amendment, capable of improving chemical properties, enhancing nutrient dynamics, and supporting yield increases across diverse crops and agroecosystems. Its capacity to raise soil pH, boost CEC, retain essential nutrients, and contribute stable OC directly strengthens soil fertility and NUE, while simultaneously alleviating issues such as soil acidity, nutrient losses, and heavy metal toxicity. These improvements are manifested in higher crop productivity, superior produce quality, and increased resilience of farming systems to climate variability and environmental stresses. Beyond benefits to soil and crops, biochar advances environmental sustainability by sequestering C, lowering GHG emissions, and reducing reliance on synthetic fertilizers, positioning it as a strategic component of climate-smart agriculture. However, the magnitude of biochar's agronomic and economic benefits is influenced by feedstock type, pyrolysis conditions, soil properties, and application rates, emphasizing the need for site-specific management strategies. Future research should prioritize cost-effective biochar production methods, the integration of biochar with organic and inorganic amendments, and long-term field trials to elucidate its cumulative impacts on soil health, productivity, and farm profitability. With adequate research, policy support, and farmer engagement, biochar holds substantial potential as a key tool for sustainable soil management, food security and CC mitigation.

References

Abagandura GO, Chintala R, Sandhu SS, Kumar S, Schumacher TE (2019) Effects of biochar and manure applications on soil carbon dioxide, methane, and nitrous oxide fluxes from two different soils. J Environ Qual 48(6):1664–1674. https://doi.org/10.2134/jeq2018.10.0374

Abdallah AM, Jat HS, Choudhary M, Abdelaty EF, Sharma PC, Jat ML (2021) Conservation agriculture effects on soil WHC and water-saving varied with management practices and

agroecological conditions: A review. Agronomy 11(9):1681. https://doi.org/10.3390/agronomy11091681
Abid F, Naz R, Asif T (2023) Biochar for crop protection from soil borne diseases. In: Sustainable agriculture reviews 61: biochar to improve crop production and decrease plant stress under a changing climate. Springer International Publishing, Cham, pp 231–246. https://doi.org/10.1007/978-3-031-26983-7_10
Adekiya AO, Ogunbode TO, Esan VI, Adedokun O, Olatubi IV, Ayegboyin MH (2025) Short term effects of biochar on soil chemical properties, growth, yield, quality, and shelf life of tomato. Sci Rep 15(1):24965. https://doi.org/10.1038/s41598-025-10411-5
Adhikari S, Moon E, Timms W (2024) Identifying biochar production variables to maximise exchangeable cations and increase nutrient availability in soils. J Clean Prod 446:141454. https://doi.org/10.1016/j.jclepro.2024.141454
Arif M, Ilyas M, Riaz M, Ali K, Shah K, Haq IU, Fahad S (2017) Biochar improves phosphorus use efficiency of organic-inorganic fertilizers, maize-wheat productivity and soil quality in a low fertility alkaline soil. Field Crop Res 214:25–37. https://doi.org/10.1016/j.fcr.2017.08.018
Arshad U (2024) Biochar application: a sustainable approach for mitigating biotic and abiotic stresses in plants. Integr Plant Biotechnol 2:77–98. https://doi.org/10.55627/pbiotech.002.01.0972
Arwenyo B, Varco JJ, Dygert A, Brown S, Pittman CU, Mlsna T (2023) Contribution of modified P-enriched biochar on pH buffering capacity of acidic soil. J Environ Manag 339:117863. https://doi.org/10.1016/j.jenvman.2023.117863
Asio VB, Asio LG (2021) Properties, fertility status, and sustainable management of degraded tropical soils. Sustainable soil fertility management. Nova Science Publishers Inc., New York, pp 1–31
Asirifi I, Werner S, Heinze S, Saba CK, Lawson IY, Marschner B (2021) Short-term effect of biochar on microbial biomass, respiration and enzymatic activities in wastewater irrigated soils in urban agroecosystems of the West African savannah. Agronomy 11(2):271. https://doi.org/10.3390/agronomy11020271
Bai SH, Omidvar N, Gallart M, Kämper W, Tahmasbian I, Farrar MB et al (2022) Combined effects of biochar and fertilizer applications on yield: a review and meta-analysis. Sci Total Environ 808:152073. https://doi.org/10.1016/j.scitotenv.2021.152073
Bakshi S, Banik C, Laird DA (2020) Estimating the organic oxygen content of biochar. Sci Rep 10(1):13082. https://doi.org/10.1038/s41598-020-69798-y
Bedassa M (2020) Soil acid management using biochar: review. Int J Agric Sci Food Technol 6:211–217. https://doi.org/10.17352/2455-815X.000076
Biederman LA, Harpole WS (2013) Biochar and its effects on plant productivity and nutrient cycling: a meta-analysis. GCB Bioenergy 5(2):202–214. https://doi.org/10.1111/gcbb.12037
Blackwell P, Riethmuller G, Collins M (2012) Biochar application to soil. In: Biochar for environmental management. Routledge, pp 239–258
Blanco-Canqui H (2017) Biochar and soil physical properties. Soil Sci Soc Am J 81(4):687–711. https://doi.org/10.2136/sssaj2017.01.0017
Brockhoff SR, Christians NE, Killorn RJ, Horton R, Davis DD (2010) Physical and mineral-nutrition properties of sand-based turfgrass root zones amended with biochar. Agron J 102(6):1627–1631. https://doi.org/10.2134/agronj2010.0188
Castaldi S, Riondino M, Baronti S, Esposito FR, Marzaioli R, Rutigliano FA et al (2011) Impact of biochar application to a Mediterranean wheat crop on soil microbial activity and greenhouse gas fluxes. Chemosphere 85(9):1464–1471. https://doi.org/10.1016/j.chemosphere.2011.08.031
Cayuela ML, Van Zwieten L, Singh BP, Jeffery S, Roig A, Sánchez-Monedero MA (2014) Biochar's role in mitigating soil nitrous oxide emissions: a review and meta-analysis. Agric Ecosyst Environ 191:5–16. https://doi.org/10.1016/j.agee.2013.10.009
Ch'ng HY, Haruna AO, Majid NMNA, Jalloh MB (2019) Improving soil phosphorus availability and yield of *Zea mays* L. using biochar and compost derived from agro-industrial wastes. Ital J Agron 14(1):1107. https://doi.org/10.4081/ija.2019.1107

Chen D, Wang C, Shen J, Li Y, Wu J (2018) Response of CH_4 emissions to straw and biochar applications in double-rice cropping systems: insights from observations and modeling. Environ Pollut 235:95–103. https://doi.org/10.1016/j.envpol.2017.12.041

Chen W, Meng J, Han X, Lan Y, Zhang W (2019) Past, present, and future of biochar. Biochar 1(1):75–87. https://doi.org/10.1007/s42773-019-00008-3

Cheraghi M, Mousavi SM, Zarebanadkouki M (2023) Functions of rhizosheath on facilitating the uptake of water and nutrients under drought stress: a review. Plant Soil 491(1):239–263. https://doi.org/10.1007/s11104-023-06126-z

Chintala R, Mollinedo J, Schumacher TE, Malo DD, Julson JL (2014) Effect of biochar on chemical properties of acidic soil. Arch Agron Soil Sci 60(3):393–404. https://doi.org/10.1080/03650340.2013.789870

Clough TJ, Condron LM, Kammann C, Müller C (2013) A review of biochar and soil nitrogen dynamics. Agronomy 3(2):275–293. https://doi.org/10.3390/agronomy3020275

Colantoni A, Evic N, Lord R, Retschitzegger S, Proto AR, Gallucci F, Monarca D (2016) Characterization of biochars produced from pyrolysis of pelletized agricultural residues. Renew Sust Energ Rev 64:187–194. https://doi.org/10.1016/j.rser.2016.06.003

Cornelissen G, Martinsen V, Shitumbanuma V, Alling V, Breedveld GD, Rutherford DW et al (2013) Biochar effect on maize yield and soil characteristics in five conservation farming sites in Zambia. Agronomy 3(2):256–274. https://doi.org/10.3390/agronomy3020256

Dai Z, Wang Y, Muhammad N, Yu X, Xiao K, Meng J et al (2014) The effects and mechanisms of soil acidity changes, following incorporation of biochars in three soils differing in initial pH. Soil Sci Soc Am J 78(5):1606–1614. https://doi.org/10.2136/sssaj2013.08.0340

Dai Z, Xiong X, Zhu H, Xu H, Leng P, Li J et al (2021) Association of biochar properties with changes in soil bacterial, fungal and fauna communities and nutrient cycling processes. Biochar 3(3):239–254. https://doi.org/10.1007/s42773-021-00099-x

Das SK, Ghosh GK, Avasthe R (2022) Valorizing biomass to engineered biochar and its impact on soil, plant, water, and microbial dynamics: a review. Biomass Convers Biorefinery 12(9):4183–4199. https://doi.org/10.1007/s13399-020-00836-5

Dzvene AR, Chiduza C (2024) Application of biofertilizers for enhancing beneficial microbiomes in push–pull cropping systems: a review. Bacteria 3(4):271–286. https://doi.org/10.3390/bacteria3040018

Eduah JO, Nartey EK, Abekoe MK, Henriksen SW, Andersen MN (2020) Mechanism of orthophosphate (PO4-P) adsorption onto different biochars. Environ Technol Innovation 17:100572. https://doi.org/10.1016/j.eti.2019.100572

Farhangi-Abriz S, Torabian S, Qin R, Noulas C, Lu Y, Gao S (2021) Biochar effects on yield of cereal and legume crops using meta-analysis. Sci Total Environ 775:145869. https://doi.org/10.1016/j.scitotenv.2021.145869

Feng Y, Xu Y, Yu Y, Xie Z, Lin X (2012) Mechanisms of biochar decreasing methane emission from Chinese paddy soils. Soil Biol Biochem 46:80–88. https://doi.org/10.1016/j.soilbio.2011.11.016

Fidel RB, Laird DA, Thompson ML, Lawrinenko M (2017) Characterization and quantification of biochar alkalinity. Chemosphere 167:367–373. https://doi.org/10.1016/j.chemosphere.2016.09.151

Fridahl M, Haikola S, Rogers PM, Hansson A (2020) Biochar deployment drivers and barriers in least developed countries. In: Leal Filho W, Luetz J, Ayal D (eds) Handbook of climate change management: research, leadership, transformation. Springer, Cham, pp 1–30. https://doi.org/10.1007/978-3-030-22759-3_324-1

Garg S, Rumjit NP, Arora P, Ahmed S (2025) Biochar-based solutions for sustainable dye treatment and agricultural waste management. J Mater Cycles Waste Manag 27:1–19

Gasim MF, Choong ZY, Koo PL, Low SC, Abdurahman MH, Ho YC et al (2022) Application of biochar as functional material for remediation of organic pollutants in water: an overview. Catalysts 12(2):210. https://doi.org/10.3390/catal12020210

Gaunt JL, Lehmann J (2008) Energy balance and emissions associated with biochar sequestration and pyrolysis bioenergy production. Environ Sci Technol 42(11):4152–4158. https://doi.org/10.1021/es071361i

Gharred J, Derbali W, Derbali I, Badri M, Abdelly C, Slama I, Koyro HW (2022) Impact of biochar application at water shortage on biochemical and physiological processes in *Medicago ciliaris*. Plants 11(18):2411. https://doi.org/10.3390/plants11182411

Glaser B, Lehmann J, Zech W (2002) Ameliorating physical and chemical properties of highly weathered soils in the tropics with charcoal—a review. Biol Fertil Soils 35(4):219–230. https://doi.org/10.1007/s00374-002-0466-4

Guenet B, Gabrielle B, Chenu C, Arrouays D, Balesdent J, Bernoux M et al (2021) Can N_2O emissions offset the benefits from soil organic carbon storage? Glob Chang Biol 27(2):237–256. https://doi.org/10.1111/gcb.15342

Haque ANA, Uddin MK, Sulaiman MF, Amin AM, Hossain M, Solaiman ZM, Mosharrof M (2022) Rice growth performance, nutrient use efficiency and changes in soil properties influenced by biochar under alternate wetting and drying irrigation. Sustainability 14(13):7977. https://doi.org/10.3390/su14137977

Hardie M, Clothier B, Bound S, Oliver G, Close D (2014) Does biochar influence soil physical properties and soil water availability? Plant Soil 376(1):347–361. https://doi.org/10.1007/s11104-013-1980-x

Hardie MA, Oliver G, Clothier BE, Bound SA, Green SA, Close DC (2015) Effect of biochar on nutrient leaching in a young apple orchard. J Environ Qual 44(4):1273–1282. https://doi.org/10.2134/jeq2015.02.0068

He Y, Wu Z, Zhao T, Yang H, Ali W, Chen J (2022) Different plant species exhibit contrasting root traits and penetration to variation in soil bulk density of clayey red soil. Agron J 114(1):867–877. https://doi.org/10.1002/agj2.20972

Hossain MZ, Bahar MM, Sarkar B, Donne SW, Ok YS, Palansooriya KN et al (2020) Biochar and its importance on nutrient dynamics in soil and plant. Biochar 2(4):379–420. https://doi.org/10.1007/s42773-020-00065-z

Huang D, Chen N, Lin Y, Ge C, Wang X, Wang D et al (2023) Pyrogenic carbon accelerates iron cycling and hydroxyl radical production during redox fluctuations of paddy soils. Biochar 5(1):38. https://doi.org/10.1007/s42773-023-00236-8

Huang J, Fisher PR, Argo WR, Jeong KY, Altland J (2025) Macronutrient solubility in response to the pH of soilless container substrates. J Soil Sci Plant Nutr 25:1–17. https://doi.org/10.1007/s42729-025-02331-0

Huong TTT, Van Hoang N, Ngoc Toan V, Tong NX, Quan TA, Thu VK (2020) Initial results of using biochar derived from spent coffee grounds to remove pollutants from livestock wastewater in Vietnam. In: Proceedings of the international conference on innovations for sustainable and responsible mining: ISRM 2020-volume 2. Springer International Publishing, Cham, pp 305–325. https://doi.org/10.1007/978-3-030-60269-7_16

Hussien Ibrahim ME, Adam Ali AY, Zhou G, Ibrahim Elsiddig AM, Zhu G, Ahmed Nimir NE, Ahmad I (2020) Biochar application affects forage sorghum under salinity stress. Chilean J Agric Res 80(3):317–325. https://doi.org/10.4067/S0718-58392020000300317

Ibrahim MM, Zhang H, Guo L, Chen Y, Heiling M, Zhou B, Mao Y (2021) Biochar interaction with chemical fertilizer regulates soil organic carbon mineralization and the abundance of key C-cycling-related bacteria in rhizosphere soil. Eur J Soil Biol 106:103350. https://doi.org/10.1016/j.ejsobi.2021.103350

Ippolito JA, Cui L, Kammann C, Wrage-Mönnig N, Estavillo JM, Fuertes-Mendizabal T et al (2020) Feedstock choice, pyrolysis temperature and type influence biochar characteristics: a comprehensive meta-data analysis review. Biochar 2(4):421–438. https://doi.org/10.1007/s42773-020-00067-x

Jaafar NM, Clode PL, Abbott LK (2015) Soil microbial responses to biochars varying in particle size, surface and pore properties. Pedosphere 25(5):770–780. https://doi.org/10.1016/S1002-0160(15)300588

Jeffery S, Verheijen FG, van der Velde M, Bastos AC (2011) A quantitative review of the effects of biochar application to soils on crop productivity using meta-analysis. Agric Ecosyst Environ 144(1):175–187. https://doi.org/10.1016/j.agee.2011.08.015

Kammann C, Ippolito J, Hagemann N, Borchard N, Cayuela ML, Estavillo JM et al (2017) Biochar as a tool to reduce the agricultural greenhouse-gas burden–knowns, unknowns and future research needs. J Environ Eng Landsc Manag 25(2):114–139. https://doi.org/10.3846/16486897.2017.1319375

Kamyab H, SaberiKamarposhti M, Hashim H, Yusuf M (2024) Carbon dynamics in agricultural greenhouse gas emissions and removals: a comprehensive review. Carbon Lett 34(1):265–289. https://doi.org/10.1007/s42823-023-00647-4

Karhu K, Mattila T, Bergström I, Regina K (2011) Biochar addition to agricultural soil increased CH_4 uptake and WHC–results from a short-term pilot field study. Agric Ecosyst Environ 140(1–2):309–313. https://doi.org/10.1016/j.agee.2010.12.005

Karim AA, Kumar M, Singh E, Kumar A, Kumar S, Ray A, Dhal NK (2022) Enrichment of primary macronutrients in biochar for sustainable agriculture: a review. Crit Rev Environ Sci Technol 52(9):1449–1490. https://doi.org/10.1080/10643389.2020.1859271

Kelly CN, Benjamin J, Calderon FC, Mikha MM, Rutherford DW, Rostad CE (2017) Incorporation of biochar carbon into stable soil aggregates: the role of clay mineralogy and other soil characteristics. Pedosphere 27(4):694–704. https://doi.org/10.1016/S1002-0160(17)60399-0

Khan MT, Aleinikovienė J, Butkevičienė LM (2024) Innovative organic fertilizers and cover crops: perspectives for sustainable agriculture in the era of climate change and organic agriculture. Agronomy 14(12):2871. https://doi.org/10.3390/agronomy14122871

Kimetu JM, Lehmann J, Ngoze SO, Mugendi DN, Kinyangi JM, Riha S et al (2008) Reversibility of soil productivity decline with organic matter of differing quality along a degradation gradient. Ecosystems 11(5):726–739. https://doi.org/10.1007/s10021-008-9154-z

Kobua CK, Wang YM, Jou YT (2025) Exploring the roles of plant growth-promoting rhizobacteria (PGPR) and alternate wetting and drying (AWD) in sustainable Rice cultivation. Soil Systems 9(2):61. https://doi.org/10.3390/soilsystems9020061

Kumar SS, Mahale AG, Patil AC (2020) Mitigation of climate change through approached agriculture-soil carbon sequestration (a review). Curr J Appl Sci Technol 39(33):47–64. https://doi.org/10.9734/CJAST/2020/v39i3331017

Kumar SS, Wani OA, Mir SA, Babu S, Sharma V, Chesti MUH et al (2022) Soil carbon dynamics in the temperate Himalayas: impact of land use management. Front Environ Sci 10:1009660. https://doi.org/10.3389/fenvs.2022.1009660

Kumar SS, Wani OA, Malik AR, Kumar S, Patel R (2025) Biochar for food security and environmental sustainability under current CC scenario. In: Agricultural diversification for sustainable food production. Springer Nature Singapore, Singapore, pp 59–88. https://doi.org/10.1007/978-981-97-7517-0_3

Kuryntseva P, Karamova K, Galitskaya P, Selivanovskaya S, Evtugyn G (2023) Biochar functions in soil depending on feedstock and pyrolyzation properties with particular emphasis on biological properties. Agriculture 13(10):2003. https://doi.org/10.3390/agriculture13102003

Laird DA, Fleming P, Davis DD, Horton R, Wang B, Karlen DL (2010) Impact of biochar amendments on the quality of a typical midwestern agricultural soil. Geoderma 158(3–4):443–449. https://doi.org/10.1016/j.geoderma.2010.05.013

Lal R (2013) Soil carbon management and climate change. Carbon Management 4(4):439–462. https://doi.org/10.4155/cmt.13.31

Lee YL, Ahmed OH, Wahid SA, Aziz AB, Z. F. (2021) Biochar tablets with and without embedded fertilizer on the soil chemical characteristics and nutrient use efficiency of *Zea mays*. Sustainability 13(9):4878. https://doi.org/10.3390/su13094878

Lehmann J, Rillig MC, Thies J, Masiello CA, Hockaday WC, Crowley D (2011) Biochar effects on soil biota—a review. Soil Biol Biochem 43(9):1812–1836. https://doi.org/10.1016/j.soilbio.2011.04.022

Leng L, Xiong Q, Yang L, Li H, Zhou Y, Zhang W, Huang H (2021) An overview on engineering the surface area and porosity of biochar. Sci Total Environ 763:144204. https://doi.org/10.1016/j.scitotenv.2020.144204

Li L, Zhang YJ, Novak A, Yang Y, Wang J (2021) Role of biochar in improving sandy soil water retention and resilience to drought. Water 13(4):407. https://doi.org/10.3390/w13040407

Liu Y, Li H, Hu T, Mahmoud A, Li J, Zhu R, Jing P (2022) A quantitative review of the effects of biochar application on rice yield and nitrogen use efficiency in paddy fields: A meta-analysis. Sci Total Environ 830:154792. https://doi.org/10.1016/j.scitotenv.2022.154792

López-Cano I, Cayuela ML, Mondini C, Takaya CA, Ross AB, Sánchez-Monedero MA (2018) Suitability of different agricultural and urban organic wastes as feedstocks for the production of biochar—part 1: physicochemical characterisation. Sustainability 10(7):2265. https://doi.org/10.3390/su10072265

Lorenz K, Lal R (2014) Biochar application to soil for climate change mitigation by soil organic carbon sequestration. J Plant Nutr Soil Sci 177(5):651–670. https://doi.org/10.1002/jpln.201400058

Lulu M, Lemma B, Hidoto L, Melese A (2022) Effect of biochar, farmyard manure, and lime on soil properties, and on growth and nutrient uptake of wheat on acidic soils in Southern Ethiopia. S Afr J Plant Soil 39(2):112–122

Mahala DM, Maheshwari HS, Yadav RK, Prabina BJ, Bharti A, Reddy KK, Ramesh A (2020) Microbial transformation of nutrients in soil: an overview. In: Rhizosphere microbes: soil and plant functions. Springer, pp 175–211. https://doi.org/10.1007/978-981-15-9154-9_7

Mathews JA (2008) Carbon-negative biofuels. Energy Policy 36(3):940–945. https://doi.org/10.1016/j.enpol.2007.11.029

Mesele SA, Soremi PA, Ayamba BE, Aderolu J, Olayiwola VA (2024) Organic carbon and nutrient enrichment in clay-rich calcareous soils in self-regenerating fallows in humid tropical agroecology. Arab J Geosci 17(12):313. https://doi.org/10.1007/s12517-024-12122-z

Miao X, Chen X, Wu W, Lin D, Yang K (2023) Topological defects strengthened nonradical oxidation performance of biochar catalyzed peroxydisulfate system. Biochar 5(1):40. https://doi.org/10.1007/s42773-023-00243-9

Mosharrof M, Uddin MK, Mia S, Sulaiman MF, Shamsuzzaman SM, Haque ANA (2022) Influence of rice husk biochar and lime in reducing phosphorus application rate in acid soil: a field trial with maize. Sustainability 14(12):7418. https://doi.org/10.3390/su14127418

Mukherjee A, Lal R, Zimmerman AR (2014) Effects of biochar and other amendments on the physical properties and greenhouse gas emissions of an artificially degraded soil. Sci Total Environ 487:26–36. https://doi.org/10.1016/j.scitotenv.2014.03.141

Murtaza G, Ahmed Z, Usman M, Tariq W, Ullah Z, Shareef M, Ditta A (2021) Biochar induced modifications in soil properties and its impacts on crop growth and production. J Plant Nutr 44(11):1677–1691. https://doi.org/10.1080/01904167.2021.1871746

Murtaza G, Ahmed Z, Usman M, Li Y, Tariq A, Rizwan M (2023) Effects of biotic and abiotic aging techniques on physiochemical and molecular characteristics of biochar and their impacts on environment and agriculture: a review. J Soil Sci Plant Nutr 23(2):1535–1564. https://doi.org/10.1007/s42729-023-01201-x

Naeem MA, Khalid M, Aon M, Abbas G, Amjad M, Murtaza B, Ahmad N (2018) Combined application of biochar with compost and fertilizer improves soil properties and grain yield of maize. J Plant Nutr 41(1):112–122. https://doi.org/10.1080/01904167.2017.1381734

Nagabovanalli Basavarajappa P, Shruthi, Lingappa M, Kadalli GG, Goudra Mahadevappa S (2021) Nutrient requirement and use efficiency of rice (*Oryza sativa* L.) as influenced by graded levels of customized fertilizer. J Plant Nutr 44(19):2897–2911. https://doi.org/10.1080/01904167.2021.1927081

Nan Q, Wang C, Wang H, Yi Q, Wu W (2020) Mitigating methane emission via annual biochar amendment pyrolyzed with rice straw from the same paddy field. Sci Total Environ 746:141351. https://doi.org/10.1016/j.scitotenv.2020.141351

Nelson NO, Agudelo SC, Yuan W, Gan J (2011) Nitrogen and phosphorus availability in biochar-amended soils. Soil Sci 176(5):218–226. https://doi.org/10.1097/SS.0b013e3182171eac

Novak JM, Lima I, Xing B, Gaskin JW, Steiner C, Das KC et al (2009) Characterization of designer biochar produced at different temperatures and their effects on a loamy sand. Ann Environ Sci 3:195–206

Obia A, Mulder J, Martinsen V, Cornelissen G, Børresen T (2016) In situ effects of biochar on aggregation, water retention and porosity in light-textured tropical soils. Soil Tillage Res 155:35–44. https://doi.org/10.1016/j.still.2015.08.002

Oguntunde PG, Fosu M, Ajayi AE, Van De Giesen N (2004) Effects of charcoal production on maize yield, chemical properties and texture of soil. Biol Fertil Soils 39(4):295–299. https://doi.org/10.1007/s00374-003-0707-1

Oladele S, Adeyemo A, Awodun M, Ajayi A, Fasina A (2019) Effects of biochar and nitrogen fertilizer on soil physicochemical properties, nitrogen use efficiency and upland rice (*Oryza sativa*) yield grown on an Alfisol in southwestern Nigeria. Int J Recycl Org Waste Agric 8(3):295–308. https://doi.org/10.1007/s40093-019-0251-0

Palansooriya KN, Wong JTF, Hashimoto Y, Huang L, Rinklebe J, Chang SX et al (2019) Response of microbial communities to biochar-amended soils: a critical review. Biochar 1(1):3–22. https://doi.org/10.1007/s42773-019-00009-2

Pan SY, Dong CD, Su JF, Wang PY, Chen CW, Chang JS et al (2021) The role of biochar in regulating the carbon, phosphorus, and nitrogen cycles exemplified by soil systems. Sustainability 13(10):5612. https://doi.org/10.3390/su13105612

Panhwar QA, Naher UA, Shamshuddin J, Ismail MR (2020) Effects of biochar and ground magnesium limestone application, with or without bio-fertilizer addition, on biochemical properties of an acid sulfate soil and rice yield. Agronomy 10(8):1100. https://doi.org/10.3390/agronomy10081100

Partey ST, Preziosi RF, Robson GD (2014) Short-term interactive effects of biochar, green manure, and inorganic fertilizer on soil properties and agronomic characteristics of maize. Agric Res 3(2):128–136. https://doi.org/10.1007/s40003-014-0102-1

Paz-Ferreiro J, Álvarez-Calvo ML, Figueiredo CCD, Mendez AM, Gascó G (2020) Effect of biochar and hydrochar on forms of aluminium in an acidic soil. Appl Sci 10(21):7843. https://doi.org/10.3390/app10217843

Peng Y, Chen Q, Guan CY, Yang X, Jiang X, Wei M et al (2023) Metal oxide modified biochars for fertile soil management: effects on soil phosphorus transformation, enzyme activity, microbe community, and plant growth. Environ Res 231:116258. https://doi.org/10.1016/j.envres.2023.116258

Qi L, Ma Z, Chang SX, Zhou P, Huang R, Wang Y et al (2021) Biochar decreases methanogenic archaea abundance and methane emissions in a flooded paddy soil. Sci Total Environ 752:141958. https://doi.org/10.1016/j.scitotenv.2020.141958

Quin PR, Cowie AL, Flavel RJ, Keen BP, Macdonald LM, Morris SG et al (2014) Oil mallee biochar improves soil structural properties—a study with X-ray micro-CT. Agric Ecosyst Environ 191:142–149. https://doi.org/10.1016/j.agee.2014.03.022

Rahman MA, Lee SH, Ji HC, Kabir AH, Jones CS, Lee KW (2018) Importance of mineral nutrition for mitigating aluminum toxicity in plants on acidic soils: current status and opportunities. Int J Mol Sci 19(10):3073. https://doi.org/10.3390/ijms19103073

Riaz MU, Ayub MA, Khalid H, Ul Haq MA, Rasul A, Ur Rehman MZ, Ali S (2020) Fate of micronutrients in alkaline soils. In: Resources use efficiency in agriculture. Springer Singapore, Singapore, pp 577–613. https://doi.org/10.1007/978-981-15-6953-1_16

Rivelli AR, Libutti A (2022) Effect of biochar and inorganic or organic fertilizer co-application on soil properties, plant growth and nutrient content in Swiss chard. Agronomy 12(9):2089

Rondon M, Ramirez JA, Lehmann J (2005) Charcoal additions reduce net emissions of greenhouse gases to the atmosphere. In: Proceedings of the 3rd USDA symposium on greenhouse gases and carbon sequestration in agriculture and forestry, vol 208. USDA, Baltimore, pp 21–24

Sajjadi B, Chen WY, Egiebor NO (2019) A comprehensive review on physical activation of biochar for energy and environmental applications. Rev Chem Eng 35(6):735–776

Sasmita KD, Anas I, Anwar S, Yahya S, Djajakirana G (2017) Application of biochar and organic fertilizer on acid soil as growing medium for cacao (*Theobroma cacao* L.) seedlings. Int J Sci Basic Appl Res 36(5):261–273

Shackley S, Sohi S, Brownsort P, Carter S, Cook J, Cunningham C et al (2010) An assessment of the benefits and issues associated with the application of biochar to soil. Department for Environment, Food and Rural Affairs, UK Government, London

Sharma P, Ali S, Biswas JK (2025) Application of biochar for soil erosion control and environmental management: implications for achieving sustainable development goals. Discover Soil 2(1):1–19. https://doi.org/10.1007/s44378-025-00065-0

Shetty R, Vidya CSN, Prakash NB, Lux A, Vaculík M (2021) Aluminum toxicity in plants and its possible mitigation in acid soils by biochar: a review. Sci Total Environ 765:142744. https://doi.org/10.1016/j.scitotenv.2020.142744

Shi RY, Ni N, Nkoh JN, Li JY, Xu RK, Qian W (2019) Beneficial dual role of biochars in inhibiting soil acidification resulting from nitrification. Chemosphere 234:43–51. https://doi.org/10.1016/j.chemosphere.2019.06.030

Singh BP, Cowie AL, Smernik RJ (2012) Biochar carbon stability in a clayey soil as a function of feedstock and pyrolysis temperature. Environ Sci Technol 46(21):11770–11778. https://doi.org/10.1021/es302545b

Singh H, Northup BK, Rice CW, Prasad PV (2022) Biochar applications influence soil physical and chemical properties, microbial diversity, and crop productivity: a meta-analysis. Biochar 4(1):8. https://doi.org/10.1007/s42773-022-00138-1

Singh S, Luthra N, Mandal S, Kushwaha DP, Pathak SO, Datta D et al (2023) Distinct behavior of biochar modulating biogeochemistry of salt-affected and acidic soil: a review. J Soil Sci Plant Nutr 23(3):2981–2997. https://doi.org/10.1007/s42729-023-01370-9

Spokas KA, Reicosky DC (2009) Impacts of sixteen different biochars on soil greenhouse gas production. Ann Environ Sci 3:179–193

Stewart CE, Zheng J, Botte J, Cotrufo MF (2013) Co-generated fast pyrolysis biochar mitigates green-house gas emissions and increases carbon sequestration in temperate soils. GCB Bioenergy 5(2):153–164. https://doi.org/10.1111/gcbb.12001

Sujeeun L, Thomas SC (2023) Biochar mitigates allelopathic effects in temperate trees. Ecol Appl 33(4):e2832. https://doi.org/10.1002/eap.2832

Sun F, Lu S (2014) Biochars improve aggregate stability, water retention, and pore space properties of clayey soil. J Plant Nutr Soil Sci 177(1):26–33. https://doi.org/10.1002/jpln.201200639

Syuhada AB, Shamshuddin J, Fauziah CI, Rosenani AB, Arifin A (2016) Biochar as soil amendment: impact on chemical properties and corn nutrient uptake in a podzol. Can J Soil Sci 96(4):400–412. https://doi.org/10.1139/cjss-2015-0044

Takala B, Teshale E, Adugna B (2025) Integrated use of biochar and lime enhances soil properties and maize yield in acidic soil of Jimma zone, Southwestern Ethiopia. Sci J Anal Chem 13(1):1–8. https://doi.org/10.11648/j.sjac.20251301.11

Thies JE, Rillig MC, Graber ER (2015) Biochar effects on the abundance, activity and diversity of the soil biota. In: Biochar for environmental management. Routledge, pp 327–389

Tomczyk A, Sokołowska Z, Boguta P (2020) Biochar physicochemical properties: pyrolysis temperature and feedstock kind effects. Rev Environ Sci Biotechnol 19(1):191–215. https://doi.org/10.1007/s11157-020-09523-3

Tusar HM, Uddin MK, Mia S, Suhi AA, Wahid SBA, Kasim S, Anwar F (2023) Biochar-acid soil interactions—a review. Sustainability 15(18):13366. https://doi.org/10.3390/su151813366

Ullah I, Khan S, Nawab J, Khan MA, Jehan S (2022) Hardwood modified and unmodified biochar amendments used for saline alkali soil remediation: phosphorus availability and its plant uptake. Arab J Geosci 15(9):910. https://doi.org/10.1007/s12517-022-10157-8

Vaccari FP, Baronti S, Lugato E, Genesio L, Castaldi S, Fornasier F, Miglietta F (2011) Biochar as a strategy to sequester carbon and increase yield in durum wheat. Eur J Agron 34(4):231–238. https://doi.org/10.1016/j.eja.2011.01.006

Van Zwieten LLVZ, Kimber S, Morris S, Chan KY, Downie A, Rust J et al (2010) Effects of biochar from slow pyrolysis of papermill waste on agronomic performance and soil fertility. Plant Soil 327(1):235–246. https://doi.org/10.1007/s11104-009-0050-x

Van Zwieten L, Kammann C, Cayuela ML, Singh BP, Joseph S, Kimber S et al (2015) Biochar effects on nitrous oxide and methane emissions from soil. In: Biochar for environmental management. Routledge, pp 489–520

Verheijen FG, Zhuravel A, Silva FC, Amaro A, Ben-Hur M, Keizer JJ (2019) The influence of biochar particle size and concentration on bulk density and maximum WHC of sandy vs sandy loam soil in a column experiment. Geoderma 347:194–202. https://doi.org/10.1016/j.geoderma.2019.03.044

Wang J, Zhang M, Xiong Z, Liu P, Pan G (2011) Effects of biochar addition on N_2O and CO_2 emissions from two paddy soils. Biol Fertil Soils 47(8):887–896. https://doi.org/10.1007/s00374-011-0595-8

Wang Z, Zheng H, Luo Y, Deng X, Herbert S, Xing B (2013) Characterization and influence of biochars on nitrous oxide emission from agricultural soil. Environ Pollut 174:289–296. https://doi.org/10.1016/j.envpol.2012.12.003

Wang Z, Guo H, Shen F, Yang G, Zhang Y, Zeng Y et al (2015) Biochar produced from oak sawdust by Lanthanum (La)-involved pyrolysis for adsorption of ammonium (NH_4^+), nitrate (NO_3^-), and phosphate (PO_4^{3-}). Chemosphere 119:646–653. https://doi.org/10.1016/j.chemosphere.2014.07.084

Wang L, O'Connor D, Rinklebe J, Ok YS, Tsang DC, Shen Z, Hou D (2020) Biochar aging: mechanisms, physicochemical changes, assessment, and implications for field applications. Environ Sci Technol 54(23):14797–14814. https://doi.org/10.1021/acs.est.0c04033

Wani OA, Akhter F, Kumar SS, Babu S, Kanth RH, Mir SA, Rathore SS (2023) Mitigating soil erosion through biomass-derived biochar: exploring the influence of feedstock types and pyrolysis temperature. Land 12(12):2111. https://doi.org/10.3390/land12122111

Waqar M, Habib-ur-Rahman M, Hasnain MU, Iqbal S, Ghaffar A, Iqbal R, Sabagh AE (2022) Effect of slow release nitrogenous fertilizers and biochar on growth, physiology, yield, and nitrogen use efficiency of sunflower under arid climate. Environ Sci Pollut Res 29(35):52520–52533. https://doi.org/10.1007/s11356-022-19289-6

Whalen JK, Ejack L, Gul S, Castro LL (2024) Characteristics of biochar: nutrient properties. In: Biochar for environmental management. Routledge, pp 183–207. https://doi.org/10.4324/9781003297673

Wong JW, Ogbonnaya UO (2021) Biochar porosity: a nature-based dependent parameter to deliver microorganisms to soils for land restoration. Environ Sci Pollut Res 28(34):46894–46909. https://doi.org/10.1007/s11356-021-14803-8

Xiao X, Chen B, Chen Z, Zhu L, Schnoor JL (2018) Insight into multiple and multilevel structures of biochars and their potential environmental applications: a critical review. Environ Sci Technol 52(9):5027–5047. https://doi.org/10.1021/acs.est.7b06487

Xu RK, Zhao AZ, Yuan JH, Jiang J (2012) pH buffering capacity of acid soils from tropical and subtropical regions of China as influenced by incorporation of crop straw biochars. J Soils Sediments 12(4):494–502. https://doi.org/10.1007/s11368-012-0483-3

Xu W, Xu H, Delgado-Baquerizo M, Gundale MJ, Zou X, Ruan H (2023) Global meta-analysis reveals positive effects of biochar on soil microbial diversity. Geoderma 436:116528. https://doi.org/10.1016/j.geoderma.2023.116528

Zhang A, Cui L, Pan G, Li L, Hussain Q, Zhang X et al (2010) Effect of biochar amendment on yield and methane and nitrous oxide emissions from a rice paddy from Tai Lake plain, China. Agric Ecosyst Environ 139(4):469–475. https://doi.org/10.1016/j.agee.2010.09.003

Zhang A, Bian R, Pan G, Cui L, Hussain Q, Li L et al (2012) Effects of biochar amendment on soil quality, crop yield and greenhouse gas emission in a Chinese rice paddy: a field study of 2 consecutive rice growing cycles. Field Crop Res 127:153–160. https://doi.org/10.1016/j.fcr.2011.11.020

Zimmerman AR (2010) Abiotic and microbial oxidation of laboratory-produced black carbon (biochar). Environ Sci Technol 44(4):1295–1301. https://doi.org/10.1021/es903140c

Biochar and Soil Biological Properties: Yield Enhancement Through Modulation of Soil Biological Properties

Tamanna Sharma, Vikas Sharma, Vivak M. Arya, Nikhil Thakur, and Owais Ali Wani

Abstract

Sustainable crop production and minimising the negative environmental effects of excessive chemical fertiliser use—which causes soil degradation, nutrient leaching, and a decline in microbial diversity—are the two main challenges facing modern agriculture. Biochar, a carbon-rich byproduct of pyrolysis, has become a promising soil supplement in response to these problems because of its capacity to improve soil health, facilitate nutrient cycling, and aid in climate mitigation. The addition of biochar to soils increases the diversity, abundance, and activity of soil microorganisms. In particular, microbial activity is increased when suitable habitats and substrates for their metabolic processes are provided. This supports a number of ecologically significant processes, including enzyme activity, element cycling, soil purification, and plant productivity. However, its interactions with rhizosphere dynamics have not been thoroughly studied. This chapter summarises the most recent research on the short- and long-term effects of biochar, with a focus on nutrient dynamics, enzyme activity, and rhizosphere properties. It will also provides insights about the function of biochar in raising agricultural yields, the mechanisms underlying it, and the different obstacles and information gaps in putting the biochar approach into practice.

T. Sharma · V. Sharma · V. M. Arya
Division of Soil Science and Agricultural Chemistry, Sher-e-Kashmir University of Agricultural Sciences and Technology of Jammu, Jammu, Jammu and Kashmir, India

N. Thakur (✉)
Division of Fruit Science, Sher-e-Kashmir University of Agricultural Sciences and Technology of Jammu, Jammu, Jammu and Kashmir, India

O. A. Wani
Division of Agronomy, Faculty of Agriculture, SKUAST-Kashmir, Wadura, Sopore, Jammu and Kashmir, India

O. A. Wani et al. (eds.), *Biochar for Sustainable Soil Health and Crop Production*, https://doi.org/10.1007/978-981-92-2230-8_5

Keywords

Biochar · Pyrolysis · Microbial · Rhizosphere · Sustainable

1 Introduction

The ever-increasing pressure on farms to enhance food production has severely impacted their ability to generate agricultural goods at a reasonable environmental cost. According to estimates, if current trends continue, 840 million people will suffer from hunger by 2030 (FAO et al. 2021). Some argue that by 2050, food production would need to double or more to meet the demands of a growing global population that is becoming increasingly wealthy (Food Security Information Network 2017). However, many lands that are currently in use or have previously been used for food production are agronomically deteriorated. Reduced soil fertility and greater environmental sensitivity to farming caused by poor soils, poor management, or both have continuously reduced yields on these lands. Many former agricultural tracts are no longer suited for cultivation, and numerous have been abandoned. Degraded lands are frequently the result of diminished soil fertility caused by intense management, ineffective soil conservation measures, and climate change. Numerous studies have demonstrated the negative effects of intensive annual crop production on the soil environment; in fact, the current revival of interest in regenerative agriculture (Giller et al. 2021) places an emphasis on soil health restoration (Schreefel et al. 2020). Some specific elements of intensive annual crop production that contribute to soil degradation include regular disturbance events such as tillage, a lack of consistent year-round plant cover, an absence of continuous deep roots systems and crop functional diversity, and imbalanced nutrient use. Furthermore, climate change induced increased global warming and altering precipitation dynamics—has aggravated soil degradation in areas where soils are particularly vulnerable to flooding and drought (Intergovernmental Panel on Climate Change 2019). Erosion losses have been notably severe—up to 1% of topsoil is lost each year in several places (Mosier et al. 2021) as a result of tillage, overgrazing, and the increasing frequency of extreme climate events that exacerbate both wind and water erosion. The ongoing use of degraded lands for crop production necessitates ever-increasing management interventions to ensure high-yielding food production. In this scenario, increased land degradation poses an additional threat to agriculture's environmental integrity by accelerating soil carbon (C), nitrogen (N), and phosphorus (P) losses. The loss of carbon and essential nutrients from the agricultural system causes land to struggle to grow nutritious food for human consumption, and losses will only worsen as management strengthens to restore lost fertility, resulting in a positive, downward spiralling feedback cycle. Innovative and eco-friendly methods are needed to improve plant productivity in adverse conditions such as nutrient deficiency, contamination, degradation, aridity, and salinity. Biochar can reduce pollution, alleviate salinity, and improve soil conditions for cultivation in diverse soil types (Jalal et al. 2020). Biochar is typically manufactured from waste materials including agricultural waste, animal dung, and forest leftovers.

These feedstocks can create biochar, transforming trash into a valuable commodity (Awad et al. 2018). Biochar is described as a solid substance produced by thermo-chemical carbonisation under oxygen-limited circumstances (Yu et al. 2017). It is mostly formed during the pyrolysis of organic matter at temperatures ranging from 300 to 700 °C and low oxygen levels (Kuryntseva et al. 2023). Other pyrolysis products include liquid fuel and pyrolysis gas (Tang et al. 2022). Biochar has gained popularity in the agroindustry and environmental sciences over the last decade due to its potential applications as a soil conditioner and fertiliser, fodder additive, wastewater cleaning agent, renewable energy source, and low-cost sorbent for the removal of heavy metals and organic pollutants. Biochar enhances soil structure and fertility, increases water holding capacity, and serves as a valuable supply of micro-elements and nutrients for plants, livestock, and poultry (An et al. 2022). The use of biochar reduces methane (CH_4) and nitrous oxide (N_2O) emissions (Huang et al. 2019), increases soil microbial activity, and promotes plant germination. It is critical that biochar provides benefits through amendments in practically all agricultural circumstances, including soil type and climate (Kuryntseva et al. 2022). Unlike many organic fertilisers, biochar carbon (C) is rather stable and mineralises slowly with the release of carbon dioxide (CO_2) (Yin et al. 2022). This makes biochar significant in green sustainability policies, carbon sequestration, yield improvement, and greenhouse gas emissions reduction. In natural biomass degradation processes, approximately 90%–95% of carbon is returned to the environment after a few years (or sooner), whereas pyrolysis of bioresidues results in C sequestration and a 50% decrease in CO_2 emissions (Haris et al. 2021). The manufacturing process, soil conditions, and application type all have a significant impact on biochar response and behavior (Arif et al. 2020). Biochar is gaining popularity as a sustainable approach to combat climate change and improve soil health (Shirzad et al. 2023). Further, numerous studies shows that intensive farming can promote soil deterioration (Basu et al. 2021), lower soil biodiversity, reduce soil organic matter contents, leading to nutrient deficiencies in soil that affect soil fertility (Jin et al. 2022), and reduce agricultural output and environmental stability (Pylro et al. 2020). Under such conditions, soil microbial communities have the ability to act as a restorative agent, improving soil health and quality traits as well as soil-mediated ecosystem services that have been harmed by intensive farming (Saccá et al. 2017). Soil biological fertility refers to soil's biological ability to provide the necessary nutrients to plants, humans, and animals to complete their lifecycles while preserving and strengthening soil's physical and chemical properties (Adedeji et al. 2020). Soil biological fertility is determined by the interactions between soil bacteria and the rhizosphere. The use of biochar improves soil biological characteristics significantly (Taheri et al. 2024). Biochar promotes microbial diversity and activity by creating a habitat for microorganisms (Hossain et al. 2020). The porous structure provides shelter and colonisation areas for beneficial microbes including nitrogen-fixing bacteria and mycorrhizal fungi, which are essential for nutrient cycling and plant health. Biochar can boost soil microbial activity by providing labile carbon and adjusting pH and moisture levels (Sharma and Thakur 2023). Enhanced microbial activity helps decompose organic matter, mineralise nutrients, and reduce soil-borne diseases

(Zaheer et al. 2021). Biochar promotes root development and plant growth by improving nutrient availability, reducing phytotoxicity, and increasing water retention. This chapter examines how biochar affects soil biological characteristics and how these changes correlate with increased yields.

2 Biochar: Properties and Mechanisms of Action

2.1 Physical and Chemical Properties

Biochar performance in various applications is mostly determined by its biochar characteristics (Fig. 1). The surface area and porosity of biochar have a crucial role in determining the quantity and quality of available active compounds, boosting biochar features such as capillary exchange capacity, water holding capacity, and adsorption capacity (Weber and Quicker 2018).

(a) **Porosity**: Pyrolysis of biomass causes water losses and the release of volatile compounds from the carbon matrix, resulting in the production of biochar pores and rudimentary porse (Tomczyk et al. 2020). Biochar contains pores ranging from nanometres to tens of micrometres. The International Union of Pure and Applied Chemistry (IUPAC) classifies pores into three types: micropores (less than 2 nm), mesopores (2–50 nm), and macropores (more than 50 nm) (Fiore et al. 2018). Pore size distribution is a key indicator of structural variability of biochar. Each type of pore serves a specific function in interacting with contaminants and microbes. Macropores transport adsorbates to micro- and

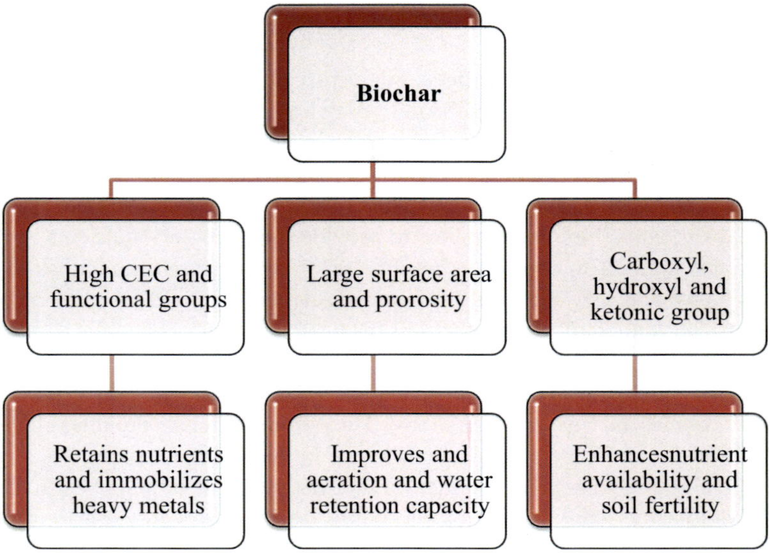

Fig. 1 Biochar as a soil conditioner

mesopore areas, provide habitat for microorganisms, and improve soil aeration and hydrology (Fig. 1). Mesopores are generated when micropores widen during heating, as determined by surface area (Xiong et al. 2017). Chemically reactive surfaces with various exterior functional groups can adsorb microorganisms depending on feedstock and pyrolysis conditions. Leng et al. (2019) found that microbial colonisation in pores is influenced by gaseous, nutritional, and aqueous content flux. Surface molecules, including cell wall components, extracellular polymeric substances (binding proteins, lipoproteins, surface-associated enzymes, and exopolysaccharides), hydrophobicity, and van der Waals attractive forces, all play a role in microbial attachment. Biochar surfaces differ in ionic strength and specificity for microbial attraction based on their ash content, ferric, calcium, magnesium, and manganate-associated components. Metallic salts in biochar increase cell surface hydrophobicity by binding to anionic functional groups in microbial lipopolysaccharide or lipoprotein outer membranes. The porous nature of biochar facilitates pore filling during sorption processes. The remarkable sorption capabilities of biochar has led to its widespread use in environmental applications, including wastewater treatment, soil remediation, and gas pollution absorption. Oliveira et al. (2017) found high sorption efficiency for both inorganic and organic contaminants, including heavy metals, N, P, CO_2, SO_2, and H_2S. Different compounds require different pore sizes for sorption due to adsorbate accessibility and kinetic diameter differences. Mesopores play a significant role in the sorption of macromolecules with kinetic diameters larger than 2 nm. The biochar pore network should be designed based on the adsorbate's kinetic diameter. However, the geometry of biochar pores can significantly impact application performance, including hydrogen storage capacity (Leng et al. 2021).

(b) **Surface Area**: The porous structure, high internal surface area, and ability to absorb organic chemicals, gases, and nutrients (Weber and Quicker 2018) of biochar make it ideal for microbial colonisation. Large surface area correlates with other biochar features, such as cation exchange capacity and water retention capacity (Fig. 1), making it necessary for various applications. Biochars from pyrolysis typically have a large surface area, whereas hydrothermal carbonisation residues have a low surface area (Chen et al. 2018). Chemically reactive surfaces with various exterior functional groups can adsorb microorganisms depending on feedstock and pyrolysis conditions. Morillo and Villaverde (2017) found that microbial colonisation in pores is influenced by gaseous, nutritional, and aqueous content flux. Microorganisms often migrate passively through water flow, making pore space properties like surface or directional connectedness important for microbial colonisation (Liu et al. 2018b). Biochar surface area varies significantly with pyrolysis temperature. Previous research (Liu et al. 2017) found that higher temperatures lead to increased pore volume and surface area. Biochar can have decreased surface area and porosity when created at temperatures higher than 550 °C. Jin et al. (2016) found that producing biochar from biosolids at 600 °C resulted in a lower surface area (5.99 m^2g^{-1}) than at 550 °C (8.45 m^2g^{-1}). The content of biochar feedstock, in addition to the pyrolysis temperature, significantly impacts its qualities. Plant biochar, such as oakwood, corn stover, and pine needle, has a higher surface area (112–642 m^2g^{-1}) as compared to swine manure (4.11 m^2g^{-1}), pig manure (3.32–20.5 m^2g^{-1}), and biosolid biochar (50.9–94.2 m^2g^{-1}) (Li et al. 2017). Biomass with high lignin content (e.g., bamboo and coconut shell) creates biochar with high surface area and macroporous structures, while biomass with high cellulose content (e.g., husks) primarily produces biochar with low surface area and microporous structures. Biochars vary in porosity and surface area, leading to different adsorption reactions for contaminants in soils (Zhu et al. 2017).

(c) **pH**: Biochar is mostly found to be alkaline. Biochar pH values varied from slightly acidic (≈6.5) to strongly alkaline (≈11.5) depending on feedstock and residence time interval (Munera-Echeverri et al. 2018). The pH value of biochar is crucial for agricultural uses such as soil amendment. Pyrolysis chars have unique features that set them apart from hydrothermal carbonisation chars. The latter process produces organic acids, making hydrochars acidic. Biomass is often slightly acidic or mildly basic, with pH levels ranging from 5 to 7.5 (Leng et al. 2021). Pyrolysis removes acidic functional groups, including carboxyl, hydroxyl, and formyl groups. The remaining solid gets more basic when functional groups are released. Biochar with higher pH can expedite the breakdown of organophosphorus and carbamate insecticides in soil via alkali catalysis (Braghiroli et al. 2020). Biochar pH, like porosity and surface area, was controlled by pyrolysis temperature and feedstock. Biochar often has an alkaline pH that rises with higher pyrolysis temperatures (Tan et al. 2021). However, there are certain exceptions depending on the raw material (Li et al. 2017). For instance, biochar made from wastewater sludge and wheat straw at a low temperature (400 °C) was acidic, with a pH range of 4.87–6.11 (Qin et al. 2018). Previous research found a significant correlation between pH and pyrolysis temperature in biochars made from diverse biomass, including biosolids, agricultural residues, and livestock manure (Zhang et al. 2015). Higher temperatures leads to increasing ash content, which causes higher pH of biochar (Jin et al. 2016). Wystalska et al. (2022) found that as temperature climbed from 300 to 700 °C, the pH of biochar increased along with the amount of basic cations and carbonates. Li et al. (2017) found that reducing acidic functional groups like carboxyl groups present on biochars at higher temperatures resulted in higher pH levels.

(d) **Cation Exchange Capacity (CEC)**: Cation exchange capacity (CEC) refers to the ability of a material to adsorb exchangeable cations, such as Ca^{2+}, Mg^{2+}, K^+, Na^+ and NH^{4+}. Negative surface charges attract cations, indicating soil fertility. Plants and microorganisms mostly absorb nutrients in their ionic form (Antonangelo et al. 2024). Surface structure and surface area play a direct role in cation exchange capacity (Panwar and Pawar 2022). Functional groups provide surface charges, while surface area facilitates their accessibility. Pyrolysis provides polar organic moieties such as carboxylic acids, ketones, and aldehydes due to the depolymerisation of cellulose and lignin, which increases the negative charge density on the surface of biochar. Biochars produced at low temperatures have the highest cation exchange capacities due to the combination of charged surface functional groups and surface area. This is because the surface area increases significantly compared to the feedstock, but the structure retains enough functional groups to provide negative charges. Biochars from various feedstocks (oak, pine, grass) generated at varied pH levels (1.5–7.5) showed the highest average CEC (51.9 ± 15.3 cmol C/kg) at the lowest production temperature of 250 °C (Paetsch et al. 2018). High-ash feedstocks have the capability to produce biochar with a higher CEC than wood and other low-ash wastes. During pyrolysis, elements such as Mg, K, Na, and P can catalyse the synthesis of surface oxygen groups such as carboxylic acids, lactones, and phenols, increasing biochars' negative charge density (Banik et al. 2018). As the pyrolysis temperature rises, the loss of O and H on the charcoal surface causes a significant fall in biochar CEC (Domingues et al. 2017). For example, at high temperatures, the quantity of carboxylic and phenolic groups diminishes, which is why CEC is inversely proportional to pyrolysis temperature (Veiga et al. 2017). Anaerobically digested bagasse residue produced biochar with greater CEC and negative surface charge when compared to undigested bagasse biochar (Hernandez-Charpak et al. 2022). Algae biochar has a higher CEC (29–41 cmol/kg) and extractable Ca, Mg, and K levels (27–485 cmol/kg) compared to other terrestrial biomass biochars (Wang and Wang 2019).

2.2 Types of Feedstocks and Pyrolysis Conditions

Biochar is usually produced from pyrolysis of plant biomass, animal and food residues, forestry wastes, crop residues (nut shells, fruit pits, bagasse, straw) (Haris et al. 2021), algae (Yu et al. 2017), manures (Qi et al. 2020), biosolids and sewage sludge formed during wastewater treatment (Xue et al. 2019). The elemental composition of biochar (primarily C, N, K, P, Ca, and Mg) is determined by the feedstock and pyrolysis conditions. The utilization of animal manures frequently leads in elevated potassium levels. Mineral components of biochar range from 1% (grasses) to around 20% (wastewater sludge) (Xu et al. 2018a). The C content of the biomass used as biochar feedstock is determined by the source, pre-treatment conditions, storage time, moisture content, and other parameters, some of which can be considered prior to thermal treatment to unify the pyrolysis parameters and final properties (Table 1). Plant residues particularly forest and grass residues, are mostly composed of cellulose, lignin, and hemicellulose (lignocellulosic biomass) (Dhyani and Bhaskar 2018). Cellulose is found in all plant cell walls. It is composed of repeated D-glucose units that interact with one another via hydrogen bonds, imparting mechanical strength and chemical stability. Hemicellulose, a less polymerised analogue of cellulose, contains glucose, galactose, mannose, xylose, arabinose, and glucuronic acid units (50–200 monomers per molecule) (Krylova and Zaitchenko 2018). Lignin forms the outer layer of cellulose fibres and holds hemicellulose and cellulose together within the cell wall. The lignin percentage ranges from 16 (soft

Table 1 Effect of pyrolysis temperature on biochar properties

Property	Low Temperature (250–400 °C)	Medium Temperature (400–550 °C)	High Temperature (>550 °C)	Effect on Biochar
Surface Area and Porosity	Lower porosity with dominant micropores	Increased pore development as micropores are converted to mesopores	Peak surface area them decline at very high temperature	Enhances adsorption capacity and microbial habitats
pH	Slightly acidic to neutral	Neutral to slightly alkaline	Strongly alkaline	Higher ash content, basic cations and carbonates
Cation Exchange Capacity (CEC)	High	Moderate	Low	Quantity of carboxylic and phenolic groups diminishes leading to lower CEC

wood) to 40% (hard wood). In addition to the three primary components, organic matter in biomass contains some extractives. Organic solvents can extract them from biomass without causing hydrolysis (Roy and Dias 2017). Urban garbage contains more lignin and cellulose than wood or agriculture leftovers, but it is still much lower. Biosolids are also high in nitrogen, largely in organic form (up to 6% of dry weight), and phosphorus (up to 3% of dry weight), however these can be lost in combustion and pyrolysis (P loss is less than N loss) (Ma et al. 2017). Biochar produced from biosolids have a high ash concentration, complicating the use of this characteristic for biochar yield estimations.

The technological process of biomass pyrolysis begins with extra drying of the substrate if its moisture content is higher than optimum (He et al. 2018). The following heating eliminates the natural volatile constituents of the raw material. Following that, the biopolymers in biomass decompose, resulting in successive increment in C content and the emission of permanent gases (carbon dioxide, nitrogen oxides, ammonia, and methane) as well as volatile organic compounds of disintegration (Kuryntseva et al. 2023). The final product can be condensed into liquid gasoline. The solid residue is biochar. The reaction routes for producing gaseous and liquid pyrolysis products compete with one another and differ depending on pyrolysis parameters (residence duration, heating rate, pyrolysis temperature) and substrate qualities. Thus, rapid heating enhances the output of pyrolysis gases and fuel, whereas slow and prolonged pyrolysis is better suited for predominant carbonisation and highest biochar yield. The pyrolysis characteristics of feedstock have a direct impact on the biochar properties. Hard wood can be heated to a final carbonisation level of 1000 °C (Murtaza et al. 2022). In the meantime, biomass from agricultural waste and other sources with comparatively low ash melting temperatures is typically burned to no more than 700 °C (Godwin et al. 2019). Fast pyrolysis is primarily used to produce liquid fuel: volatile products are condensed through quick cooling to avoid polymerisation of byproducts. Fast pyrolysis may result in partial biomass conversion, leaving up to 9% hydrocarbons (mainly in the form of cellulosic and hemicellulosic fractions) in the final product. Flash pyrolysis (Ho et al. 2018) is accomplished by igniting the press biomass under high pressure (1–2 MPa). The processes heat the feedstock to 300–600 °C for approximately 30 min. The production of biochar rises with temperature and pressure (Hoekman et al. 2017).

Biochar has a direct impact on physical and chemical soil attributes, but biological properties are primarily influenced indirectly. The effects are mostly associated with sources of feedstock and pyrolysis conditions (Table 2). Although the beneficial impacts of biochar on soil fertility and agriculture ecosystems is primarily due to pH alterations in acidic soils, increased nutrient retention, and aeration, biochar amendments have been shown to have a substantial effect on microbial communities and soil animals (Glazunova et al. 2018). Changes in microbial composition and abundance are also associated with improved soil structure and nutrient availability as a result of biochar application. All of these factors influence plant development and crop quality, resulting in increased rhizodeposit input into the soil. Soil biota variety is frequently seen to be a crucial component in soil functioning (Mohamed et al. 2017), and organic amendments derived from biochar are important factors influencing trophic relationships in soil (Igalavithana et al. 2017). Microbial abundance increases significantly with biochar amendments, regardless of feedstock or soil type. The mycorrhizal response measured by root colonisation rose twice after a 2-year exposure of 0.6–6 t/ha of biochar to Eucalyptus wood (Ren et al. 2023). The protective impact of biochar particles, sorption of signalling molecules, detoxification of allelochemicals, and modification of soil physical qualities were suggested as possible explanations for biochar's favourable effect on extraradical mycelium (Zilberman et al. 2023). In addition, biochar has been shown to increase spore germination. The microbial reaction to biochar addition is also dependent on cell adherence to the particle surface. Biofilm development has a greater impact on bacterial abundance than fungal abundance. Hydrophobic attraction and electrostatic forces can both help cells adhere to biochar particles. The mineralisation of the biochar, as well as the greater salt concentration, promote adhesion. The size of the pores, which should be at least 2–5 times larger than the cell size, limits the cell's inclusion. In addition to abundance, biochar addition considerably alters microbial community composition compared to untreated soils of the same mineralogy (Prasad et al. 2020). Because of the increased nutrient availability, reduced viscosity promotes the development of plant roots and rhizobia. In addition, the altered soil allows invertebrates to move more easily, creating pores and affecting the predator/prey ratio. Furthermore, the microbial community might actively respond to the addition of biochar. Higher microbial diversity may result in increased mineralisation or oxidation of the biochar itself, as has been demonstrated for organic carbons of non-pyrolysis origin. In most cases, an increase in microbial biomass stimulates these processes (Jie et al. 2017).

Table 2 Physicochemical properties of biochars from different feedstocks and their potential effects on soil biology

Biochar Feedstock	pH	Surface Area (m^2/g)	Cation Exchange Capacity ($cmol^+ kg^{-1}$)	C (%)	H (%)	O (%)	N (%)	Effect on biological property	References
Corn straw	9.97	9.97	304	57.33	1.46	4.87	0.71	Increased soil microbial biomass especially Gram positive bacteria	Tang et al. (2019)
Corn cob	8.94	33.4	–	77.4	3.25	9.29	1.24	Resulted in eight fold increase in culturable fungal colonies	Maaz et al. (2021)
Wheat straw	10.4	8.92	–	55.0	5.90		1.10	Enhanced bacterial gene abundance and reduced fungal gene abundance	Hu et al. (2021)
Willow leaves	9.1	62	–	54.6	2.64	6.72	1.0	Increased total bacteria by 28%, Gram negative bacteria by 27% and actinobacteria by 62%	Wu et al. (2025)
Rice Husk	7.8	–	11	38	2.1	59	0.9	Reduced the abundance and diversity of indigenous bacteria	Liu et al. (2018a, b)
Poultry litter	8.7	3.9		51.1	3.79	15.6	4.45	Increased relative abundance of the phyla *Actinobacteria* and arbuscular mycorrhizal colonization	Deinert et al. (2024)
Poplar wood	9	1.8	–	67.3	4.42	–	0.78	Relative abundance of several of the dominant fungal phyla and genera (*Humicola* and *Pseudallescheria*) increased with the addition of biochar	Jin et al. (2023)
Oak and hickory hardwood sawdust	6.8	–	–	66.2			0.32	Little biochar C was available for microbial degradation BC did not improve total N, NH_4-N, NO_3-N, or available P in soil	Elzobair et al. (2016)
Pine needle		30	–	67.00	6.70	–	2.80	Promoted phosphate solubilizing bacteria, enhanced the abundance of bacterial families that can degrade more recalcitrant C com pounds, decreased bacterial plant pathogens	Saxena et al. (2024)
Wood biochar	6.45	–	–	35.47	–	–	2.17	A shift towards a bacterial dominated microbial community along with a lower fungi to bacteria ratio in the soil	Gao and DeLuca (2018)

2.3 Interaction with Soil Components

Biochar (BC) is a widely studied material because of its ability to enhance soil properties and act as an adsorbent (Moreira et al. 2017). BC has a wide surface area, micropore volume, high CEC, and rich surface functional groups, making it an excellent amendment for remediating contaminated soils and increasing soil health (de Caprariis et al. 2017). Biochar has a large specific surface area along with numerous oxygen-containing functional groups on the surface, which improves soil fixation and heavy metal adsorption while also purifying the agricultural environment (Yavari et al. 2017), reducing soil toxicity, and promoting crop growth through enhanced functional group activity. The use of biochar as a soil conditioner has the ability to improve soil aeration, soil structure, water holding capacity, microbial biomass carbon, enzymatic activities, soil organic carbon (SOC) content (García-Ruiz et al. 2019), and nutrient retention and availability (Zhao et al. 2018), potentially reducing nutrient leaching from the soil surface to groundwater. It is extremely likely that the surface area, porosity, and ion exchange capacity of various BCs may alter their ability to absorb and possibly release organic matter (OM) or nutrients over time (Purakayastha et al. 2019). The large surface areas oxygen-containing functional groups, and CECs of biochar allows it to retain, stabilise, and inactivate harmful heavy metals, as well as lower phytotoxicity and bioavailability of heavy metal ions in the soil environment (Ma et al. 2022). Because of its wide surface area and fenestrated pore structure, biochar can adsorb retained inorganic N ions by cation exchange after application to soil, reducing N loss and runoff while also enhancing water holding capacity (Zhang et al. 2017). Furthermore, the leaching effects of biochar on soil nutrients varied significantly depending on soil type. The leaching effect after biochar application was more substantial on sandy loam soil than on clay planted with rice (Nguyen et al. 2017), and biochar regulated nitrate nitrogen loss better on sandy loam soil than on control clay, which could be connected to soil characteristics. Furthermore, when applied to soil, biochar promotes N retention and biomass uptake, hence boosting the nitrogen cycle, soil mineral quality, and N fertilizer utilisation (Luo et al. 2023). Biochar inputs change soil physicochemical parameters such as pH, capacity weight, CEC, water retention, and bioactivity, all of which affect nutrient retention. Biochar can greatly increase soil resin soluble P and inorganic soluble P, allowing for a nutrient increase in soil accessible P, and the drop in P content is attributable to plant uptake rather than conversion to a more stable form (Nair et al. 2017).

Various functional groups present on the surface of biochar, such as carboxyl (− COOH), hydroxyl (− OH), quinone (= O−O =), ester (C = O−O−C), and ketone (= O), have a high affinity for adsorbing nutrient ions like nitrate and phosphate (Gęca et al. 2023). These ions can raise essential nutrients for microorganisms, enhancing their value-added tolerance and survival (Nguyen et al. 2022). Some investigations have demonstrated that porous biochar can provide a habitat and biological barrier for extraneous microorganisms, reducing microbial impact from the soil environment (Mukherjee et al. 2022). Biochar generally offers a favourable habitat for heterotrophic microbial communities by serving as a surface for microbial adhesion and improving water retention; this biochar shelter protects dominant microbes from predators and unfavourable conditions, such as chemical toxicity and unfavourable growth environments (Zhou et al. 2020). Functional groups present on the surface of biochar, particularly oxygenated carboxylates, have been demonstrated to attract nutrients and inorganic ions, supplying nutrition to microorganisms (Bolan et al. 2023). Furthermore, biochar contains insoluble particles such as organic and inorganic debris, which can be destroyed or absorbed by microorganisms during microbial growth, altering the structure of biochar and influencing microbial community expansion (Zhang et al. 2021b). Biochar additionally possesses essential elements such as minerals, which can be used as fertiliser to gradually release nutrients into the soil and improve soil fertility, hence fostering microbial growth (Cai et al. 2019). Furthermore, it has been demonstrated that modifying biochar can enhance adsorption sites by changing surface characteristics, shape, composition, and adding oxygen-containing functional groups to improve microbial adhesion to the biochar surface (Wu et al. 2022). Biochar may also serve as a carrier for beneficial bacteria, creating a favourable microbial environment and substrate, and increasing microbial metabolic activity, all of which improve microbial function and assist essential ecological processes such as element cycling, soil decontamination, and plant growth.

3 Soil Biological Properties and Their Role in Agroecosystems

3.1 Soil Microbial Biomass

Soils are host to around 25% of the world's living species (Korneykova and Nikitin 2021). Bioresource function of soil is crucial for maintaining biodiversity and quantity of soil-associated organisms (Bhargava et al. 2017). Soil biodiversity contributes to mineralisation of organic compounds, carbon sequestration, biophilic element cycles, plant nutrition, and health (Jacoby et al. 2017). Reduced biodiversity and abundance of organisms (soil bioresource function) slows soil formation and decomposition, reducing fertility, buffer capacity, and potential to self-restore after degradation (Tibbett et al. 2020). Soil health is essential for the sustainability and productivity of agricultural systems, forest ecosystems, and the environment (Bargali et al. 2019). To completely understand the complicated balance of soil ecosystems, multiple parameters must be considered. Soil microbial biomass is a key indicator of these characteristics (Padalia et al. 2022). "Soil microbial biomass" refers to the living elements of soil, such as bacteria, fungi, protozoa, and other microscopic creatures. Microorganisms contribute significantly to the Earth's biomass, despite their small size (Pant et al. 2017). Their relevance derives from their adaptability, ubiquity, and critical involvement in ecosystem functioning. Microorganisms play key roles in nutrient cycling, organic matter breakdown, disease prevention, and soil structure preservation (Awasthi et al. 2022). Microbial activities promote soil aggregation, improve structure, and reduce erosion, making them a potential indicator of soil health. A broad and robust microbial population helps soil withstand environmental pressures. Enzymatic degradation of organic materials is primarily carried out by microbes. They break down complex molecules into simpler forms, releasing nutrients from dead and decaying stuff (Bargali et al. 2019). Microbial species play a crucial role in the nitrogen cycle, fixing atmospheric nitrogen for plant use. Nitrogen-fixing bacteria, such as *Rhizobium*, work in tandem with leguminous plants to increase soil fertility. Microbes form symbiotic connections with plants, animals, and other species. Interactions often result in better resilience to environmental stressors, pathogen defence, and nutrition exchange (Mourya et al. 2019). Microbes can be found at various trophic levels in ecosystems. In some cases, they can produce organic molecules through photosynthesis or chemosynthesis, acting as primary producers (Pandey et al. 2024). Microbial predation limits population growth and inhibits the spread of other bacteria. Soil bacteria play an important part in bioremediation too. Microbes have exceptional ability to breakdown contaminants like hydrocarbons, insecticides, and heavy metals. Bioremediation uses microbial metabolic processes to reduce environmental contamination, providing a sustainable solution for ecosystem restoration (Nikitin et al. 2022). Microbial activities affect greenhouse gas emissions, such as CO_2, CH_4, and N_2O. Understanding the involvement of microbes in these processes is vital for understanding the impact on global climate patterns. Microbial biomass has a crucial role in soil fertility, which is part of the complex ecosystem that supports life on Earth. Microorganisms have a crucial role in soil fertility, nutrient cycling, plant and soil health, and ultimately world food production (Manral et al. 2022). Microorganisms stimulate the decomposition process, releasing important elements including nitrogen, phosphorus, and potassium. Plants require these nutrients for growth and development, which directly impacts soil fertility. Microbial biomass contributes significantly to nutrient cycling. Plants and bacteria cohabit in a symbiotic relationship, such as mycorrhizal fungus on plant roots. Nitrogen-fixing bacteria like Rhizobium and Azotobacter can convert atmospheric nitrogen into plant-usable form, contributing to soil fertility and reducing the need for synthetic fertilizers (Karki et al. 2021). Microorganisms have a crucial role in soil stability and structure (Pandey et al. 2023). Microbial secretions produce extracellular polymeric compounds (EPS), which promote soil aggregation and porosity. This improves water infiltration, aeration, and root penetration, resulting in an ideal environment for plant growth (Shahi et al. 2023). Microbial biomass enhances soil structure, cycles nutrients, and prevents plant diseases. Soil microbes that produce antibiotics or compete for resources can reduce disease incidence and promote plant health. Thus, measuring soil microbial biomass is important for understanding soil health, nutrient cycling, and ecosystem function. Sustainable land management prioritises soil biodiversity protection and minimises disruptions to microbial ecosystems, ensuring long-term soil fertility and global food security.

3.2 Microbial Diversity and Community Structure

Soils include a vast range of microorganisms that play an important part in ecological processes (Guerra et al. 2021). It was estimated that one gramme of surface soil includes billions of bacterial cells, 200 million fungal hyphae, thousands of protists, and trillions of viruses (Sokol et al. 2022). Belowground soils serve as a seed bank for microbial colonisation in crops (Trivedi et al. 2020). Plant hosts provide micro-habitats for microbes, including the rhizosphere, phyllosphere, and endosphere (Knights et al. 2021). Plants interact with a wide range of microbes in the rhizosphere micro-habitat, including bacteria, fungus, and viruses. Many useful microbes, such as bacteria and fungi, live in the soil and provide ideal circumstances for plant development (Fig. 2) (Sansinenea 2021). These bacteria' favourable interactions with plants include nutrient supply, plant growth stimulation, phytohormone production, phytopathogen biocontrol, soil structure improvement, inorganic compound bioaccumulation, and bioremediation of metal-contaminated soils (Ortiz and Sansinenea 2021).

Bacteria, sometimes known as enzyme bags or fertiliser soluble bags, have a fundamental structure (Philippot et al. 2024). Soil bacteria play an important role in

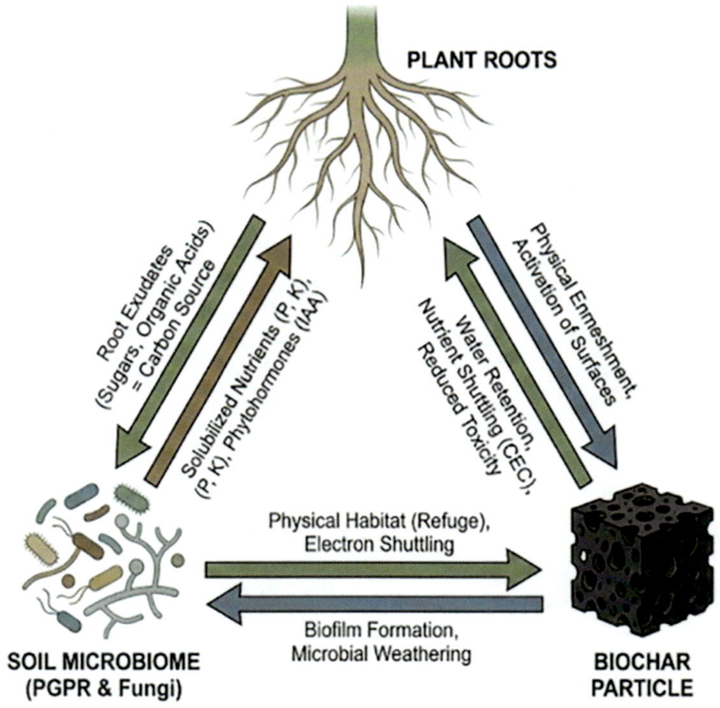

Fig. 2 Biochar mediated microbial and root interactions

crop processing by providing nutrients, promoting plant growth, regulating pathogens, improving soil composition, and facilitating bioaccumulation and inorganic leaching (Shah et al. 2021). Bacteria have a higher nitrogen content (10%–30% N, 3–10 C:N ratio) as compared to other microorganisms (Philippot et al. 2024). "Rhizobia" are believed to cause nodule formation at the roots of leguminous plants (and occasionally stalks). The "bacteroid" helps in nitrogen fixation from the atmosphere within nodules, resulting in ammonia as a fixed nitrogen supply.

This symbiotic connection provides a unique environment for bacteria while also providing plants with a specialised supply of nitrogen. Bacteria including *Azotobacter*, *Azospirillum*, *Rhizobium*, *MesoRhizobium*, and *SinoRhizobium* are known to improve plant growth (Sumbul et al. 2020). *Azospirillum* species are endophytic diazotrophs that colonise the surface and interior of roots to fix nitrogen, known as bio-nitric nitrogen fixation globally. Cyanobacteria, a type of photoautotrophic gram-negative bacteria found in many soils, play an important role in soil formation and fertility (Pandey et al. 2020). Their impact on soil quality and plant resistance to abiotic and biotic stressors, including diseases, is significant. Cyanobacteria provide a sustainable source of solubilised organic matter, which is mineralised by soil microorganisms and contributes to crop growth (Kultschar and Llewellyn 2018). Cyanobacteria can boost N levels and microorganisms in semiarid soils (Jacoby et al. 2017). Actinomycetes are gram-positive aerobic bacteria renowned for producing both substrate and aerial mycelium. They associate with non-leguminous plants and fix nitrogen, making it available to the host and nearby plants (Hozzein et al. 2019). They play a critical role in organic matter cycling and suppress plant pathogens in the rhizosphere. Further, they also aids in the decomposition of complex polymer combinations in dead plant, animal, and fungal material and produces extracellular enzymes which assist in crop production (Bhatti et al. 2017). Actinomycetes in agricultural crops improve soil fertility by recycling organic matter and phosphorus (Hozzein et al. 2019).

Another group known as Fungi are very prolific in soil due to their adaptability to changing environments. Fungi help stabilise and decompose soil organic materials (Shah et al. 2021). Fungi convert dead organic materials into CO_2, organic acids, and biomass. Fungal species accumulate hazardous metals like cadmium, copper, mercury, lead, and zinc in their fruiting bodies, making them effective biosorbents. Arbuscular Mycorrhizae Fungi symbionts are recognised as important microbiological components in the development of primary biogeochemical cycles (P, N, and C). Mycorrhizal fungi and plant roots form a mutually beneficial symbiosis (Shaikh et al. 2025). Mycorrhiza fungi are classified as either ectomycorrhiza or endomycorrhiza. VAM fungi can absorb and transport macro and micronutrients (N, P, K, Ca, S, Cu, and Zn) from soil to plants. Mycorrhizal hyphae can break down large biomolecules into smaller molecules, such as N or P (Verma et al. 2021). Hyphae can increase root surface area and absorb nutrients up to 12 cm away from the root surface (Fall et al. 2022). AMF plays an important function in increasing soil P availability. Indeed, it is a P activator that could accelerate the process to change P into bio-available forms via a range of chemical reactions and biological interactions (Zhu et al. 2018). AMF can also convert inorganic phosphate into soluble forms via acidification, chelation, exchange reactions, and the formation of organic acids, H^+ and metabolites (Yadav et al. 2020). It has been proven that AMF's metabolic activities produce alkaline phosphatases, which split soil substrates and make phosphate available (Rodrigues et al. 2021). Furthermore, the organic acid generated by AMF converts insoluble mineral phosphate to soluble form (Rana et al. 2020). AM fungi in plants boost chlorophyll levels, disease resistance, parasite tolerance, water stress, salinity tolerance, and heavy metal toxicity.

Further, the soil microbiome can also influence soil aggregation indirectly, such as through mycorrhizal fungus, root-associated fungi, viruses, or bacteria that affect plants and plant communities. Bacteria form macroaggregates and microaggregates (less than 250 μm), while fungi primarily form macroaggregates (Totsche et al. 2024). Fungi play a key role in soil aggregation due to their filamentous nature, which allows for particle enmeshment and physical entanglement (Fig. 2). They also produce binding agents and amphiphilic proteins that can modulate surface polarity, known as hydrophobins. A meta-analysis found that arbuscular mycorrhizal fungi, symbionts of most land plants, significantly improve soil aggregation (Lehmann et al. 2021a). Extracellular polymeric substances (EPSs), such as polysaccharides, proteins, and lipids, can act as binding agents for aggregation, including bacteria (Costa et al. 2018). Cyanobacteria in biocrust communities at the soil surface produce EPSs, which can help stabilise soil (Chamizo et al. 2018). Microbial interactions with clay minerals become increasingly essential as the scale of soil particles decreases (Cuadros 2017). Bacteria and fungi build biofilms on mineral and organic surfaces, altering mineral characteristics through weathering. Adsorption of microbial cells or cell products can lead to the development of microaggregates (Olagoke et al. 2022). Soil aggregation is influenced by diverse microbial groups at different scales, including the smallest, intermediate, and largest.

3.3 Soil Enzyme Activities

Soil enzymes are another essential indicator in the ecology of soil microorganisms because they govern the direction of biochemical changes in the soil. They engage in circulation of elements and play a crucial role in humus synthesis and breakdown, mineral substance release and supply to plants, molecular nitrogen fixation, xenobiotic detoxification, nitrification, and denitrification. Extracellular enzymes released by soil microorganisms play an important part in the breakdown of organic matter as well as the global cycles of C, N & P and they also serve as soil health and fertility indicators. Soil enzymes accelerate the decomposition of organic crop residue by increasing the contact area between substrate and water during hydrolysis, releasing plant nutrients into the soil solution (Dotaniya et al. 2021). The rhizosphere is a unique ecosystem where microorganisms are selectively enriched due to root activity and exudates. Enzyme activity is more prominent in the rhizosphere zone, influencing mineral biogeochemistry and nourishing soil-plant ecosystems (Datt and Singh 2019). The following subsections will cover some of the most important microbial enzymes:

(a) **β-Glucosidases**: BG is a prevalent and major enzyme in soils, where it catalyses the hydrolysis of different glucose derivatives. They are exceedingly diversified enzymes because of the large variety of glycosidic linkages in their substrates (Sun et al. 2022). β-Glucosidase breaks down the β (1 → 4) glycosidic bonds of cellobioses and cellotrioses to produce glucose. They selectively catalyse the final reaction by hydrolysing cellobiose residues, as well as the hydrolysis of maltose and cellobiose (Wei et al. 2024). These activities, finally yield glucose as an end product (Wang et al. 2023). Plant residues and litter contain considerable levels of cellulose, which must be broken down by the soil microbial community. β-glucosidase breaks down low molecular weight carbohydrates in soil organic matter. The majority of soil bacteria generate the β-glucosidase enzyme, which is becoming an important indicator of microbial activity, akin to soil Dehydrogenase activity (Dotaniya et al. 2021). Unlike dehydrogenase, this enzyme is located outside of the cell. β-glucosidase releases glucose during organic matter decomposition, influencing the proliferation of soil microbes that prefer glucose as a carbon source. Soil health, physicochemical attributes (e.g. pH), and crop management techniques (e.g. fertiliser application, organic residues) all have a significant impact on its stability and seasonal variability. BG activity is intimately related to soil organic matter (SOM), biological activity, and carbon cycling. Some studies reported that BG activity reduces dramatically when soil pH increases, especially in paddy soil, where pH values range from 4.3/4.5 to 7.4/8.5 (Adetunji et al. 2017). It has also been demonstrated that soils rich in organic material with a high C:N ratio and lignified roots have lower BG activity and slower organic matter breakdown (Krolicka et al. 2018). On the other hand, soils rich in easily decomposable organic matter, have higher levels of BG activity. As a result, adding soil organic wastes including biosolids, manure, urban sludge, and poultry litter boosts the activity of this enzyme in the soil (Stirling et al. 2017). BGs are important enzymes in soil that predominantly catalyse the hydrolysis of glucose derivatives, impacting cellulose degradation, C cycling, and maintaining the availability of C sources for the soil microbiota (Meena et al. 2017). However, their activity is greatly influenced by soil pH and organic matter concentration. Soil β-glucosidase activity is a reliable predictor of the impact of crop management strategies on soil health. It can help improve and quantify ecological changes.

(b) **Urease**: Nitrogen input into the soil is critical to sustain optimal crop productivity. Crop residues utilised as organic inputs are typically low in nitrogen due to high C:N ratios. Chemical fertilisation provides nitrogen in the form of urea ($CO(NH_2)_2$), which plants cannot absorb (Lincoln and More 2017). Plants require ammonia or ammonium ions, which must be released through degradation. Soil microbes release enzymes that break down urea to produce ammonium and CO_2. This raises soil pH and causes nitrogen to be lost to the atmosphere via NH_3 volatilisation (Huang et al. 2021). Urease is produced by various microbial groups in soil, including fungi, bacteria, yeast, algae, and plant roots. This soil enzyme is found in both extracellular and intracellular forms in microorganisms. The strength of the soil arose when the Urease activity was between 4.4 and 9.5 mM hydrolysed urea per minute, demonstrating that maintaining Urease activity is critical for efficient biocementation (Gao et al. 2020). Moreover, urease activity influences the rate of nitrogenous nutrient introduction into the soil and can be utilised to estimate N uptake or loss (Nim and Wong 2019). Because its activity increases with organic fertilisation and declines with soil tillage, it has been widely employed to assess changes in soil quality caused by management (De Deyn and Kooistra 2021). Several factors influence the stability of urease, including temperature, moisture content, microbial community, and soil physical and chemical properties (Vaskevicius et al. 2023). Urease also catalyses the hydrolysis of hydroxyurea and semicarbazide, with Ni^{2+} ions acting as a cofactor (Adetunji et al. 2017). Urease is important in soil ecosystems because it plays a function in the nitrogen cycle, specifically in hydrolysing urea to release ammonium ions and altering soil pH. Urease activity, which varies depending on factors such as organic fertilisation, soil tillage, and environmental conditions, is an important indication of soil quality and nitrogen management in agricultural and ecological scenarios.

(c) **Acid/Alkaline Phosphatases**: Phosphorus is a crucial nutrient for plant growth, but its availability is limited to only 15%–20%, despite ample sources in the soil. Even with chemical fertilisers, over 80% remains immobile in the soil, making it difficult for plants to absorb during crop growth. Phosphotase is crucial for the mineralisation of organic P, particularly in tropical climates (Zhang et al. 2023). Enzymes called APs catalyse the hydrolysis of phosphomonoesters and, in certain situations, phosphodiesters, resulting in the release of phosphate. Because plants only consume inorganic P and a major amount of soil P is organically bound, mineralisation of this organic component can have a significant impact on plant nutrition (Park et al. 2022). P transformation-related AP hydrolyses the ester bonds in organic P in soil, promoting P conversion and hence influencing plant P availability. When there is a deficiency of P in the soil, plant roots and microorganisms enhance AP secretion to improve phosphate solubilisation and remobilization, which influences the plant's ability to withstand P-stressed circumstances (Zheng et al. 2022). Ps are often triggered when soil P levels are low. The most thoroughly studied soil AP is phosphomonoesterase, which plays an important role in the biogeochemical cycle of P. It catalyses the hydrolysis of phosphate monoesters, producing free phosphate, which is required for biological absorption (Adetunji et al. 2017). This enzyme targets low molecular weight P molecules containing monoester linkages, such as nucleotides, sugar phosphates, and polyphosphates. Its activity is dependent on soil pH, and it can function in both acidic and alkaline circumstances depending on the ideal pH (Chen et al. 2019b). As a result, acid phosphatase (AcP) activity is more common in acidic soils, whereas alkaline phosphatase (AlP) activity is predominant in alkaline soils. The soil pH has a considerable impact on the rate of AP synthesis, release, and stability (Kunito et al. 2022). It was also discovered that APs can be suppressed in organically amended soils, although mineral fertilisation increases enzyme activity. It was also reported that lead and other heavy metals in soil inhibited AP action (Rahul et al. 2022). AP is essential for converting organic P into a form useable by plants, and its activity is regulated by soil pH, environmental stresses, and heavy metals. This enzyme's functionality directly affects plant growth and soil health by adjusting to changing environmental conditions.

(d) **Dehydrogenase Activity (DHA)**: DHA activity is often employed to assess biological activity in soils as it reflects the total oxidative metabolic activities of soil microbes (Ujjainiya et al. 2022). DHA is an intracellular enzyme of the oxidoreductase family. DHA is known for oxidising SOM by transferring protons and electrons from acceptors to substrates. These mechanisms are part of the soil microorganism respiration pathways, and DHA activity is substantially influenced by the metabolic status of the soil biota and has a considerable correlation with soil biomass C (Daunoras et al. 2024). They transport hydrogen ions and electrons, facilitating cellular oxidation and reduction reactions. Dehydrogenases, which oxidise various substrates to produce reductants, play a crucial role in respiration and electron transport. These dehydrogenases convert H^+ ions to either the nicotinamide adenine dinucleotide or its phosphate (Fang et al. 2022). Dehydrogenases oxidise soil organic materials by transferring protons and electrons to inorganic acceptors. After nutrient modifications, DHA activity increases as microbial populations grow (Zaborowska et al. 2017). Because these reactions are part of soil microorganisms' respiration pathways, studying the activity of the DHA enzyme in soil is significant because it can indicate the soil's ability to support biochemical processes required for soil fertility and health.

3.4 Soil Fauna

Soil invertebrates are creatures that live continuously or temporarily in soil or on its surface. They include all stages of development. Soil invertebrates play a crucial role in soil quality. They are classified as microfauna (e.g., nematodes), mesofauna (e.g., mites, colembola, enchytraeids, diplopoda), and macrofauna (e.g., isopods, earthworms, diptera, and coleoptera). Soil fauna influences the physical and chemical composition and accessibility of organic materials (El-Naggar et al. 2019). The most important groups of soil invertebrates along with their functions are given below:

Nematodes Nematodes are one of the most diverse and abundant animal phyla. They have high genetic diversity and morphological plasticity, allowing them to live in a wide variety of environments. Nematodes are the most abundant multicellular creatures that feed and reproduce in the water layer around and within soil aggregates. However, a tiny number of plant parasite organisms spend a considerable portion of their lifespan in roots. Healthy soils often feature a large and diverse population of free-living nematodes in complex food webs with long chains and feedback loops, as well as a low number of herbivores (Martin et al. 2022). Nematodes, as major components of soil food webs, play a critical part in all soil processes and ecosystem activities (Lazarova et al. 2021). They play a role in C transformation and nutrient cycling, which originate from active plant roots ("direct pathway") and dead plant wastes ("indirect pathway"). Nematodes promotes plant growth and primary productivity, C and nitrogen cycling, and pest and disease control, including entomopathogenic nematodes (EPN) (Weisser et al. 2017). Nematodes also indirectly contribute to the breakdown of soil organic matter (Manning et al. 2018). The majority of free-living nematodes feed on bacteria and fungi, and their impact can be both direct and indirect, since they excrete organic substances into soil and modify the size, composition, and activity of the microbial community (Sechi et al. 2017). Microbial feeders can enhance the competitive ability and dispersion of beneficial microbiota by specifically grazing on harmful soil microorganisms (Bokhorst et al. 2017) or by transferring beneficial bacteria to the rhizosphere as vectors (Poeydebat et al. 2017) and vice versa.

Earthworms Earthworms play a crucial function in agricultural ecosystems. Earthworms play a crucial role as soil "engineers" (Ghosh 2021). These organisms can create organomineral formations with specific physical, chemical, and microbiological properties through movement. Earthworms contribute significantly to the breakdown and decomposition of organic matter, releasing and recycling nutrients. Earthworms consume more surface organic materials than any other soil organisms combined. These elements are excreted in the form of casts, which include water-soluble nutrients that plants can easily absorb (Murariu 2024). Earthworms move crop residues, plant litter, and partially degraded materials to the sub-surface layer, where they consume and fragment them. Earthworms are one of the major decomposers of organic matter. Earthworms deposit their faeces, known as cast, on the soil surface, either inside their burrows or in open places beneath the soil surface (Bastida et al. 2020). Earthworms also boost soil mineralisation and nitrogen levels through superior nitrification in their casts. Earthworm biomass allows for direct nitrogen bypass in terrestrial ecosystems. *L. Terrestris* in forest in England is projected to return up to 60–70 kg nitrogen per ha in the form of dead tissue to the soil over a year (Edwards and Arancon 2022). Earthworm tissues degrade quickly, allowing nitrogen to be easily mineralised. The presence of nitrogen-fixing bacteria in earthworm guts and casts improves nitrogen fixation compared to soil, increasing the activity of the nitrogenase enzyme (Ezeokoli et al. 2021). Earthworms help in improving soil aeration, infiltration, structure, nutrient cycling, water movement, and plant growth. They obtain their sustenance from microorganisms that reside on organic matter and in soil. Earthworms create tubular pathways or burrows as they feed through the soil (Akhila and Entoori 2022). These burrows can last for a long time in the soil. Earthworm burrows enhance soil porosity, allowing more air and water to enter. Increased porosity reduces bulk density improving and promotes root growth.

Microarthropods Soil microarthropods, which include springtails (Collem bola) and mites (Acari), play a crucial role in various soil activities. Microarthropod interactions in soil provide a link between smaller and larger scale activities due to their intermediate size (Innocenti and Sabatini 2018). Microarthropods interact with soil and other biota, affecting plants both indirectly and directly. Microarthropods have a significant impact on plant health by influencing soil organic matter decomposition, nutrient cycling, pathogen density, and plant-soil feedbacks (Neher and Barbercheck 2019). Microarthropods can affect SOM-plant interactions in several ways. Plant residue fragmentation alters the physical structure of particulate soil organic matter (Minasny and McBratney 2018). Translocating plant remains from the soil surface alters the vertical distribution of organic matter in soil (Potapov 2022). Microarthropod excreta contains high levels of nitrogen and other nutrients, which can mobilise nutrients and impact soil microbial activity (Winck et al. 2020). Mites play a crucial part in the micro- and macro-elements circuit, responsible for the growth of primary producers. Microbiota and mites play a crucial role in accelerating the biodegradation and decomposition of organic compounds in ecosystems, working in tandem with abiotic forces. Collembola have a significant part in decomposition. Larger species promote mineralisation, while smaller species contribute to soil humidity (Jernigan et al. 2023). Certain species in agroecosystems can speed up decomposition, impact microbial activity, and affect nutrient cycles. Phytophagous species affect the distribution of mobile metals, such as potassium, while detritivores speed up the mineralisation of less mobile elements (Santorufo et al. 2023). It can be concluded that the microarthropods affect plants via interacting with microbial communities, which influence organic matter decomposition, nutrient cycling, and plant-pathogen dynamics in soils.

4 Modulation of Soil Biological Properties by Biochar

4.1 Biochar as a Microbial Habitat

Biochar is one of the most desirable materials for use as an inoculum carrier. Biochar manufacturing is more economically profitable than other carriers due to the wide availability and inexpensive cost of feedstock, as well as its high sterility (Zheng et al. 2019). Several studies have shown that biochar can be used as a carrier for microbial inoculants under a variety of climatic and environmental circumstances, increasing soil productivity and stimulating plant growth and nutrient uptake (Fig. 3) (Egamberdieva et al. 2018). These beneficial bacteria play important roles in carbon and nutrient cycle, as well as soil biochemical processes (Chuaphasuk and Prapagdee 2019). Furthermore, biochar-based carriers loaded with microbes have demonstrated significant potential in the elimination of numerous environmental pollutants (Zhu et al. 2022). Soil physicochemical properties can impact microbial community structure and function by changing their habitats, either positively or adversely. Improvements in soil characteristics, including pH, water retention capacity, and nutrient availability (Li et al. 2019a, b, c), may stimulate microbial

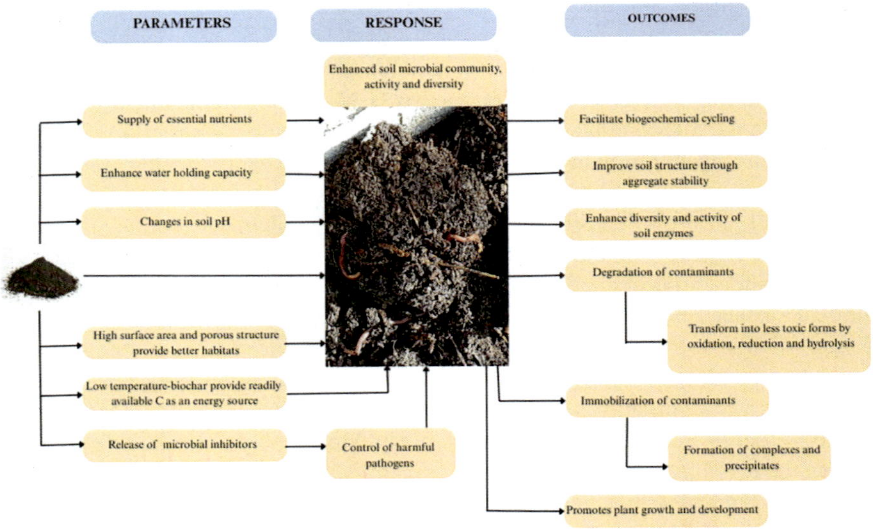

Fig. 3 Conceptual framework illustrating the role of biochar in modulating soil biological properties and enhancing plant growth

development. Biochar's nutrient composition and concentration depend on feedstock and production conditions, including temperature and retention time. Increasing production temperature leads to higher biochar pH, total surface area, carbon-to-oxygen ratio (C:O), and C:N, despite decreasing dissolved organic matter and C content (Xu et al. 2018c). Biochar supports microbial communities by supplying labile carbon substrates for decomposition. The porous structure and large surface area of biochar provide an ideal environment for soil biota (Ye et al. 2017). The specific area and porosity of biochar vary based on feedstock type and production temperature. It was found that high temperatures accelerate the volatilisation of organic matter and water during thermal decomposition, resulting in larger pores (Rahman et al. 2018). These pores support a variety of soil biota, including bacteria (0.3–3 mm), ectomycorrhizal fungi, ericoid mycorrhizal fungi, and arbuscular mycorrhizal fungi (2–80 mm) (Wong et al. 2019). Bacteria and fungal hyphae (<16 μm) colonise charcoal pores, providing protection from predators like mites and worms. AMF has also been shown to colonise biochar pores. Mesopores and micropores hold water and soluble chemicals used in microbial metabolism, such as acids, alcohols, aldehydes, ketones, and sugars (Li et al. 2018). Biochar may influence the growth and activity of bacteria that benefit from its physical attributes. Further, surface area of biochar plays a crucial role in promoting soil microbial growth. It typically exceeds feedstock materials and soil levels. Soybean stover biochar generated at 300 and 700 °C had surface areas of 5.61 and 420.33 $m^2\ g^{-1}$, while pine needles biochar produced at the same temperatures had 4.09 and 390.52 $m^2\ g^{-1}$, respectively (Irfan et al. 2019). Increasing porosity and surface area during pyrolysis can facilitate microbial colonisation, depending on the temperature and feedstock used. Biochar has a higher specific surface area and porosity, leading to better water retention for microbial growth in various soil types (Wong et al. 2017).

Biochars can significantly influence soil microbial abundance and activity by modifying soil chemical characteristics, including pH. A sufficient amount of biochar can be added to the soil. Biochar's surface contains negatively charged hydroxyl, carboxyl, and phenolic groups, making it alkaline (Chen et al. 2017b). Carbonates, bicarbonates, and silicates in biochar can bind with hydrogen ions in soil water. Reducing H^+ concentration leads to a rise in pH value. Biochar treatment typically increases soil pH (Wong et al. 2017). Luo et al. (2017) reported that addition of bamboo stick biochar to sandy loam soil boosted the microbial population. The alkaline pH, EC, and K^+ levels enhance soil conditions for microbial colonisation (Chen et al. 2017a). Biochar treatment can increase soil nutrient concentration, perhaps promoting biota activity and growth. Luo et al. (2018) found that applying maize straw-derived biochar at 30 t ha^{-1} to a silt loam soil boosted soil organic carbon, total nitrogen, and total phosphorus levels, boosting nutrient availability. They also reported that there was an increase in the relative abundance of AMF and theAMF/saprotrophic fungi.

4.2 Biochar Effects on Soil Microbial Community Composition

Soil microbial diversity and activity are strongly linked to organic carbon content, soil nutrient cycle, and plant productivity (Fig. 3) (Song et al. 2018). The effect of biochar inclusion is determined by soil type, biochar quality and dosage, as well as the activity of soil microorganisms (Farkas et al. 2020). Several studies have shown that biochar can improve soil microbial diversity (Table 3) (Liu et al. 2019; Xu et al. 2018a, b, c). Three probable mechanisms include that leads to increase are: (i) The relatively large surface area and porous nature of biochar can provide a habitat for soil microbes, allowing them to live while protecting them from predation and exposure to adverse circumstances (Tang et al. 2020). (ii) The addition of biochar to the soil may alter the physico-chemical properties of the soil, such as improving the soil cation exchange capacity (CEC), soil pore space, and soil water content, which may have a positive impact on soil microbial diversity (Li et al. 2019a, b, c). (iii) The inclusion of biochar would enhance the growth and reproduction of soil microorganisms (Li et al. 2021) because it affects the pH and adds organic carbon, nitrogen, phosphorous, and potassium to the soil, which will boost soil microbial diversity (Chen et al. 2018). Added nutrients may stimulate nutrient cycling in the soil, hence increasing soil microbial diversity. It was discovered that incorporating biochar in soil could improve the diversity of soil bacteria. The increase in soil bacterial diversity following biochar application could be attributed to the many organic compounds included in biochar, whilst the porous structure of biochar provides a constant supply of nutrients and increases the ecological habitat for microorganisms (Kolton et al. 2017). When biochar is applied at a specific concentration, it alters the pH, nutrient concentration, water holding capacity, and other physicochemical characteristics of the soil, promoting the development and reproduction of soil bacteria

Table 3 Impact of biochar application on soil biological properties and crop yield in selected field studies

Biochar Type and Dose	Soil Type	Effect on Microbial Biomass	Effect of Yield	References
Eupatorium plant biochar @ 15 t ha^{-1}	Silty loam Inceptisol	Enhance microbial count and microbial diversity (Acidobacteria and Proteobacteria by 2% and 1.7%)	50% increase in maize yield and 96% increase in mustard yield	Pandit et al. (2018)
Wheat straw biochar @ 10 t ha^{-1}	Ultisol (Red soils)	Increased soil microbial biomass C, N, and enzymatic activities	30% increase in rapeseed yield was observed	Jin et al. (2019)
Cotton stalk biochar @ 5 t ha^{-1}	Fluventic Haplocambic (silty loam calcareous soil)	Enhanced microbial communities of bacteria and fungi	Increase grain yield of wheat by 11%	Khan et al. (2021)
Biochar of Organic residues (digestates produced from a biogas plant) @ 17.1 t ha^{-1}	Gleyic Plaggic Umbrisol (Sandy soils)	Soil biomass carbon concentration increased by 4%	Increase crop yields by 33%–37% in corn	Davys et al. (2023)
Rice husk biochar @ 15 t ha^{-1}	Sandy loam soils	Microbial biomass C:N:P ratio increased from 242.12:18.30:11.49 μg/g to 302.30:18.81:11.75 μg/g	Increase yields of tomato by 28%	Adebajo et al. (2022)
Peanut shell biochar @ 101.25 t ha^{-1}	Solonchak (Clay loam soils)	Increased catalase and alkaline phosphatase activity by 20%–54.76% and 17.54%–21.13%	Increased yields of rice by 7.07%	Yao et al. (2021)
Corn stalks biochar@ 40 t ha^{-1}	Mollisols	Enhanced activities of phosphotase by 70%, sucrose by 56% and catalase by 30%	Increase maize yield by 37%	Yan et al. (2022)
Rice husks biochar @ 12 t ha^{-1}	Isohyperthermic Typic Oxic-Paleustalf Alfisol (sandy clay loam)	Increased the activities of urease activity by 40%, invertase activity increased by 9% and catalase activity by 11%	Increase grain and straw yield by 46% and 47% of rain-fed upland rice	Oladele (2019)
Maize straw biochar @ 5 t ha^{-1}	Albic Luvisols (loamy sand)	Increased the community and diversity of soil bacteria (*Firmicutes*, *Proteobacteria*, and *Acidobacteria*)	Increase the yield of maize crop by 20%	Wolna-Maruwka et al. (2021)

(continued)

Table 3 (continued)

Biochar Type and Dose	Soil Type	Effect on Microbial Biomass	Effect of Yield	References
Maize straw biochar @ 75 t ha^{-1}	Vertisol	Increased the abundances of bacterial diversity (*Micrococcaceae*, *Gaiellale*, *Sphingomonadaceae*, and *Nocardioidaceae*)	Increase wheat and maize yields by 8.4 and 8.9%	Chen et al. (2022)

and thereby expanding their diversity (Zhang et al. 2018a, b). These soil bacteria are more mobile and can easily enter the pores of biochar, or that they are more adaptable and can use more nutrients and minerals (He et al. 2021). The fungal population reduced following biochar addition because, unlike bacteria, soil fungal communities are more stable and less variable (de Vries et al. 2018). Fungal mycelia can be adsorbed onto the granular structure of soil, resulting in lower mobility than bacteria. A lower absorption of nutrients and mineral elements brought by biochar may also reduce reproductive advantage for fungi leading to the decrease in fungal dominance.

Biochar promotes soil fertility by directly providing nutrients including nitrogen, phosphate, and potassium (Fig. 3). Biochar alters the composition, biomass, and enzyme activity of microorganism communities, especially those capable of nitrogen transformation. The mechanisms involve the provision of refractory and metabolisable carbon, mineral ions, sorption events, pH changes, and physical features such as pore shape and surface area (Seneviratne et al. 2017). As these mechanisms vary, the nitrogen cycle produces more or fewer substrates and products. The impact of biochar on fixation rates is mostly due to its provision of two types of carbon: metabolisable and recalcitrant, both of which have been shown to influence the number of nitrogen-fixing organisms. Biochar addition alters the abundance of the Bradyrhizobiaceae and Hyphomicrobiaceae families, increasing relative abundance by up to 11% when compared to the control (Rittl et al. 2021). Root nodulation can alter the rate of biological fixation, and the presence of biochar can affect nodulation rate, development, and nitrogenase activity. For example, pyrolysis of bamboo improved root nodulation by 243%, resulting in greater soybean growth (Wang et al. 2018). The primary mechanisms of influence include the availability of metabolisable carbon, which influences microorganism growth. For example, because the process of ammonification is mediated by microorganisms (typically *Bacillus* spp., *Proteus* spp., and *Pseudomonas* spp.) that rely on nutrient, energy, and water resources, as well as communication mechanisms, biochar can influence the rate and outcome of ammonification via its effect on soil structure and WHC, the provision and sorption of toxins, signalling compounds, nutrients, and energy sources (that is, metabolisable carbon) (Lataf et al. 2022).

Biochar amendment can directly provide P to soil, increasing its availability (Xu et al. 2019). The impact of biochar on soil P status depends on its characteristics, dosage, amended soil type, and agricultural methodologies followed (Zhu et al.

2018). Biochar amendment boosts soil nutrient levels and protects microorganisms. Low-temperature biochar amendment contains partially degraded nucleic acid and phospholipid, which can be used as a source of P for microorganisms that absorb it. Soil P-solubilizing bacteria can liberate inorganic P from biochar, increasing soil P availability by speeding organic P mineralisation and activating insoluble mineral P (Alori et al. 2017). For example, After 80 days of incubation with 3% 600 °C *Cunnighamia lanceolata* leaf biochar, the abundance of *Burkholderia-Para-burkholderia* and *Planctomyces* increased by 123% and 436% in an acidic red loam soil, respectively. After 300 °C biochar incorporation, the abundance increased by 959% and 91%, respectively. Zhou et al. (2020) found that amending soil raised total P content by 211 and 85 mg kg^{-1} and accessible P content by 20.9 and 5.86 mg kg^{-1}. P-solubilizing bacteria secrete organic acids such as lactic, malic, and gluconic acids, which can dissolve apatite (Zheng et al. 2019). Inoculating Aspergillus niger into a sterilised 1:1 mixture of 480 °C holm oak biochar and P rock increased citric, gluconic, and oxalic acid content by 97.8%, 128.6%, and 230.5%, respectively, and increased P solubility by threefold compared to the control group (Wei et al. 2020). Biochar promotes plant growth, microbial colonisation, and enhances alkaline phosphatase activity in the rhizosphere (Wang et al. 2019b). Yang and Lu (2021) found that adding 2% water hyacinth-derived biochar to red soil enhanced microbial biomass and alkaline phosphatase activity by 218% and 22.8%, respectively. Alkaline phosphomonoesterase activity increased from 100 to 175 mg p-nitrophenol kg^{-1} h^{-1} and 150 to 250 mg p-nitrophenol kg^{-1} h^{-1} in silt loam and clay soils amended with 1.5% swine manure derived biochar, respectively (Kong et al. 2021).

4.3 Impact on Enzyme Activities

Soil microorganisms use secreted enzymes to assist or catalyse several biogeochemical processes in nature. Soil enzymes play a crucial role in microbial activity and nutrient cycling, making them essential indicators of soil quality (Jaiswal et al. 2018). Biochar additions can improve degraded soil quality by boosting biological attributes, including enzyme activity (Table 3). Evaluating changes in dehydrogenase activity in biochar-amended/unamended soils can indicate the degree of recovery of damaged soils (Johnson et al. 2017). The inclusion of different forms of biochar to soils increased dehydrogenase activity under various situations (Irfan et al. 2019). Biochar-treated soils have higher dehydrogenase activities due to the inclusion of labile organic matter and the high volatile matter content of the biochars (Chen et al. 2017a). Pandey et al. (2016) found that adding charcoal boosted soil enzyme activity: dehydrogenase by 27% and urease activity from 7.4% to 39%. Enzyme activity varied depending on microbial populations. It was found that increasing the colocalization efficiency of microorganisms and carbon on biochar lowered the need for enzyme production (Nguyen et al. 2023). Further, the amount of biochar used correlates with the point of change in enzyme activity. Additionally, β-cellobiohydrolase and β-glucosidase play crucial roles in the carbon cycle and soil cellulose breakdown. Biochar treatment at a rate of 12 t/ha resulted in sweet glucose and low enzyme levels at maximum speed (Qiu et al. 2018).

The porous structure of biochar supports soil organisms and microorganisms, enhances biological activity and diversity (Palansooriya et al. 2019) and promotes soil aggregate formation. Biochar application significantly raised microbial biomass and the activities of urease, alkaline phosphatase, dehydrogenase, and N-acquiring enzymes. Alkaline phosphatase is produced by a variety of microorganisms, plants, and even soil flora (Alharbi et al. 2024). The enzyme hydrolyses organic phosphates, releasing inorganic phosphate that plants can absorb. This process improves nutrient availability and plant development, both of which are critical for increasing soil productivity (Timofeeva et al. 2022). Application of compost biochar was reportedly found to increase the activity of alkaline phosphatase (Zaid et al. 2024). However, the magnitude of these increases varied depending on soil properties, biochar characteristics, feedstock, pyrolysis temperature, and application rate. Biochar application can boost microbial and enzyme activities in soil, known as the priming effect (Zhang et al. 2019). Although biochar contains less labile carbon than recalcitrant carbon, it has been shown to stimulate short-term microbial growth and enzyme activities when added to the soil. Biochar's surfaces and pores create a habitat for microorganisms and stimulate the passage of air, water, and nutrients in soil, leading to increased microbial activity (Rahmanian and Khadem 2024). Biochar pores provide shelter for soil microorganisms (bacteria and fungus), leading to increased microbial biomass and enzyme activity (Pokharel et al. 2020). The dark colour of biochar can trap heat and increase soil temperature, potentially promoting microbial growth and enzyme activity.

4.4 Interactions with Organic Matter

As a popular soil supplement, biochar has been demonstrated to have a considerable effect on the mineralisation of soil organic carbon (SOC), also known as the priming effect (PE) (Gaudel et al. 2022). PE is a short-term alteration in soil organic matter (SOM) breakdown caused by pulsed or continuous inputs of nutrients. The PE plays a critical role in soil carbon and nutrient cycling, as well as triggering CO_2 fluxes into the atmosphere. The direction and intensity of PE are primarily influenced by the intrinsic features of biochar and soil, the time scale, and the interaction of biochar and soil (Fu et al. 2022). Positive PE occurs when exogenous organic substrate additions increase soil microbial activity, accelerating the loss of native SOC. Biochar induces positive priming through direct effects such as increased microbial activity and enzyme synthesis. Biochar has several benefits, including enhanced mineralisable carbon, microbial nutrient mining (e.g., N and P), and indirect effects such as improving acidity and promoting microbial activity (El-Naggar et al. 2019), alleviating nutrient constraints (de la Rosa et al. 2018), and improving microbial habitat. Soil microorganisms can occasionally switch their choice from native SOC to readily available fresh organic substrates (substrate switching), resulting in negative PE or slowing SOC decomposition (Pan et al. 2021). Changes in soil physicochemical and microbiological characteristics following biochar application can affect the quantity and direction of PE. Wang et al. (2016) found that biochar slowed SOC decomposition by 3.8% compared to soil without biochar.

This increment was because of the ability of biochar to contribute towards extra C storage in soil, which, if maintained for an extended period of time, can result in substantial increases in SOC stocks under certain conditions (Fang et al. 2018a). The binding of soil organic and mineral components to the vast surface area of biochar improves inter-particular cohesion, which is the first step in the aggregate formation and stabilisation process. Biochar increases soil aggregation, thereby, protecting the native SOC from microbial degradation (Weng et al. 2017). Other potential mechanisms for biochar-induced negative priming include substrate switching and dilution effects (microbes preferring to breakdown labile biochar C and other easily mineralisable substrates over native SOC due to biochar). The dilution effect is similar to substrate switching, but labile biochar is not used preferentially by microbes. Instead, there is a temporary increase in total labile C in soil (from both biochar and native SOC) after biochar amendment, resulting in a decrease in SOC mineralisation for microbes. Adding 350 °C sugar maple (*Acer saccharum*) biochar to a Typic Fragiudept may result in a short-term decrease in native SOC mineralisation due to substrate shift (Joseph et al. 2021).

Pyrogenic materials have a high sorption capacity for many natural organic chemicals, including recalcitrant OM (Brown et al. 2024) and labile and water-soluble SOM fractions, reducing their availability to bacteria. Biochar particles can encapsulate SOM, making it physically unavailable to microorganisms, or provide sorptive protection, making SOM less accessible. The slow diffusion rate within biochar pores may limit sorption, postponing its effects (Cui et al. 2017). Biochar's surface and pore networks may become clogged with high molecular weight organic molecules in soil, limiting its ability to absorb dissolved organic matter over time (Raya-Moreno et al. 2017). Biochar's sorptive characteristics may lead to decreased SOC mineralisation, particularly in its more resistant fractions. Biochar may facilitate interactions between soil minerals and organic matter through ligand exchange or cation bridging. Dissolved OM and colloidal particles of biochar can create stable soil aggregates that limit accessibility of microorganisms to SOC (Du et al. 2017). When applying biochar to improve carbon storage in agroecosystems, it's important to evaluate the impact on soil-resident microorganisms, especially those involved in soil carbon cycling. Understanding the interactions between biochar and SOC(-plant) is crucial for predicting the impact of biochar additions on soil carbon stocks throughout time. These rhizodeposits can prime and serve as a source of SOC, and biochar can influence newly added plant-derived C to the soil. When plants were present, a 13C-depleted hardwood charcoal (450 °C) in a subtropical pasture on a Rhodic Ferralsol started positive priming up to 0.15 Mg C ha^{-1} over 62 days and switched to negative priming after 188 days. In addition to decreasing priming brought on by plant C input, biochar increases SOC by 6%–16% by soil aggregation processes that stabilise additional C (i.e., rhizodepos it) (Ventura et al. 2019).

Negative priming can be caused directly by (1) substrate switching, in which microbes may preferentially utilise the readily mineralisable carbon from biochar to temporarily replace the use of SOC (De Ciucies et al. 2018); (2) a dilution effect of substrates, in which the addition of biochar temporarily decreases the mineralisation of the more readily mineralisable carbon in soil; (4) improved organo-mineral protection and stable aggregation, which slows down the mineralisation of SOC within the organo-mineral complexes (Fang et al. 2018b) and (5) suppression of microbial activity by polyaromatic toxic compounds (Zhang et al. 2018a). In comparison to the unamended soils, biochar amendments decreased the metabolic quotient by 12%–21% (i.e., respiration rate CO_2-C per unit of microbial biomass C), improved C-use efficiency (Zhou et al. 2017), increased soil microbial biomass (Li et al. 2020), and decreased the activities of soil enzymes linked to C cycling by 6% (Zhang et al. 2019). The latter was attributed to superior microbial habitats and the mitigation of environmental stresses, including acid soil constraints. In the short term, substrate switching (Weng et al. 2018) and dilution (De Ciucies et al. 2018) are found to be the primary causes of negative priming, with adsorption becoming increasingly significant after a few weeks.

5 Biochar-Induced Yield Enhancement: Biological Pathways

5.1 Improved Nutrient Mineralization and Availability

Amendments like biochar also improved the physical characteristics of the soils, including bulk density and several aspects of soil structure (Blanco-Canqui 2017). Furthermore, by creating environments with adequate aeration, water, and nutrients, biochar amendments may improve the soil's ability to hold water (Razzaghi et al. 2020) and alter the composition, abundance, and activities of microbial communities in soils. The addition of biochar to soil may modify important soil processes as OC formation, carbon mineralisation, and nutrient transformations as a result of these alterations in microbial populations (Ullah et al. 2020).

The composition, biomass, and enzyme activity of communities of microorganisms, particularly those with the capacity to convert nitrogen, all are impacted by biochar (Table 2). The mechanisms include the supply of mineral ions, metabolisable and recalcitrant carbon, sorption phenomena, pH changes, and physical characteristics including surface area and pore structure (Tan et al. 2023). The nitrogen cycle produces more or fewer substrates and products as a result of these processes changing. Fixation is primarily influenced by either physical protection against fungal grazers or the encouragement of mycorrhizal root colonisation through the availability of carbohydrates that make up the metabolisable fraction of carbon given by biochar (Mosa et al. 2023). Because biochar absorbs nitrogen and decreases its availability in soil, it may have a significant impact on nodulation (Wang et al. 2017). It absorbs contaminants that hinder the growth of microorganisms, such as fixers, but it can also disrupt signal pathways, which may stop nodule formation.

Although this effect may wane with time, biochar can change the pH of soil in a way that benefits soil fixers (Lataf et al. 2022). Nitrogen-fixing organisms have been reported to grow when metabolisable and recalcitrant forms of carbon are provided (Lee et al. 2021). Any pH shift caused by biochar has the potential to either increase nutrient availability (and thus leaching) or decrease nutritional availability (Tomczyk et al. 2020). Furthermore, enhanced microbial proliferation and nitrogen immobilisation may be caused by the metabolisable carbon fraction (Zhu et al. 2017). The microorganisms (usually Bacillus, Proteus, and Pseudomonas species) that promote the ammonification process depend on nutrient, energy, and water resources as well as communication mechanisms. Biochar can affect the rate and outcome of ammonification by influencing soil structure and WHC, as well as the provision and sorption of toxins, signalling compounds, nutrients, and energy sources (metabolisable carbon). Recalcitrant organic sources of nitrogen may mineralise if there is a nitrogen deficit in that soil (Rivka et al. 2018). The kind and quantity of mineralising microorganisms might be impacted by any pH shift (Mao et al. 2019). Mineralisation may be restricted by sorption of substrates (NH_4^+) (Hagemann et al. 2016), while nitrifying populations may benefit from sorption of contaminants (Fawzy et al. 2021).

Both biotic and abiotic conditions influence tranformation of nutrients. The ash fraction of biochar contains P. When biochar was added to the soils, plants absorbed more P. Through a mechanism of anion and cation exchange that interacts with P, biochar can alter the availability of P to plants (Heidari et al. 2020). Because of its liming impact, poultry-based biochar can raise the pH of soil. It has been discovered that biofertilizers significantly contribute to the solubilisation of unavailable P and the enhancement of P uptake by plants (Nandini and Prakasha 2022). Biochar can also serve as a promoter to create a favourable habitat for microorganisms that can mineralise and solubilise both organic and inorganic P so that plants can use them. Soil microorganisms may be impacted by the sorption effects of P using biochar (Li et al. 2019a). The porous structure of biochar provides favourable habitat to microbes, potentially protecting bacteria from protozoa and other predators (Rafique et al. 2017). This potential of biochar could have a more effective impact on the transformation of nutrients. Biochar usually comprises of ash, labile carbon, and recalcitrant carbon. While labile C is one that mineralises easily, recalcitrant C controls the biochar's chemical stability (Wang et al. 2021). The mineral elements (macro and micronutrients) that make up the ash component are important resources for biological activity in the soil food web. As a result, by boosting the microbial population in the soil environment, it supports numerous significant physical, chemical, and biological processes in the soil (Mamatha and Basavarajappa 2024). Phosphorus is an essential element that gets fixed with calcium, iron, and aluminium in soils and become unavailable to plants. However, it has been determined that phosphate solubilising bacteria (PSB) are a crucial biological component in boosting plant access to soil P (Wang et al. 2025). In order to increase plant uptake and productivity, biochar can also serve as a soil conditioner because of its strong potential to create an environment that is conducive to microorganisms mineralising and solubilising both organic and inorganic forms of P (Lu et al. 2023). Because of its high surface area and nutrient content, biochar offers a good environment for PSB colonisation and activity. Phosphate-solubilizing bacteria can solubilise phosphate and potentially provide mangrove plants with soluble phosphorous due to their capacity for phosphatase catalysis (Ali et al. 2020). PSBs have a stable habitat in biochar, which shields them from environmental challenges and increases their lifespan in soil.

5.2 Disease Suppression

The soil environment is made more conducive to plant growth by a combination of enhancements made to the soil by biochar. The efficiency of a biochar as a management technique is linked to its capacity to modify the makeup of soil microbial communities, which affects the growth of beneficial microorganisms that will shield plants and soil from diseases (Akanmu et al. 2020), as well as the growth of fungi belonging to the Trichoderma genus (De Medeiros et al. 2020). By suppressing plant diseases, biochar works with a variety of soil microorganisms to promote plant growth. An environmentally appropriate substitute for treating plant diseases is biochar. The addition of biochar to soil changes root exudates, nutrient availability and cycling, and soil characteristics, all of which have an impact on plant pathogenic populations in the soil (Mansoor et al. 2021). Additionally, adding biochar to soil causes systemic resistance, activates active oxygen species, and triggers a stress hormone response, all of which make host plants more resistant to infections (Rasool et al. 2021). The plant parasitic nematode population in the soil is also influenced by changes in microbial communities, root exudates, and the induction of defence response in host plants. The addition of biochar to soil not only helps to reduce pollution, but its porosity allows microorganisms like bacteria and mycorrhizae to have enough surface area to support them and improve their metabolic processes (Silva et al. 2020). Depolymerisation, decarboxylation, aromatisation, and decarboxylation of O-alkyl carbons, lignin, hemicellulose, and cellulose are among the complicated chemical processes that often occur during the pyrolysis process of organic (lignocellulosic) materials. The action of *Fusarium oxysporum* and *Rhizoctonia solani* in solanaceous crops is inhibited by pyrolysis, which also produces a number of other volatile organic compounds, including furan, terpenes, benzaldehyde, 2,3-butadione, phenol, acetonitrile, and pyrazine (Wang et al. 2019a). The porous structure of animal bone biochar has been found to colonise bacterial populations, including *Bacillus pumilus*, *Streptomyces pseudovenezuelae*, and *Pseudomonas chlororaphis*, as reported by Hui (2021) and Liu et al. (2023). They have utilised this biochar to control tomato seedling disease caused by *Pythium aphanidermatum* and *Fusarium oxysporumf.*sp. *lycopersici*.

In addition to the fact that pores serve as microhabitats, biochar can also promote microbial activity as a result of alterations in abiotic conditions that may provide certain populations with a competitive edge over others (e.g., modulating pH or changing the presence and availability of toxic compounds). Furthermore, microorganisms can use biochar as a source of mineral nutrients or energy (Iacomino et al. 2022). *Trichoderma* sp., one of the most researched antagonistic fungi, suppresses disease through a variety of strategies, such as mycoparasitism, competition, antibiosis, and the induction of systemic resistance in plants (Silva et al. 2017). Trichoderma sp. activity is increased when biochar is added to soil, which lowers the incidence of diseases. Disease suppression is associated with slight changes in soil enzymatic activity, which are sensitive indicators of changes in the soil ecosystem. Co-application of compost and biochar had a synergistic effect on controlling soil-borne diseases and increasing the activity of beneficial soil microbial populations, such as plant growth promoting rhizobacteria (PGPRs), arbuscular mycorrhizal fungi, and other bio-control agents (Ojha et al. 2018). The majority of the various microbial communities found in soil are bacteria, including PGPR. In addition to enhancing plant nutrient availability, PGPR such as *Bacillus subtilis*, *Pseudomonas fluorescens*, *Burkholderiaphytofirmans*, and *Azospirillum* spp. significantly reduce plant diseases and other abiotic stressors (Wang et al. 2018). By creating systemic resistance through metabolic pathways involving ethylene or jasmonic acid (JA), PGPR inhibits foliar pathogens (Attia et al. 2020). Thus, taking into account a balanced application of biological antagonists and soil organic additives offers a novel platform to control both soil-borne and airborne pathogens (Panth et al. 2020).

5.3 Enhanced Symbiotic Associations

BNF is a metabolic process that occurs in root nodules and represents a symbiotic relationship between rhizobacteria and legumes in the rhizosphere and provides access to nitrogen dioxide from the atmosphere. The BNF of legume plants is greatly influenced by the C content of the soil, which controls microbial activity and improves soil physical conditions (Torabian et al. 2019). One useful method of boosting stable carbon forms in the rhizosphere is to incorporate biochar into the soil. Biochars are C-rich soil additives that affect the physicochemical characteristics of soil in a variety of ways (Swagathnath et al. 2019). In addition to improving overall soil conditions and nutrient availability, biochar also has an impact on BNF and nodulation (Ghassemi-Golezani and Farhangi-Abriz 2019). Macil et al. (2020) revealed that biochar enhanced the number of nodules and their weights in three chickpea cultivars. Nonetheless, a number of studies have demonstrated that adding biochar to the soil raises BNF in a variety of leguminous plants, including soybean, faba bean (Ma et al. 2019), clover, and common bean (Yu et al. 2018). Biochar boosts biological nitrogen fixation (Semida et al. 2019), gram-positive, gram-negative, and actinomycetes, as well as N_2-fixing bacteria in mutualists linked to different plant types that form N_2-fixing nodules on roots or free-living soil bacteria (*Azotobacter* sp. or *Azospirillum* sp.) (Semida et al. 2019). Applying biochar to cold-season legumes enhanced their nodule count, nodule weight, BNF, and plant biomass by 23.8%, 24.5%, 57.9%, and 19.8%, respectively (Farhangi-Abriz et al. 2021). There are three established ways that biochar can promote nodulation and development in legumes. First, biochar improves the chemical conditions of the soil, including pH, CEC, and organic C content; second, it increases nutrient bioavailability in rhizosphere and the uptake of nutrients by legume plants; and third, it fosters the rhizosphere's symbiotic activities (Zhang et al. 2021b).

There is a lot of research being done on the affects of biochar on soil–plant–microorganism system. Microorganisms are crucial to the dynamics of soil nutrients, and because of their extreme sensitivity to minor, transient changes in their habitat, they can be employed as bioindicators (Dai et al. 2021). A class of soil microorganisms known as arbuscular mycorrhizal fungi (AMF) develops symbiotic associations with the roots of the majority of plants, improving the vital supply of nutrients and water in exchange for fixed carbon from the host plants (Rosenthal and Gunn 2025). AMF creates a network of mycelium that connects plant roots to the surrounding soil microhabitats and enters the root's cortical cells to create arbuscules that increase the root system's zone of capture (Luginbuehl and Oldroyd 2017). Biochar has been demonstrated to affect soil microbiology, particularly AMF (Jacott et al. 2017). The pyrolysis conditions, the biotic and abiotic conditions of the soil environment (Joseph et al. 2021), or the chemical and structural properties of the pyrolysed feedstock, interactions between biochar and AMF in the soil environment have been found to range from advantageous to detrimental. Applying biochar to soils can enhanced the CEC, porosity, functional groups, and soil structural aggregates (SSA). The creation and redistribution of soil microaggregates may result from changes and increases in the functional groups present in the soil environment (Tomczyk et al. 2020). It is well known that soil aggregation enhances the physical properties of the soil, which support stability, aeration, and water retention (Halder et al. 2024). This improves the environment for microbial activity. Like the effects of changing functional groups, increased SSA, CEC, and porosity brought on by the addition of biochar may have an impact on the quantity, connectivity, and size of pores (also known as macropores) in the soil environment in addition to aggregate forms (Wood et al. 2024). Variations in the macropore conditions can affect the amount of space available for the growth of microbial populations as well as the movement and content of air and moisture, either favourably or unfavourably (Gujre et al. 2021). Fungal spore formation in the soil environment is accelerated by this initial release of chemicals retained by biochar, which also alters the chemical composition of the soil solution and, consequently, the pH and CEC of the soil (Askeland et al. 2019). The chemicals that are released from the surface of the biochar increase its surface area and porosity, which makes it possible to store cations, anions, nutrients, and minerals (Schreiter et al. 2020). Microbial activity around biochar particles tends to rise as a result of this chemical accumulation on the surface. Although the particles of biochar and AMF do not directly interact, biochar functions as a storage medium to help retain nutrients in the soil environment until plants can reach them through root expansions made possible by the mycelium network (Neuberger et al. 2024). Therefore, more AMF mycelium would grow to access the nutrients if the soil contained a higher density of biochar particles that were conditioned to hold available nutrients. Biochar additions have been shown to increase spore release by a factor of ten (Feng et al. 2021). Furthermore, according to experiments, biochar boosts soil temperatures by an average of 2 °C and lessens night-to-day temperature variations, both of which promote the growth of AMF spores (Sun et al. 2021). The growth of AMF spore activity is closely correlated with enhanced soil porosity and carbon availability, which are achieved through tilling biochar into soils (Neuberger et al. 2024). Biochar raises the amount of dissolved organic carbon, cations, and anions in the soil solution, which enhances electrical conductivity and promotes the growth of spores (Joseph et al. 2021). Addition of biochar to soils has significantly altered the colonisation of AMF roots (Sarfraz et al. 2019). The addition of biochar is has improved soil nutrient profiles, favpourable habitat for spores, and additional root surface area for growth, all of which account for these increases in AMF colonisation (Neuberger et al. 2024). Higher rates of AMF colonisation have been observed in the roots of wheat, soy (*Glycine max*), maize (*Zea mays*), and other legumes in soils amended with biochar. Incorporation of biochar to soil has been shown to increase the size and complexity of microbial co-occurrence networks by altering the structure of soil and encouraging the formation of more AMF hyphae (Yan et al. 2021).

5.4 Improved Soil Structure via Biological Agents

Soil aggregates are larger structural units that are formed when soil particles aggregate due to physical, chemical, or biological processes. These structural units are crucial for soil stability, water retention, aeration, nutrient cycling, and the typical growth and metabolism of soil microorganisms. It has been shown that the production of proteins by microbial communities, such as hydrophobic and globular proteins, improves the stability of microaggregates in soil (Yudina and Kuzyakov 2023). It has also been demonstrated that the exra-polymeric substances released by soil microorganisms strengthens the connections between soil minerals, organic matter, and microorganisms, improving the cohesiveness of soil macro-aggregates and aggregate stability as a whole (Sun et al. 2021). Mucilage and hyphae created at the interface between soil particles and biochar, promotes greater microbial activity that adhere to microaggregates and form macroaggregates.

A favourable soil structure can be created by utilising cationic bridges to connect the base cations found in biochar with clay and organic particles (Hansen et al. 2017). The interactions of multivalent ions of biochar with negatively charged surface functional groups on soil minerals and SOM may be advantageous for aggregation (Zheng et al. 2018). Multivalent ions like Fe^{3+} have bridging effects that can improve sorption of soil organic matter to clay minerals (Juriga and Šimanský 2018). Basic cations, which can serve as an interface between the soil's mineral particles and the biochar particles. The stability of soil aggregates is influenced more by hydrophobic than hydrophilic organic matter components (Ramezanzadeh et al. 2023). Because of its highly aromatic C structure, biochar can assist in binding native SOM, increase soil aggregates' resistance to water, and make aggregates more resilient to physical disturbances like wet-dry cycles (Polláková et al. 2018). By altering the ionic composition of the soil solution, biochar can also affect soil aggregation. The improvements in soil aggregation brought about by biochar could be due to a number of factors. By serving as a feedstock for the synthesis of extracellular polymeric compounds that serve as cementing agents for soil aggregates, biochar can also have a major impact on the microbial communities in the soil (Šimanský 2017). Earthworms can impact the soil's aggregate stability in two ways: (1) by mechanically binding soil to biochar particles, and (2) by promoting the growth of fungi through excreted casts. Because they create polysaccharides that bind soil particles together, fungal hyphae are thought to be important stabilising agents for soil aggregates (Rahman et al. 2017). The merging of smaller particles nearly activates soil bacteria, but the joining of larger particles activates soil fungi.

6 Case Studies and Meta-Analyses

6.1 Meta-Analysis of Biochar Effects on Soil Biology and Yield

(i) The ability of biochar to provide these advantages is closely related to its impacts on soil microbial and enzyme activity. Quantitative reviews employing the meta-analysis approach are useful to critically analyse biochar's impacts on microbial and enzymatic activities on a worldwide scale, especially given the current spike in biochar amendment research that incorporate assessment of soil microbial and enzymatic activities. Using 964 data points from 72 publications, Pokharel et al. (2020) conducted a global meta-analysis to evaluate the impact of biochar on soil microbial biomass C and nitrogen (N), as well as the activity of 12 enzymes. They also identified important parameters influencing those soil microbial properties. They observed that the type of soil, the properties of the biochar, and the enzyme significantly alter the influence of biochar on enzyme activities. With no noticeable negative impacts on these enzymes, biochar significantly raised the activity of urease, alkaline phosphatase, dehydrogenase, and microbial biomass C (MBC) by 21.7%, 23.1%, 25.4%, and 19.8%, respectively. Further, the application of biochar was more successful in boosting MBC and enzyme activity in soils with fine textures (such as clay, clay loam, and silt clay), low pH (<6.5), TC (<20 g kg^{-1}), and TN (<2 g kg^{-1}). MBC as well as urease and dehydrogenase activities were elevated by biochars synthesised at pyrolysis temperatures between 350 and 550 °C, with a high pH (>10) and low C/N ratio (<50). The pyrolysis temperature, feedstock type, and biochar properties (such as pH and C/N ratio) related to the pyrolysis conditions and feedstock used also had an impact on the effects of biochar on enhancing the activities of urease, alkaline phosphatase, dehydrogenase, and N-acq enzymes as well as MBC, but those factors had no discernible effect on the activities of other enzymes. With no significant change in urease activities across all biochar application rates, urease activities were raised by 25.9% by biochar made at high pyrolysis temperatures (>550 °C), 23.9% by biochar derived from manure, and 33.5% by biochar with high pH (>10). However, biochars made from wood and crop waste and smaller rates of biochar application (<3%) greatly enhanced alkaline phosphatase activities. Biochars made at low pyrolysis temperatures (<350 °C) and with low C/N ratios (<50) enhanced dehydrogenase activities; however, they were unaffected by feedstock types, pH ranges, or application rates of biochar. Applications of biochar resulted in a 23.3% increase in N cycling enzyme activity while having no effect on C or P cycling enzyme activity. The application of biochar also significantly raised MBC by 21.7%, although MBN was unaffected. The observed rise in this enzyme activity may be the result of increased resource availability, such as labile organic C, or improved reaction kinetics brought about by the addition of biochar to raise the pH of the soil matrix. Applying biochar to soil has also been referred to as the priming effect, which increases the microbial and enzyme activities in the soil (Pukalchik et al. 2018). The addition of biochar to soil has been shown to stimulate short-term microbial growth and enzyme activity, despite the fact that the amount of labile carbon in the biochar is often considerably less than the amount of recalcitrant carbon. The surfaces and pores of biochar provides a stable habitat for microbes, which also improves the flow of nutrients, water, and air through the soil matrix, potentially boosting microbial activity and abundance (Chen et al. 2019b). The growth in microbial biomass and the activity of enzymes released by soil microorganisms (bacteria and fungi) has also been linked to the protection of these microorganisms from grazers or competitors on biochar pores (Lopes et al. 2021). Furthermore, the black colour of biochar traps heat, raising the soil temperature and perhaps accelerating enzyme activity and microbial development. To evaluate the impact of biochar on raising soil temperature, which in turn influenced soil microbial and enzyme activity, more research is necessary. Alkaline phosphatase activity raised in soils treated with biochar, indicating that (i) microbial demand for P increased and (ii) soil P supply for microbial development became constrained (Nguyen et al. 2017). The addition of labile organic C through the application of biochar increased dehydrogenase activity, which is thought to be an excellent indicator of metabolic activity. The application of biochar did not significantly alter the oxidative enzymes that mediate the deterioration of phenolic compounds using oxygen (though phenol oxidase tended to decrease slightly). This suggests that biochar is not essential for the key ecosystem functions of lignin degradation, humification of xenobiotic chemicals containing aromatic rings, and released dissolved organic C (Zhang et al. 2018a). Since fungi are the primary producers of phenol oxidase, the declining trend of this enzyme activity may be connected to the decline in fungal biomass brought on by the addition of biochar, which raises the pH of the soil. When external N is added to the soil (for example, by adding biochar), microorganisms must produce more N cycling enzymes to meet the growing microbial demand of N because the N in the biochar is typically not readily available for microbial consumption (Azeem et al. 2019). The addition of biochar increases the activity of the N cycling enzyme, implying that the microbes in these soils are N-limited. This could be because biochars have high C/N ratios, which can promote N immobilisation in the soil.

(ii) Conducted a meta analysis on the impacts of biochar on biological parameters of soil and the role of microbial diversity in boosting crop yields. They considered about 110 datasets for crop production, 36 datasets for fungal diversity and 85 datasets for bacterial diversity both defined by the Shannon index. The meta-analysis comprised of 59 published studies after screening. Wood, crop waste, straw, peanut hulls, pine chips, acacia bark, animal manure, and chicken manure were among the feedstocks utilised to make biochar. In the various experiments, the pyrolysis temperature utilised to create biochar varied from 200 to 900 °C, with a median of 500 °C. The pyrolysis time in the chosen investigations varied from less than a minute (3 s) to 4300 min, with a median of 60 min. Sandy loam soils accounted for the majority of the studies, with clay and loam soils coming in second and third. The Shannon index was utilised to evaluate the diversity of microorganisms, including both bacteria and fungi. There were several studies describing changes in fungal diversity, but there were enough studies describing bacterial diversity for a robust meta-analysis. In comparison to the control, biochar treatments considerably enhanced bacterial diversity; however, this increase was only observed in laboratory (55%) and field (38%) investigations. Bacterial diversity was considerably ($P < 0.05$) enhanced by biochar derived from herbaceous feedstocks, whereas responses to biochars derived from wood (11%), manure (18%), or lignocellulosic waste (18%) did not differ significantly from the control. The use of biochar considerably ($P < 0.05$) boosted the bacterial diversity in medium-textured soils (31%) and coarse-textured soils (24%), but not in fine-textured soils (18%). When compared to the control, the application of biochar significantly ($P < 0.0001$) increased fungal diversity. Both field and greenhouse research showed considerably (P 0.05) higher fungal diversity when biochar was treated, but not laboratory studies did. In greenhouse experiments, the rise in fungal diversity was more pronounced (43%). While wood biochar had no effect, biochars made from lignocellulosic waste and herbaceous feedstocks considerably ($P < 0.05$) enhanced fungal diversity (63% and 25%, respectively) as compared to the control. Fungal diversity was significantly higher ($P < 0.05$) in fine-textured soils (26%), but not in medium- or coarse-textured soils, as compared to changes in bacterial diversity. The levels of labile carbon and micro- or macronutrients that biochar provides to the microbiological communities in soils are responsible for this improvement (Xu et al. 2016). Furthermore, applying biochar increases the porosity of the soil, which shields some microbial organisms from predators and boosts microbial diversity and biomass (Palansooriya et al. 2019). Biochar applications had larger effects on bacterial diversity in medium and coarse-textured soils than in fine-textured soils. Fungal diversity, on the other hand, did not increase because soil textures did not affect the Shannon index for fungal diversity. Because fungal hyphae are affixed to soil particles, fungal communities are often less mobile and more stable than bacterial colonies. The use of biochar made at higher pyrolytic temperatures reduced the diversity of both bacteria and fungi (Muhammad et al. 2018). The availability of both macro- and micro-pores may be the cause of this impact. Only a limited percentage of microbial communities can colonise the large percentage of micropores (less than 2 nm in diameter) in biochar made with high pyrolytic temperatures, which reduces the diversity of microorganisms overall (Schlatter et al. 2018). Biochar made at lower pyrolytic temperatures contains volatile chemicals that are bioavailable and unstable, which can increase the diversity of bacteria and fungi. High application rates of biochar may have the following negative effects on microbial diversity: (i) they disrupt the microenvironment for microbial growth (Li et al. 2020) which leads to selection within microbial populations and reduces microbial diversity; (ii) they introduce toxic components (Muhammad et al. 2018); or (iii) higher C:N ratio of biochars limit carbon metabolism by microbial communities (Liu et al. 2022).

A variety of crops, including wheat, maize, canola, barley, rice, sorghum, tomato, groundnut, faba bean, turnip, and peanuts, had yields reported by the studies that comprised up the meta-analysis. Applications of biochar considerably ($P < 0.0001$) raised crop yield in comparison to the control. Biochars made from lignocellulosic waste (35%) and herbaceous feedstocks (53%) showed significant ($P < 0.05$) improvements in crop yield among feedstock types, but wood biochar (17%) had no substantial influence on yields. The most effective biochar for raising crop productivity was made at pyrolysis temperatures between 400 and 500 °C, but biochar made at higher temperatures—particularly beyond 600 °C-decreased crop yield (Li et al. 2019a). Water and dissolved minerals were tightly retained by biochar made at high pyrolytic temperatures, which reduced the amount of water and minerals accessible for plant growth (Pallarés et al. 2018) resulting in the decrease in crop productivity at elevated temparatures.

6.2 Empirical Evidence from Field Trials

(a) **Cereal Crops**

Field experiments on cereal crops showed that using biochar significantly increased yield. The studies in Brasília, Brazil, used biosolid-derived biochar generated at pyrolysis temperatures of 300 and 500 °C, with different pyrolysis durations. Corn yields rose dramatically, with a 58% rise at 300 °C over 5 years and a 47% increase at 500 °C over the same period (Chagas et al. 2021). In a shorter 2-year research, maize yields increased by 50% at 300 °C and 54% at 500 °C (Faria et al. 2018). The persistent yield improvements over numerous years indicate that BDB has long-term effects for soil fertility and structure. The higher yield gains at lower temperatures could be attributed to enhanced nutrient preservation in biochar, which improves soil nutrient availability.

Khan et al. (2022) investigated the effect of Acacia wood biochar on wheat growth yield in calcareous soil of Khyber Pakhtunkhwa, Pakistan. It was revealed that the application of biochar had significantly improved the yields. The nutritive character of BC, enriches the soil with long-term vital nutrients, decreases nutrient leaching and eventually increases soil fertility, which could be the cause of increased production under biochar treated plots. Additionally, BC enhanced the soil's organic matter (OM) mineralisation, which benefited crop development and, eventually, yield.

(b) **Oilseed crops**

A study was conducted by You et al. (2019) in the North China Plain, where biosolid-derived biochar generated at higher temperatures (700–850 °C) was used on peanut crops. A large yield increase of 60.43% was reported over a 2-year period. The high pyrolysis temperatures most likely resulted in biochar with a large surface area and porosity, which improved soil water retention and nutrient adsorption (Li et al. 2019c).

(c) **Legume crops**

In Delaware, USA, biosolid-derived biochar was pyrolysed at 300 °C for 660 min and sprayed at 1% and 2% w/w to various crops such as mungbean (Junior and Guo 2023). Yield improvements varied greatly according to the application rate. For mungbean, a 1% treatment rate reduced yield by 31%, whereas a 2% rate boosted yield by 49%. The adverse impact for mungbean at the lower rate could be attributed to initial soil nitrogen imbalances or phytotoxicity caused by incomplete biochar maturation. However, higher application rates generally had a beneficial effect, demonstrating that enough biochar amendment can greatly improve soil characteristics and crop yields.

(d) **Vegetable crops**

Vegetable crops exhibited a significant increase in yield after biochar addition. In a study conducted by Lei et al. (2024) it was found that biochar derived from wood, straw, shell wastes, and mixed materials also shows promise in increasing tomato yield. The TSS and VC in fruits can be significantly raised at the same time by using biochar made from wood and straw. Because of its vast surface area and porous structure, wood and straw biochar can absorb a variety of chemicals, including organic compounds and minerals with a wide range of molecular sizes and chemical characteristics. The wood and straw-based biochar had a relatively high carbon-to-nitrogen (C/N) ratio, which may have provided a favourable environment for the growth of fungal proliferation (Zheng et al. 2018). Furthermore, compared to other biochar, wood biochar had a substantially greater concentration of ferric ions, which improved soil structure (Blanco-Canqui 2017) and SOM adsorption onto clay minerals.

7 Synergistic Approaches and Future Directions

7.1 Co-application of Biochar with Microbial Inoculants

Soil ecosystems are composed of up of microorganisms such bacteria, fungi, viruses, archaea, and protists (Pattnaik et al. 2021) and they are essential to the productivity and health of agricultural ecosystems (Fu et al. 2023). They are engaged in a number of functions, including cycling of nutrients, degradation of organic matter, prevention of disease, encouragement of plant growth, and the development of tolerance to stresses like drought and heavy metal pollution (Joshi et al. 2019). Climate change has the potential to disturb microbiomes and adversely affect alterations in specific functions conducted by soil microorganisms. Furthermore, it has been demonstrated that soil microbial populations are impacted by agricultural practices such heavy chemical fertiliser application and conventional tillage (Pittol et al. 2018). Therefore, the use of microbial inoculants and other biostimulants can aid in the restoration of climate-damaged microbiomes. Biochar enhances a variety of soil physicochemical characteristics, which eventually helps in the functioning of native soil microbial populations. As an instance, biochar may increase microbial biomass, abundance, and diversity by improving substrate and nutrient availability, pH (liming effect), and water holding capacity (Kolton et al. 2017). Co-application of PGPR and biochar, however, may also facilitate nodulation and enhance rhizobium symbiotic performance (Seneviratne et al. 2017). Additionally, it has been demonstrated that biochar enhances the natural nodulation of rhizobium in plants. The reason for this is that biochar improves aeration, and provides a more air to nodule bacteria, which increase the lifespan of bacteria on porous surface of biochar (Tao et al. 2019). For example, as compared to the addition of Pseudomonas sp. and/or AMF alone, the addition of biochar and *Pseudomonas* sp. enhanced root colonisation by arbuscular mycorrhizal fungi. By presumably solubilising the phosphorus from the biochar, the phosphate-solubilizing *Pseudomonas* sp. increased the amount of phosphorus available in the soil, which improved root colonisation and plant growth (Ning et al. 2019). The general abundance of specific microbial groups in soil may be increased by applying biochar and PGPR together, which would improve the soil's overall quality. Similarly, addition of biochar made from wheat and inoculating a eucalyptus plantation with *Bacillus megaterium* significantly enhanced the microbial population of soil, which in turn enhanced nutrient availability. Numerous intracellular and extracellular enzymes have been studied in order to evaluate and report on the possible positive effects of using biochar and PGPR together. Similarly, it was discovered that using *Bacillus subtilis* in conjunction with cotton-derived biochar greatly increased the invertase and catalase activities in soil compared to using biochar alone (Alharbi et al. 2024). In addition to improving bacterial richness and diversity, the combination use of biochar and microbial inoculants dramatically raised soil bacterial abundance. The organic carbon in biochar can operate as an external carbon source for bacteria, encouraging their growth and reproduction and boosting soil bacterial diversity (Nguyen et al. 2018). The diversity of soil fungus is reduced by BCM. This could be ascribed to the properties of biochar providing soil microorganisms an ideal spot to adhere and multiply, which causes the number of pathogenic fungus to decrease and the number of hostile bacteria to rise (Azizi et al. 2021).

7.2 Integrating Biochar in Regenerative Agriculture

Regenerative agriculture (RA) is a farming approach that preserves or boosts farm profitability by utilising natural processes to improve soil health, boost biological activity, improve nutrient cycling, restore landscape function, and generate food and fibre (Chabbi et al. 2017). Based on a set of guiding principles, the practitioners employ a range of techniques that combine ecological and biological processes in an effort to boost output and restore landscape functionality (Khangura et al. 2023). Restoring the natural pre-agricultural ecology and biological function is not the goal of RA; rather, it aims to enhance the health of farming systems by utilising natural ecological processes. The following are the five guiding principles of the approach: (1) Reduce soil disturbance; (2) maintain year-round soil cover; (3) prolong the life of live plants and roots in the soil; (4) integrate biodiversity; and (5) integrate animals (Voisin et al. 2024). Incorporation of organic matter to soil that is resistant to decomposition is encouraged through the use of pyrolysed organic matter, or biochar, which is produced by anoxic thermochemical reactions. With its ability to increase cation exchange capacity, retain moisture, and reduce nitrous oxide emissions, biochar is a desirable addition to soils (Lehmann et al. 2021b). Thus, biochar may improve crop yields, decrease greenhouse gas emissions from management practices, and boost soil carbon retention. Producing biochar could potentially lower atmospheric carbon dioxide emissions if the energy from pyrolysis is harnessed and used to replace fossil fuels. Lehmann et al. (2021b) estimated that the widespread use of biochar could lower greenhouse gas emissions globally by 3.4–6.3 Pg CO_2/year (0.92–1.72 PgC), with improved soil carbon absorption accounting for half of that amount. Biochar application to soils is thought to improve nutrient cycling, increase bioavailable water, increase soil organic matter, decrease bulk density, function as a liming agent, and reduce nutrient and pesticide leaching to surface and ground water (McCauley and Barlow 2023). According to Schmidt et al. (2021), Application of biochar to the soil can boost agricultural output without or significantly reducing the need for inorganic fertiliser applications. Additionally, because of its biological, nutritional, and microbiological qualities, biochar is a viable alternative for highly disturbed soils. One of the most significant phenomena for storing carbon for a longer amount of time is thought to be the recalcitrance mechanism in biochar (Nocentini et al. 2021). Slow microbial degradation and chemical transformation are also responsible for biochar's long-term carbon sink (Siegelman et al. 2019). In comparison to unamended soils, biochar amended at 2%, 5%, 10%, 20%, 40%, and 60% w/w levels, which correspond to a field application rate of 24–720 Mg ha^{-1}, has been shown to significantly suppress ambient CH_4 oxidation and N_2O production at all levels and reduce CO_2 production (Karimi et al. 2022). Biochar is therefore one of the best options for carbon sequestration in order to mitigate climate change because it can provide a significant and long-lasting carbon sink in the soil. Because of its effects on soil organic carbon content (SOC) and water retention capacity, biochar offers soil organisms a favourable habitat (Layek et al. 2022). Beneficial microorganisms can also be physically protected by changes in the nitrogen and carbon availability in soil (Tan et al. 2017). Because of these characteristics, scientists are now more interested in employing biochar as a soil amendment method to stabilise soil organic matter. Therefore, using biochar as a soil supplement has the potential to boost the productivity of conventional agriculture. Because it improves fertility, replenishes soil organic matter, and re-establishes ecosystem function, biochar is essential to regenerative agriculture. It reduces environmental footprints, closes nutrient loops, and encourages carbon sequestration. Biochar promotes long-term soil regeneration by enhancing biodiversity and soil resilience. Its use lessens reliance on outside resources, which is consistent with the regenerative principles of self-sustaining systems. As a result, biochar is essential to the shift to genuinely regenerative farming methods.

8 Challenges and Research Gaps

8.1 Inconsistent Responses Across Environments

The majority of research assessing biochar's impact on soil characteristics was carried out for shorter than 2 years. Longer-term research is crucial because the characteristics of biochar and its effects on soil qualities may alter as it ages. Long-term research will help to determine timing of biochar reapplication and will give a better knowledge of the effects of biochar over time. Compared to laboratory or greenhouse experiments, which are carried out on a smaller scale, field studies are less common. It is crucial to carry out additional field trials at multi-locations because the results of these investigations differ from those of field research because of variations in soil characteristics, weather, and environmental factors at different locations.

8.2 Mechanistic Understanding of Microbe-Biochar Interactions

The majority of the current research on biochar and its impact on the structure of soil microbial communities, particularly in contaminated soils, is empirical. Various soil types (parent material and soil texture) were treated with biochars that had varying physicochemical characteristics and application rates. After a predetermined amount of time, the enzymatic activity or soil microbial community were measured. The primary drawback of this kind of approach is the lack of knowledge regarding the mechanisms underlying the anticipated result. Based on this review and the larger body of literature, future studies that use biochar as a growth promoter of particular soil microorganisms to accomplish a desired goal (like promoting soil nutrient cycling) should employ customised biochar (proactively select biochar feedstock and production conditions).

8.3 Standardization of Biochar Quality

A sustainable utilisation approach that fully addresses spatial heterogeneity and encompasses field-to-regional dimensions within the pertinent socio-economic framework is necessary due to the variability of biochar and the variety in soil properties across spatial and temporal scales. The availability of feedstock, competition for resources, land usage, agricultural methods, and greenhouse gas (GHG) emissions are all taken into account in this context. The physicochemical characteristics of biochar are significantly influenced by varying pyrolysis durations, temperatures, and preparation materials. Uniform biochar preparation standards must be implemented because some biochar preparations are currently costly. For sustainable production and utilization of biochar, standard techniques and processes are necessary. Biochar certification is recognised as a workable strategy that serves as a crucial tool in the application of sustainability regulations. Further, the lack of replication to increase the reliability of methods and particular outcomes is a persistent barrier to rapid advancement in crucial areas of the science-policy interface. Industry (i.e., producers of biochar) will need to increase its involvement in order to achieve experimental reproducibility and to start filling the global database needed to provide sustainable regulation over pertinent time scales.

9 Conclusion

The environmental processes and functions of agricultural ecosystems are largely determined by the biological characteristics of the soil. Biochar is currently attracting more attention as a sustainable alternative in modern farming, providing a number of environmental advantages such carbon sequestration, nutrient cycling, and greenhouse gas emission reduction. Biochar has the ability to address some of the most important issues facing agricultural production, such as soil degradation, nutrient loss, and climate change, as a crucial component of sustainable soil management techniques. Application of biochar as a soil conditioner has several advantages from an agricultural perspective, including improving the physical, chemical, and biological characteristics of soils, which raises crop yields. The dynamics of nutrients, soil pollutants, and microbial activity can all be significantly altered by applying biochar to the soil. Biochar could improve soil quality, boost agroforestry and agroecosystem resilience, and strengthen their ability to adjust to changing climatic conditions. The beneficial effects of biochar amendment on soil quality and crop development indicate that biochar is a sustainable way to overcome nutrient deficiencies and boost agricultural crop productivity.

References

Adebajo SO, Oluwatobi F, Akintokun PO, Ojo AE, Akintokun AK, Gbodope IS (2022) Impacts of rice-husk biochar on soil microbial biomass and agronomic performances of tomato (Solanum lycopersicum L.). Sci Rep 12(1):1787. https://doi.org/10.1038/s41598-022-05757-z

Adedeji AA, Häggblom MM, Babalola OO (2020) Sustainable agriculture in Africa: plant growth-promoting rhizobacteria (PGPR) to the rescue. Sci African 9:e00492. https://doi.org/10.1016/j.sciaf.2020.e00492

Adetunji AT, Lewu FB, Mulidzi R, Ncube B (2017) The biological activities of beta-glucosidase, phosphatase and urease as soil quality indicators: a review. J Soil Sci Plant Nutr 17:794–807. https://doi.org/10.4067/S0718-95162017000300018

Akanmu AO, Sobowale AA, Abiala MA, Olawuyi OJ, Odebode AC (2020) Efficacy of biochar in the management of Fusarium verticillioides Sacc causing ear rot in *Zea mays* L. Biotechnol Rep:e00474. https://doi.org/10.1016/j.btre.2020.e00474

Akhila A, Entoori K (2022) Role of earthworms in soil fertility and its impact on agriculture: a review. Int J Fauna Biol Stud 9(3):55–63. https://doi.org/10.22271/23940522.2022.v9.i3a.907

Alharbi K, Hafez EM, Elhawat N, Omara AED, Rashwan E, Mohamed HH, Alshaal T, Gadow SI (2024) Revitalizing soybean plants in saline, Cd polluted soil using Si-NPs, biochar, and PGPR. Plants 13:3550. https://doi.org/10.3390/plants13243550

Ali MA, Ajaz MM, Rizwan M, Qayyum MF, Arshad M, Hussain S, Ahmad N, Qureshi MA (2020) Effect of biochar and phosphate solubilizing bacteria on growth and phosphorus uptake by maize in an Aridisol. Arab J Geosci 13(9):333. https://doi.org/10.1007/s12517-020-05326-6

Alori ET, Glick BR, Babalola OO (2017) Microbial phosphorus solubilization and its potential for use in sustainable agriculture. Front Microbiol 8:971. https://doi.org/10.3389/fmicb.2017.00971

An N, Zhang L, Liu Y, Shen S, Li N, Wu Z, Yang J, Han W, Han X (2022) Biochar application with reduced chemical fertilizers improves soil pore structure and rice productivity. Chemosphere 298:134304. https://doi.org/10.1016/j.chemosphere.2022.134304

Antonangelo JA, Culman S, Zhang H (2024) Comparative analysis and prediction of cation exchange capacity via summation: influence of biochar type and nutrient ratios. Front Soil Sci 4:1371777. https://doi.org/10.3389/fsoil.2024.1371777

Arif M, Ali S, Ilyas M, Riaz M, Akhtar K, Ali K, Wang H (2020) Enhancing phosphorus availability, soil organic carbon, maize productivity, and farm profitability through biochar and organic-inorganic fertilizers in an irrigated maize agroecosystem under semi-arid climate. Soil Use Manag. https://doi.org/10.1111/sum.12661

Askeland M, Clarke B, Paz-Ferreiro J (2019) Comparative characterization of biochars produced at three selected pyrolysis temperatures from common woody and herbaceous waste streams. PeerJ 7:e6784. https://doi.org/10.7717/peerj.6784

Attia MS, El-Sayyad GS, Abd Elkodous M, El-Batal AI (2020) The effective antagonistic potential of plant growth-promoting rhizobacteria against Alternaria solani causing early blight disease in tomato plant. Sci Hortic 266:109289. https://doi.org/10.1016/j.scienta.2020.109289

Awad YM, Lee SS, Kim KH, Ok YS, Kuzyakov Y (2018) Carbon and nitrogen mineralization and enzyme activities in soil aggregate-size classes: effects of biochar, oyster shells, and polymers. Chemosphere 198:40. https://doi.org/10.1016/j.chemosphere.2018.01.034

Awasthi P, Bargali K, Bargali SS, Jhariya MK (2022) Structure and functioning of Coriarianepalensis Wall dominated shrublands in degraded hills of Kumaun Himalaya. I. Dry matter dynamics. Land Degrad Dev 33(9):1474–1494. https://doi.org/10.1002/ldr.4235

Azeem M, Hayat R, Hussain Q, Tahir MI, Imran M, Abbas Z, Sajid M, Latif A, Irfan M (2019) Effects of biochar and NPK on soil microbial biomass and enzyme activity during 2 years of application in the arid region. Arab J Geosci 12(10):311. https://doi.org/10.1007/s12517-019-4482-1

Azizi S, Kouchaksaraei MT, Hadian J, Abad ARFN, Sanavi SAMM, Ammer C, Bader MKF (2021) Dual inoculations of arbuscular mycorrhizal fungi and plant growth-promoting rhizo-

bacteria boost drought resistance and essential oil yield of common myrtle. For Ecol Manag 497:119478. https://doi.org/10.1016/j.foreco.2021.119478
Banik C, Lawrinenko M, Bakshi S, Laird DA (2018) Impact of pyrolysis temperature and feedstock on surface charge and functional group chemistry of biochars. J Environ Qual 47:452–461. https://doi.org/10.2134/jeq2017.11.0432
Bargali SS, Padalia K, Bargali K (2019) Effects of tree fostering on soil health and microbial biomass under different land use systems in the Central Himalayas. Land Degrad Dev 30(16):1984–1998. https://doi.org/10.1002/ldr.3394
Bastida F, Eldridge DJ, Abades S, Alfaro FD, Gallardo A, García-Velázquez L, Delgado Baquerizo M (2020) Climatic vulnerabilities and ecological preferences of soil invertebrates across biomes. Mol Ecol 29(4):752–761. https://doi.org/10.1111/mec.15299
Basu A, Prasad P, Das SN, Kalam S, Sayyed RZ, Reddy MS, Enshasy HE (2021) Plant growth promoting rhizobacteria (PGPR) as green bioinoculants: recent developments, constraints, and prospects. Sustainability 13:1140. https://doi.org/10.3390/su13031140
Bhargava P, Singh AK, Goel R (2017) Microbes: bioresource in agriculture and environmental sustainability. In: Plant-microbe interactions in agro-ecological perspectives, vol 1: Fundamental mechanisms, methods and functions. Springer-Verlag, Singapore, pp 361–376. https://doi.org/10.1007/978-981-10-5813-4_18
Bhatti AA, Haq S, Bhat RA (2017) Actinomycetes benefaction role in soil and plant health. Microb Pathog 111:458–467. https://doi.org/10.1016/j.micpath.2017.09.036
Blanco-Canqui H (2017) Biochar and soil physical properties. Soil Sci Soc Am J 81(4):687–711. https://doi.org/10.2136/sssaj2017.01.0017
Bokhorst S, Kardol P, Bellingham PJ, Kooyman RM, Richardson SJ, Schmidt S, Wardle DA (2017) Responses of communities of soil organisms and plants to soil aging at two contrasting long-term chronosequences. Soil Biol Biochem 106:69–79. https://doi.org/10.1016/j.soilbio.2016.12.014
Bolan S, Hou D, Wang L, Hale L, Egamberdieva D, Tammeorg P, Li R, Wang B, Xu J, Wang T, Sun H, Padhye LP, Wang H, Siddique KHM, Rinklebe J, Kirkham MB, Bolan N (2023) The potential of biochar as a microbial carrier for agricultural and environmental applications. Sci Total Environ 886:163968. https://doi.org/10.1016/j.scitotenv.2023.163968
Braghiroli FL, Bouafif H, Neculita CM, Koubaa A (2020) Influence of pyro-gasification and activation conditions on the porosity of activated biochars: a literature review. Waste Biomass Valoriz 11(9):5079–5098. https://doi.org/10.1007/s12649-019-00797-5
Brown RW, Reed EY, Chadwick DR, Hill PW, Jones DL (2024) Agronomic amendments drive a diversity of real and apparent priming responses within a grassland soil. Soil Biol Biochem 189:109265. https://doi.org/10.1016/j.soilbio.2023.109265
Cai P, Sun X, Wu Y, Gao C, Mortimer M, Holden PA, Redmile GM, Huang Q (2019) Soil biofilms: microbial interactions, challenges, and advanced techniques for ex-situ characterization. Soil Ecol Lett 1(3):85–93. https://doi.org/10.1007/s42832-019-0017-7
Chabbi A, Lehmann J, Ciais P, Loescher HW, Cotrufo MF, Don A, SanClements M, Schipper L, Six J, Smith P et al (2017) Aligning agriculture and climate policy. Nat Clim Chang 7:307–309
Chagas JKM, de Figueiredo CC, Paz-Ferreiro J (2021) Sewage sludge biochars effects on corn response and nutrition and on soil properties in a 5-yr field experiment. Geoderma 401:115323. https://doi.org/10.1016/j.geoderma.2021.115323
Chamizo S, Mugnai G, Rossi F, Certini G, De Philippis R (2018) Cyanobacteria inoculation improves soil stability and fertility on different textured soils: gaining insights for applicability in soil restoration. Front Environ Sci. https://doi.org/10.3389/fenvs.2018.00049
Chen J, Li S, Liang C, Xu Q, Li Y, Qin H, Fuhrmann JJ (2017a) Response of microbial community structure and function to short-term biochar amendment in an intensively managed bamboo (Phyllostachys praecox) plantation soil: effect of particle size and addition rate. Sci Total Environ 574:24–33. https://doi.org/10.1016/j.scitotenv.2016.08.190
Chen Y, Liu Y, Li Y, Wu Y, Chen Y, Zeng G, Zhang J, Li H (2017b) Influence of biochar on heavy metals and microbial community during composting of river sediment with agricultural wastes. Bioresour Technol 243:347–355. https://doi.org/10.1016/j.biortech.2017.06.100

Chen J, Sun X, Zheng J, Zhang X, Liu X, Bian R, Li L, Cheng K, Zheng J, Pan G (2018) Biochar amendment changes temperature sensitivity of soil respiration and composition of microbial communities 3 years after incorporation in an organic carbon-poor dry cropland soil. Biol Fertil Soils 54:175–188. https://doi.org/10.1007/s00374-017-1253-6

Chen H, Liu J, Li D, Xiao K, Wang K (2019a) Controls on soil arylsulfatase activity at a regional scale. Eur J Soil Biol 90:9–14. https://doi.org/10.1016/j.ejsobi.2018.11.001

Chen J, Chen D, Xu Q, Fuhrmann JJ, Li L, Pan G, Li Y, Qin H, Liang C, Sun X (2019b) Organic carbon quality, composition of main microbial groups, enzyme activities, and temperature sensitivity of soil respiration of an acid paddy soil treated with biochar. Biol Fertil Soils 55(2):185–197. https://doi.org/10.1007/s00374-018-1333-2

Chen W, Li P, Li F, Xi J, Han Y (2022) Effects of tillage and biochar on soil physiochemical and microbial properties and its linkage with crop yield. Front Microbiol 13:929725. https://doi.org/10.3389/fmicb.2022.929725

Chuaphasuk C, Prapagdee B (2019) Effects of biochar-immobilized bacteria on phytoremediation of cadmium-polluted soil. Environ Sci Pollut Res 26:23679–23688. https://doi.org/10.1007/s11356-019-05661-6

Costa OYA, Raaijmakers JM, Kuramae EE (2018) Microbial extracellular polymeric substances: ecological function and impact on soil aggregation. Front Microbiol 9:1636. https://doi.org/10.3389/fmicb.2018.01636

Cuadros J (2017) Clay minerals interaction with microorganisms: a review. Clay Miner 52:235–261. https://doi.org/10.1180/claymin.2017.052.2.05

Cui J, Ge T, Kuzyakov Y, Nie M, Fang C, Tang B, Zhou C (2017) Interactions between biochar and litter priming: a three-source 14C and δ13C partitioning study. Soil Biol Biochem 104:49–58. https://doi.org/10.1016/j.soilbio.2016.10.014

Dai Z, Xiong X, Zhu H, Xu H, Leng P, Li J, Tang C, Xu J (2021) Association of biochar properties with changes in soil bacterial, fungal and fauna communities and nutrient cycling processes. Biochar 3(3):239–254. https://doi.org/10.1007/s42773-021-00099-x

Datt N, Singh D (2019) Enzymes in relation to soil biological properties and sustainability. In: Sustainable management of soil and environment. Springer, Singapore, pp 383–406. https://doi.org/10.1007/978-981-13-8832-3_11

Daunoras J, Kačergius A, Gudiukaitė R (2024) Role of soil microbiota enzymes in soil health and activity changes depending on climate change and the type of soil ecosystem. Biology 13(2):85. https://doi.org/10.3390/biology13020085

Davys D, Rayns F, Charlesworth S, Lillywhite R (2023) The effect of different biochar characteristics on soil nitrogen transformation processes: a review. Sustainability 15(23):16446. https://doi.org/10.3390/su152316446

de Caprariis B, De Filippis P, Hernandez AD, Petrucci E, Petrullo A, Scarsella M, Turchi M (2017) Pyrolysis wastewater treatment by adsorption on biochars produced by poplar biomass. J Environ Manag 197:231–238. https://doi.org/10.1016/j.jenvman.2017.04.007

De Ciucies S, Whitman T, Woolf D, Enders A, Lehmann J (2018) Priming mechanisms with additions of pyrogenic organic matter to soil. Geochim Cosmochim Acta 238:329–342. https://doi.org/10.1016/j.gca.2018.07.004

De Deyn GB, Kooistra L (2021) The role of soils in habitat creation, maintenance and restoration. Philos Trans R Soc B Biol Sci 376:20200170. https://doi.org/10.1098/rstb.2020.0170

de la Rosa JM, Rosado M, Paneque M, Miller AZ, Knicker H (2018) Effects of aging under field conditions on biochar structure and composition: implications for biochar stability in soils. Sci Total Environ 613:969–976. https://doi.org/10.1016/j.scitotenv.2017.09.124

De Medeiros EV, dos Santos MDC, da Costa DP, Duda GP, de Oliveira JB, da Silva JA, Hammecker C (2020) Effect of biochar and inoculation with Trichoderma aureoviride on melon growth and sandy Entisol quality. Aust J Crop Sci 14(6):971–977. https://doi.org/10.21475/ajcs.20.14.06.p2302

de Vries FT, Griffiths RI, Bailey M, Craig H, Girlanda M, Gweon HS, Hallin S, Kaisermann A, Keith AM, Kretzschmar M et al (2018) Soil bacterial networks are less stable under drought than fungal networks. Nat Commun 9:3033. https://doi.org/10.1038/s41467-018-05516-7

Deinert L, Hossen S, Ikoyi I, Kwapinski W, Noll M, Schmalenberger A (2024) Poultry litter biochar soil amendment affects microbial community structures, promotes phosphorus cycling and growth of barley (*Hordeum vulgare*). Eur J Soil Biol 120:103591. https://doi.org/10.1016/j.ejsobi.2023.103591

Dhyani V, Bhaskar T (2018) A comprehensive review on the pyrolysis of lignocellulosic biomass. Renew Energy 129:695–716. https://doi.org/10.1016/j.renene.2017.04.035

Domingues RR, Trugilho PF, Silva CA, Melo ICNA, Melo LCA, Magriotis ZM, Sánchez-Monedero MA (2017) Properties of biochar derived from wood and high-nutrient biomasses with the aim of agronomic and environmental benefits. PLoS One 12:e0176884. https://doi.org/10.1371/journal.pone.0176884

Dotaniya ML, Dotaniya CK, Kumar K, Doutaniya RK, Meena HM, Shirale AO, Meena MD, Meena VD, Kumar R, Meena BP, Kumawat N (2021) Type of soil pollutant and their degradation: methods and challenges. In: Handbook of nanomaterials and nanocomposites for energy and environmental applications. Springer International Publishing, Cham, pp 3103–3134. https://doi.org/10.1007/978-3-030-36268-3_65

Du ZL, Zhao JK, Wang YD, Zhang QZ (2017) Biochar addition drives soil aggregation and carbon sequestration in aggregate fractions from an intensive agricultural system. J Soils Sediments 17:581–589. https://doi.org/10.1007/s11368-015-1349-2

Edwards CA, Arancon NQ (2022) The role of earthworms in organic matter and nutrient cycles. In: Biology and ecology of earthworms, pp 233–274. https://doi.org/10.1007/978-0-387-74943-3_8

Egamberdieva D, Hua M, Reckling M, Wirth S, Bellingrath-Kimura SD (2018) Potential effects of biochar-based microbial inoculants in agriculture. Environ Sustain 1:19–24. https://doi.org/10.1007/s42398-018-0010-6

El-Naggar A, Lee SS, Rinklebe J, Farooq M, Song H, Sarmah AK, Zimmerman AR, Ahmad M, Shaheen SM, Ok YS (2019) Biochar application to low fertility soils: a review of current status, and future prospects. Geoderma 337:536–554. https://doi.org/10.1016/j.geoderma.2018.09.034

Elzobair KA, Stromberger ME, Ippolito JA, Lentz RD (2016) Contrasting effects of biochar versus manure on soil microbial communities and enzyme activities in an Aridisol. Chemosphere 142:145–152. https://doi.org/10.1016/j.chemosphere.2015.06.044

Ezeokoli OT, Oladipo OG, Bezuidenhout CC, Adeleke RA, Maboeta MS (2021) Assessing the ecosystem support function of South African coal mining soil environments using earthworm (*Eisenia andrei*) bioassays. Appl Soil Ecol 157:103771. https://doi.org/10.1016/j.apsoil.2020.103771

Fall AF, Nakabonge G, Ssekandi J, Founoune-Mboup H, Apori SO, Ndiaye A, Badji A, Ngom K (2022) Roles of arbuscular mycorrhizal fungi on soil fertility: contribution in the improvement of physical, chemical, and biological properties of the soil. Front Fungal Biol 3:723892

Fang Y, Nazaries L, Singh BK, Singh BP (2018a) Microbial mechanisms of carbon priming effects revealed during the interaction of crop residue and nutrient inputs in contrasting soils. Glob Change Biol 24(7):2775–2790. https://doi.org/10.1111/gcb.14154

Fang Y, Singh BP, Luo Y, Boersma M, Van Zwieten L (2018b) Biochar carbon dynamics in physically separated fractions and microbial use efficiency in contrasting soils under temperate pastures. Soil Biol Biochem 116:399–409. https://doi.org/10.1016/j.soilbio.2017.10.042

Fang Y, Xu W, Zhang W, Guang C, Mu W (2022) Microbial elimination of pyrethroids: specific strains and involved enzymes. Appl Microbiol Biotechnol 106:6915–6932

FAO, IFAD, UNICEF, WFP, WHO (2021) The state of food security and nutrition in the world 2021: transforming food systems for food security, improved nutrition and affordable healthy diets for all. FAO, Rome

Farhangi-Abriz S, Torabian S, Qin R, Noulas C, Lu Y, Gao S (2021) Biochar effects on yield of cereal and legume crops using meta-analysis. Sci Total Environ 775:145869. https://doi.org/10.1016/j.scitotenv.2021.145869

Faria WM, Figueiredo CCD, Coser TR, Vale AT, Schneider BG (2018) Is sewage sludge biochar capable of replacing inorganic fertilizers for corn production? Evidence from a two-year field experiment. Arch Agron Soil Sci 64:505–519

Farkas É, Feigl V, Gruiz K, Vaszita E, Fekete-Kertész I, Tolner M, Kerekes I, Pusztai E, Kari A, Uzinger N et al (2020) Long-term effects of grain husk and paper fibre sludge biochar on acidic and calcareous sandy soils—a scale-up field experiment applying a complex monitoring toolkit. Sci Total Environ 731:138988. https://doi.org/10.1080/03650340.2017.1360488

Fawzy S, Osman AI, Yang H, Doran J, Rooney DW (2021) Industrial biochar systems for atmospheric carbon removal: a review. Environ Chem Lett 19:3023. https://doi.org/10.1007/s10311-021-01210-1

Feng W, Yang F, Cen R, Liu J, Qu Z, Miao Q, Chen H (2021) Effects of straw biochar application on soil temperature, available nitrogen and growth of corn. J Environ Manag 277:111331. https://doi.org/10.1016/j.jenvman.2020.111331

Fiore S, Berruti F, Briens C (2018) Investigation of innovative and conventional pyrolysis of ligneous and herbaceous biomasses for biochar production. Biomass Bioenergy 119:381–391. https://doi.org/10.1016/j.biombioe.2018.10.010

Food Security Information Network (2017) Global report on food crises 2017. Available online at: https://www.wfp.org/publications/global-report-food-crisis-2017

Fu Y, Luo Y, Auwal M, Singh BP, Van Zwieten L, Xu J (2022) Biochar accelerates soil organic carbon mineralization via rhizodeposit-activated Actinobacteria. Biol Fertil Soils 58:565–577. https://doi.org/10.1007/s00374-022-01643-y

Fu F, Li J, Li Y, Chen W, Ding H, Xiao S (2023) Simulating the effect of climate change on soil microbial community in an Abies georgei var. smithii forest. Front Microbiol 14:1189859. https://doi.org/10.3389/fmicb.2023.1189859

Gao S, DeLuca TH (2018) Wood biochar impacts soil phosphorus dynamics and microbial communities in organically-managed croplands. Soil Biol Biochem 126:144–150. https://doi.org/10.1016/j.soilbio.2018.09.002

Gao Y, Yao D, Wang C, Wang Y, Li C (2020) *Staphylococcus* derived from desert and its solidified ability on aeolian sandy soil. Adv Civ Eng 2020:8877039. https://doi.org/10.1155/2020/8877039

García-Ruiz R, Carranza-Gallego G, Aguilera E, González De Molina M, Guzmán GI (2019) C and N mineralisation of straw of traditional and modern wheat varieties in soils of contrasting fertility. Nutr Cycl Agroecosyst 113:167–179. https://doi.org/10.1007/s10705-019-09973-4

Gaudel G, Poudel M, Mosongo PS, Xing L, Oljira AM, Zhang Y, Bizimana F, Liu B, Wang Y, Dong W, Uwamungu JY (2022) Meta-analysis of the priming effect on native soil organic carbon in response to glucose amendment across soil depths. Plant Soil 479(1):107–124. https://doi.org/10.1007/s11104-021-05168-5

Gęca M, Khalil AM, Tang M, Bhakta AK, Snoussi Y, Nowicki P, Wiśniewska M, Chehimi MM (2023) Surface treatment of biochar—methods, surface analysis and potential applications: a comprehensive review. Surfaces 6(2):179–213. https://doi.org/10.3390/surfaces6020013

Ghassemi-Golezani K, Farhangi-Abriz S (2019) Biochar alleviates fluoride toxicity and oxidative stress in safflower (*Carthamus tinctorius* L.) seedlings. Chemosphere 223:406–415. https://doi.org/10.1016/j.chemosphere.2019.02.087

Ghosh S (2021) Climate change and earthworms: a global perspective. Int J Entomol Res 6(4):167–171

Giller KE, Hijbeek R, Andersson JA, Sumberg J (2021) Regenerative agriculture: an agronomic perspective. Outlook Agric 50:13–25. https://doi.org/10.1177/0030727021998063

Glazunova DM, Kuryntseva PA, Selivanovskaya SY, Galitskaya PY (2018) Assessing the potential of using biochar as a soil conditioner. IOP Conf Ser Earth Environ Sci 107:012059. https://doi.org/10.1088/1755-1315/107/1/012059

Godwin PM, Pan Y, Xiao H, Afzal MT (2019) Progress in preparation and application of modified biochar for improving heavy metal ion removal from wastewater. J Bioresour Bioprod 4:31–42. https://doi.org/10.21967/jbb.v4i1.180

Guerra CA, Bardgett RD, Caon L, Crowther TW, Delgado Baquerizo M, Montanarella L, Navarro LM, Orgiazzi A, Singh BK, Tedersoo L, Vargas-Rojas R (2021) Tracking, targeting, and conserving soil biodiversity. Science 371:239–241. https://doi.org/10.1126/science.abd7926

Gujre N, Soni A, Rangan L, Tsang DCW, Mitra S (2021) Sustainable improvement of soil health utilizing biochar and arbuscular mycorrhizal fungi: a review. Environ Pollut 268:115549. https://doi.org/10.1016/j.envpol.2020.115549

Hagemann N, Harter J, Behrens S (2016) Elucidating the impacts of biochar applications on nitrogen cycling microbial communities. In: Ralebitso-Senior TK, Orr CH (eds) Biochar application. Elsevier Inc., Amsterdam, Chapter 7

Halder M, Islam MU, Liu S, Guo Z, Zhang Z, Peng X (2024) Organic materials quality to control soil aggregation: a meta-analysis. J Soil Sci Plant Nutr 24(2):1857–1870. https://doi.org/10.1007/s42729-024-01815-9

Hansen V, Müller-Stöver D, Imparato V, Krogh PH, Jensen LS, Dolmer A, Hauggaard-Nielsen H (2017) The effects of straw or straw-derived gasification biochar applications on soil quality and crop productivity: a farm case study. J Environ Manag 186:88–95. https://doi.org/10.1016/j.jenvman.2016.10.041

Haris M, Hamid Y, Usman M, Wang L, Saleem A, Su F, Guo J, Li Y (2021) Crop-residues derived biochar: synthesis, properties, characterization and application for the removal of trace elements in soils. J Hazard Mater 416:126212. https://doi.org/10.1016/j.jhazmat.2021.126212

He P, Liu Y, Shao L, Zhang H, Lü F (2018) Particle size dependence of the physicochemical properties of biochar. Chemosphere 212:385–392. https://doi.org/10.1016/j.chemosphere.2018.08.106

He X, Xie H, Gao D, Khashi U, Rahman M, Zhou X, Wu F (2021) Biochar and intercropping with potato–onion enhanced the growth and yield advantages of tomato by regulating the soil properties, nutrient uptake, and soil microbial community. Front Microbiol 12:695447. https://doi.org/10.3389/fmicb.2021.695447

Heidari E, Mohammadi K, Pasari B, Rokhzadi A, Sohrabi Y (2020) Combining the phosphate solubilizing microorganisms with biochar types in order to improve safflower yield and soil enzyme activity. Soil Sci Plant Nutr 66(2):255–267. https://doi.org/10.1080/00380768.2019.1704180

Hernandez-Charpak YD, Trabold TA, Lewis CL, Diaz CA (2022) Biochar-filled plastics: effect of feedstock on thermal and mechanical properties. Biomass Convers Biorefinery 12(10):4349–4360. https://doi.org/10.1007/s13399-022-02340-4

Ho SH, Zhang C, Chen WH, Shen Y, Chang JS (2018) Characterization of biomass waste torrefaction under conventional and microwave heating. Bioresour Technol 264:7–16. https://doi.org/10.1016/j.biortech.2018.05.047

Hoekman SK, Broch A, Felix L, Farthing W (2017) Hydrothermal carbonization (HTC) of loblolly pine using a continuous, reactive twin-screw extruder. Energy Convers Manag 134:247–259. https://doi.org/10.1016/j.enconman.2016.12.035

Hossain MZ, Bahar MM, Sarkar B, Donne SW, Ok YS, Palansooriya KN, Kirkham MB, Chowdhury S, Bolan N (2020) Biochar and its importance on nutrient dynamics in plant. Biochar 2:379–420. https://doi.org/10.1007/s42729-024-01784-z

Hozzein WN, Abuelsoud W, Wadaan MA, Shuikan AM, Selim S, Al Jaouni S, AbdElgawad H (2019) Exploring the potential of actinomycetes in improving soil fertility and grain quality of economically important cereals. Sci Total Environ 651:2787–2798. https://doi.org/10.1016/j.scitotenv.2018.10.048

Hu Y, Sun B, Wu S, Feng H, Gao M, Zhang B, Liu Y (2021) After-effects of straw and straw-derived biochar application on crop growth, yield, and soil properties in wheat (*Triticum aestivum* L.)-maize (*Zea mays* L.) rotations: a four-year field experiment. Sci Total Environ 780:146560. https://doi.org/10.1016/j.scitotenv.2021.146560

Huang Y, Wang C, Lin C, Zhang Y, Chen X, Tang L, Liu C, Chen Q, Onwuka MI, Song T (2019) Methane and nitrous oxide flux after biochar application in subtropical acidic paddy soils under tobacco-rice rotation. Sci Rep 9:17277. https://doi.org/10.1038/s41598-019-53044-1

Huang S, Li J, Wang J (2021) Clay minerals change the toxic effect of cadmium on the activities of leucine aminopeptidase. Adsorpt Sci Technol 2021:1024085. https://doi.org/10.1155/2021/10240085

Hui D (2021) Effects of biochar application on soil properties, plant biomass production, and soil greenhouse gas emissions: a mini-review. Agric Sci 12(03):213–236

Iacomino G, Idbella M, Laudonia S, Vinale F, Bonanomi G (2022) The suppressive effects of biochar on above-and belowground plant pathogens and pests: a review. Plants 11(22):3144. https://doi.org/10.3390/plants11223144

Igalavithana AD, Park J, Ryu C, Lee YH, Hashimoto Y, Huang L, Kwon EE, Ok YS, Lee SS (2017) Slow pyrolyzed biochars from crop residues for soil metal(loid) immobilization and microbial community abundance in contaminated agricultural soils. Chemosphere 177:157–166. https://doi.org/10.1016/j.chemosphere.2017.02.112

Innocenti G, Sabatini MA (2018) Collembola and plant pathogenic, antagonistic and arbuscular mycorrhizal fungi: a review. Bull Insectology 71:71–76

Intergovernmental Panel on Climate Change (2019) Climate change and land. In: Shukla PR, Skea J, Calvo Buendia E, Masson-Delmotte V, Pörtner H-O, Roberts DC et al (eds) Available online at: https://www.ipcc.ch/srccl/

Irfan M, Hussain Q, Khan KS, Akmal M, Ijaz SS, Hayat R, Khalid A, Azeem M, Rashid M (2019) Response of soil microbial biomass and enzymatic activity to biochar amendment in the organic carbon deficient arid soil: a 2-year field study. Arab J Geosci 12:95. https://doi.org/10.1007/s12517-019-4239-x

Jacoby R, Peukert M, Succurro A, Koprivova A, Kopriva S (2017) The role of soil microorganisms in plant mineral nutrition–current knowledge and future directions. Front Plant Sci 8:1617. https://doi.org/10.3389/fpls.2017.01617

Jacott CN, Murray JD, Ridout CJ (2017) Trade-offs in arbuscular mycorrhizal symbiosis: disease resistance, growth responses and perspectives for crop breeding. Agronomy 7(4):75. https://doi.org/10.3390/agronomy7040075

Jaiswal AK, Frenkel O, Tsechansky L, Elad Y, Graber ER (2018) Immobilization and deactivation of pathogenic enzymes and toxic metabolites by biochar: a possible mechanism involved in soilborne disease suppression. Soil Biol Biochem 121:59–66. https://doi.org/10.1016/j.soilbio.2018.03.001

Jalal F, Arif M, Akhtar K, Khan A, Naz M, Said F, Ali M (2020) Biochar integration with legume crops in summer gape synergizes nitrogen use efficiency and enhances maize yield. Agronomy 10:58. https://doi.org/10.3390/agronomy10010058

Jernigan A, Kao-Kniffin J, Pethybridge S, Wickings K (2023) Soil microarthropod effects on plant growth and development. Plant Soil 483(1):27–45. https://doi.org/10.1007/s11104-022-05766-x

Jie L, Ding Y, Ma L, Gao G, Wang Y (2017) Combination of biochar and immobilized bacteria in cypermethrin-contaminated soil remediation. Int Biodeterior Biodegrad 120:15–20. https://doi.org/10.1016/j.ibiod.2017.01.039

Jin J, Kang M, Sun K, Pan Z, Wu F, Xing B (2016) Properties of biochar-amended soils and their sorption of imidacloprid, isoproturon, and atrazine. Sci Total Environ 550:504–513. https://doi.org/10.1016/j.scitotenv.2016.01.117

Jin Z, Chen C, Chen X, Hopkins I, Zhang X, Han Z et al (2019) The crucial factors of soil fertility and rapeseed yield – a five-year field trial with biochar addition in upland red soil, China. Sci Total Environ 649:1467–1480. https://doi.org/10.1016/j.scitotenv.2018.08.412

Jin Q, Wang C, Sardans J, Vancov T, Fang Y, Wu L, Huang X, Gargallo-Garriga A, Peñuelas J, Wang W (2022) Effect of soil degradation on the carbon concentration and retention of nitrogen and phosphorus across Chinese rice paddy fields. Catena 209:105810. https://doi.org/10.1016/j.catena.2021.105810

Jin Y, Tian Y, Yang R, Li W, Liu C, Li T (2023) Poplar sawdust biochar altered community composition of dominant soil fungi but not bacteria depending on pyrolysis temperature. Forests 14(6):1114

Johnson MS, Webster C, Jassal RS, Hawthorne I, Black TA (2017) Biochar influences on soil CO_2 and CH_4 fluxes in response to wetting and drying cycles for a forest soil. Sci Rep 7:6780

Joseph S, Cowie AL, Van Zwieten L, Bolan N, Budai A, Buss W et al (2021) How biochar works, and when it doesn't: a review of mechanisms controlling soil and plant responses to biochar. GCB Bioenergy 13(11):1731–1764. https://doi.org/10.1111/gcbb.12885

Joshi H, Somduttand CP, Mundra SL (2019) Role of effective microorganisms (EM) in sustainable agriculture. Int J Curr Microbiol Appl Sci 8(3):172–181. https://doi.org/10.20546/ijcmas.2019.803.024
Junior A, Guo M (2023) Efficacy of sewage sludge derived biochar on enhancing soil health and crop productivity in strongly acidic soil. Front Soil Sci 3:1066547. https://doi.org/10.3389/fsoil.2023.1066547
Juriga M, Šimanský V (2018) Effect of biochar on soil structure—review. Acta Fytotech Zootech 21:11–19. https://doi.org/10.15414/afz.2018.21.01.11-19
Karimi M, Shirzad M, Silva JA, Rodrigues AE (2022) Biomass/biochar carbon materials for CO_2 capture and sequestration by cyclic adsorption processes: a review and prospects for future directions. J CO2 Util 57:101890. https://doi.org/10.1016/j.jcou.2022.101890
Karki H, Bargali K, Bargali SS (2021) Spatial and temporal trends in soil N-mineralization rates under the agroforestry systems in Bhabhar belt of Kumaun Himalaya, India. Agrofor Syst 95:1603–1617. https://doi.org/10.1007/s10457-021-00669-9
Khan Z, Rahman MHU, Haider G, Amir R, Ikram RM, Ahmad S, Schofield HK, Riaz B, Iqbal R, Fahad S, Datta R (2021) Chemical and biological enhancement effects of biochar on wheat growth and yield under arid field conditions. Sustainability 13(11):5890. https://doi.org/10.3390/su13115890
Khan MA, Basir A, Fahad S, Adnan M, Saleem MH, Iqbal A, Amanullah, Al-Huqail AA, Alosaimi AA, Saud S, Liu K (2022) Biochar optimizes wheat quality, yield, and nitrogen acquisition in low fertile calcareous soil treated with organic and mineral nitrogen fertilizers. Front Plant Sci 13:879788. https://doi.org/10.3389/fpls.2022.879788
Khangura R, Ferris D, Wagg C, Bowyer J (2023) Regenerative agriculture—a literature review on the practices and mechanisms used to improve soil health. Sustainability 15(3):2338. https://doi.org/10.3390/su15032338
Knights HE, Jorrin B, Haskett TL, Poole PS (2021) Deciphering bacterial mechanisms of root colonization. Environ Microbiol Rep 13:428–444. https://doi.org/10.1111/1758-2229.12934
Kolton M, Graber ER, Tsehansky L, Elad Y, Cytryn E (2017) Biochar-stimulated plant performance is strongly linked to microbial diversity and metabolic potential in the rhizosphere. New Phytol 213:1393–1404. https://doi.org/10.1111/nph.14253
Kong FX, Ling X, Iqbal B, Zhou ZG, Meng YL (2021) Soil phosphorus availability and cotton growth affected by biochar addition under two phosphorus fertilizer levels. Arch Agron Soil Sci. https://doi.org/10.1080/03650340.2021.1955355
Korneykova MV, Nikitin DA (2021) Qualitative and quantitative characteristics of the soil microbiome in the impact zone of the Kandalaksha Aluminum Smelter. Eurasian Soil Sci 54:897–906. https://doi.org/10.1134/S1064229321060089
Krolicka M, Hinz SWA, Koetsier MJ, Eggink G, Van Den Broek LAM, Boeriu CG (2018) β-N-Acetylglucosaminidase MthNAG from Myceliophthora thermophila C1, a thermostable enzyme for production of N-acetylglucosamine from chitin. Appl Microbiol Biotechnol 102:7441–7454. https://doi.org/10.1007/s00253-018-9166-3
Krylova AY, Zaitchenko VM (2018) Hydrothermal carbonization of biomass: a review. Solid Fuel Chem 52:91–103. https://doi.org/10.3103/S0361521918020076
Kultschar B, Llewellyn C (2018) Secondary metabolites in cyanobacteria. In: Secondary metabolites-sources and applications, vol 2. InTech, London, pp 23–36
Kunito T, Kurita H, Kumori M, Sakaguchi K, Nishizawa S, Fujita K, Moro H, Sawada K, Miyabara Y, Toda H et al (2022) Microbial synthesis of arylsulfatase depends on the soluble and adsorbed sulfate concentration in soils. Eur J Soil Biol 111:103418. https://doi.org/10.1016/j.ejsobi.2022.103418
Kuryntseva P, Galitskaya P, Selivanovskaya S (2022) Optimization of pyrolysis regime for chicken manure treatment and biochar production. Water Environ J 36:270–281. https://doi.org/10.1111/wej.12764
Kuryntseva P, Karamova K, Galitskaya P, Selivanovskaya S, Evtugyn G (2023) Biochar functions in soil depending on feedstock and pyrolyzation properties with particular emphasis on biological properties. Agriculture 13(10):2023. https://doi.org/10.3390/agriculture13102003

Lataf A, Jozefczak M, Vandecasteele B, Viaene J, Schreurs S, Carleer R, Yperman J, Marchal W, Cuypers A, Vandamme D (2022) The effect of pyrolysis temperature and feedstock on biochar agronomic properties. J Anal Appl Pyrolysis 168:105728. https://doi.org/10.1016/j.jaap.2022.105728

Layek J, Narzari R, Hazarika S, Das A, Rangappa K, Devi S, Balusamy A, Saha S, Mandal S, Idapuganti RG, Babu S (2022) Prospects of biochar for sustainable agriculture and carbon sequestration: an overview for Eastern Himalayas. Sustainability 14(11):6684. https://doi.org/10.3390/su14116684

Lazarova S, Coyne D, Rodriguez MG, Peteira B, Ciancio A (2021) Functional diversity of soil nematodes in relation to the impact of agriculture—a review. Diversity 13(2):64. https://doi.org/10.3390/d13020064

Lee SI, Park HJ, Jeong YJ, Seo BS, Kwak JH, Yang HI, Xu X, Tang S, Cheng W, Lim SS et al (2021) Biochar-induced reduction of N2O emission from East Asian soils under aerobic conditions: review and data analysis. Environ Pollut 291:118154. https://doi.org/10.1016/j.envpol.2021.118154

Lehmann A, Leifheit EF, Gerdawischke M, Rillig MC (2021a) Microplastics have shape- and polymer-dependent effects on soil aggregation and organic matter loss–an experimental and meta-analytical approach. Microplast Nanoplast 1(1):7. https://doi.org/10.1186/s43591-021-00007-x

Lehmann J, Cowie A, Masiello CA, Kammann C, Woolf D, Amonette JE, Cayuela ML, Camps-Arbestain M, Whitman T (2021b) Biochar in climate change mitigation. Nat Geosci 14(12):883–892. https://doi.org/10.1038/s41561-021-00852-8

Lei Y, Xu L, Wang M, Sun S, Yang Y, Xu C (2024) Effects of biochar application on tomato yield and fruit quality: a meta-analysis. Sustainability 16(15):6397. https://doi.org/10.3390/su16156397

Leng L, Huang H, Li H, Li J, Zhou W (2019) Biochar stability assessment methods: a review. Sci Total Environ 647:210–222. https://doi.org/10.1016/j.scitotenv.2018.07.402

Leng L, Xiong Q, Yang L, Li H, Zhou Y, Zhang W, Jiang S, Li H, Huang H (2021) An overview on engineering the surface area and porosity of biochar. Sci Total Environ 763:144204. https://doi.org/10.1016/j.scitotenv.2020.144204

Li H, Dong X, da Silva EB, de Oliveira LM, Chen Y, Ma LQ (2017) Mechanisms of metal sorption by biochars: biochar characteristics and modifications. Chemosphere 178:466–478. https://doi.org/10.1016/j.chemosphere.2017.03.072

Li Y, Li Y, Chang SX, Yang Y, Fu S, Jiang P, Luo Y, Yang M, Chen Z, Hu S (2018) Biochar reduces soil heterotrophic respiration in a subtropical plantation through increasing soil organic carbon recalcitrancy and decreasing carbon-degrading microbial activity. Soil Biol Biochem 122:173–185. https://doi.org/10.1016/j.soilbio.2018.04.019

Li F, Liang X, Niyungeko C, Sun T, Liu F, Arai Y (2019a) Effects of biochar amendments on soil phosphorus transformation in agricultural soils. Adv Agron 158:131–172. https://doi.org/10.1016/bs.agron.2019.07.002

Li S, Harris S, Anandhi A, Chen G (2019b) Predicting biochar properties and functions based on feedstock and pyrolysis temperature: a review and data syntheses. J Clean Prod 215:890–902. https://doi.org/10.1016/j.jclepro.2019.01.106

Li Y, Xu D, Guan Y, Yu K, Wang W (2019c) Phosphorus sorption capacity of biochars from different waste woods and bamboo. Int J Phytoremediation 21:145–151. https://doi.org/10.1080/15226514.2018.1488806

Li X, Wang T, Chang SX, Jiang X, Song Y (2020) Biochar increases soil microbial biomass but has variable effects on microbial diversity: a meta-analysis. Sci Total Environ 749:141593. https://doi.org/10.1016/j.scitotenv.2020.141593

Li S, Li Z, Feng X, Zhou F, Wang J, Li Y (2021) Effects of biochar additions on the soil chemical properties, bacterial community structure and rape growth in an acid purple soil. Plant Soil Environ 67:121–129. https://doi.org/10.17221/390/2020-PSE

Lincoln L, More SS (2017) Bacterial invertases: occurrence, production, biochemical characterization, and significance of transfructosylation. J Basic Microbiol 57:803–813. https://doi.org/10.1002/jobm.201700269
Liu Y, Yao S, Wang Y, Lu H, Brar SK, Yang S (2017) Bio- and hydrochars from rice straw and pig manure: inter-comparison. Bioresour Technol 235:332–337. https://doi.org/10.1016/j.biortech.2017.03.103
Liu Q, Zhang Y, Liu B, Amonette JE, Lin Z, Liu G, Ambus P, Xie Z (2018a) How does biochar influence soil N cycle? A meta-analysis. Plant Soil 426(1):211–225. https://doi.org/10.1007/s11104-018-3619-4
Liu Y, Lonappan L, Brar SK, Yang S (2018b) Impact of biochar amendment in agricultural soils on the sorption, desorption, and degradation of pesticides: a review. Sci Total Environ 645:60–70. https://doi.org/10.1016/j.scitotenv.2018.07.099
Liu Y, Zhu J, Gao W, Guo Z, Xue C, Pang J, Shu L (2019) Effects of biochar amendment on bacterial and fungal communities in the reclaimed soil from a mining subsidence area. Environ Sci Pollut Res 26:34368–34376. https://doi.org/10.1007/s11356-019-06567-z
Liu J, Tang Y, Bao J, Wang H, Peng F, Tan P, Chu G, Liu S (2022) A stronger rhizosphere impact on the fungal communities compared to the bacterial communities in pecan plantations. Front Microbiol 13:899801. https://doi.org/10.3389/fmicb.2022.899801
Liu Z, Zhou W, Sun Y, Peng Y, Niu J, Tan J, Wei M (2023) Biochar and its coupling with microbial inoculants for suppressing plant diseases: a review. Appl Soil Ecol 190:105025. https://doi.org/10.1016/j.apsoil.2023.105025
Lopes ÉMG, Reis MM, Frazão LA, da Mata Terra LE, Lopes EF, Dos Santos MM, Fernandes LA (2021) Biochar increases enzyme activity and total microbial quality of soil grown with sugarcane. Environ Technol Innov 21:101270. https://doi.org/10.1016/j.eti.2020.101270
Lu J, Liu S, Chen W, Meng J (2023) Study on the mechanism of biochar affecting the effectiveness of phosphate solubilizing bacteria. World J Microbiol Biotechnol 39(3):87. https://doi.org/10.1007/s11274-023-03533-3
Luginbuehl LH, Oldroyd GED (2017) Understanding the arbuscule at the heart of endomycorrhizal symbioses in plants. Curr Biol 27(17):R952–R963. https://doi.org/10.1016/j.cub.2017.06.042
Luo S, Wang S, Tian L, Li S, Li X, Shen Y, Tian C (2017) Long term biochar application influences soil microbial community and its potential roles in semiarid farmland. Appl Soil Ecol 117:10–15. https://doi.org/10.1016/j.apsoil.2017.04.024
Luo Y, Dungait JA, Zhao X, Brookes PC, Durenkamp M, Li G, Lin Q (2018) Pyrolysis temperature during biochar production alters its subsequent utilization by microorganisms in an acid arable soil. Land Degrad Dev 29:2183–2188. https://doi.org/10.1002/ldr.2846
Luo K, Xie C, Yuan X, Liu S, Chen P, Du Q, Zheng B, Fu Z, Wang X, Yong T et al (2023) Biochar and biofertilizer reduced nitrogen input and increased soybean yield in the maize soybean relay strip intercropping system. BMC Plant Biol 23:38. https://doi.org/10.1186/s12870-023-04058-5
Ma W, Du G, Li J, Fang Y, Hou L, Chen G, Ma D (2017) Supercritical water pyrolysis of sewage sludge. Waste Manag 59:371–378. https://doi.org/10.1016/j.wasman.2016.10.053
Ma H, Egamberdieva D, Wirth S, Bellingrath-Kimura SD (2019) Effect of biochar and irrigation on soybean-rhizobium symbiotic performance and soil enzymatic activity in field rhizosphere. Agronomy 9:626. https://doi.org/10.3390/agronomy9100626
Ma J, Saleem MH, Yasin G, Mumtaz S, Qureshi FF, Ali B, Ercisli S, Alhag SK, Ahmed AE, Vodnar DC et al (2022) Individual and combinatorial effects of SNP and NaHS on morpho-physio-biochemical attributes and phytoextraction of chromium through Cr-stressed spinach (*Spinacia oleracea* L.). Front Plant Sci 13:973740. https://doi.org/10.3389/fpls.2022.973740
Maaz TM, Hockaday WC, Deenik JL (2021) Biochar volatile matter and feedstock effects on soil nitrogen mineralization and soil fungal colonization. Sustainability 13(4):2018. https://doi.org/10.3390/su13042018
Macil PJ, Ogola JB, Odhiambo JJ (2020) Response of soil pH and nodulation of three chickpea genotypes to biochar and rhizobium inoculation. Commun Soil Sci Plant Anal 51:2377–2387. https://doi.org/10.1080/00103624.2020.1836204

Mamatha B, Basavarajappa M (2024) Enhancing phosphorus release in acidic soils using biochar enriched with PSB: a review. Mysore J Agric Sci 58(3):24–34

Manning P, van der Plas F, Soliveres S, Allan E, Maestre FT, Mace G, Whittingham MJ, Fischer M (2018) Redefining ecosystem multifunctionality. Nat Ecol Evol 2:427–436. https://doi.org/10.1038/s41559-017-0461-7

Manral V, Bargali K, Bargali SS, Jhariya MK, Padalia K (2022) Relationships between soil and microbial biomass properties and annual flux of nutrients in Central Himalayan forests, India. Land Degrad Dev 33(12):2014–2025. https://doi.org/10.1002/ldr.4283

Mansoor S, Kour N, Manhas S, Zahid S, Wani OA, Sharma V, Wijaya L, Alyemeni MN, Alsahli AA, El-Serehy HA, Paray BA (2021) Biochar as a tool for effective management of drought and heavy metal toxicity. Chemosphere 271:129458. https://doi.org/10.1016/j.chemosphere.2020.129458

Mao J, Zhang K, Chen B (2019) Linking hydrophobicity of biochar to the water repellency and water holding capacity of biochar-amended soil. Environ Pollut 253:779–789. https://doi.org/10.1016/j.envpol.2019.07.051

Martin T, Wade J, Singh P, Sprunger CD (2022) The integration of nematode communities into the soil biological health framework by factor analysis. Ecol Indic 136:108676. https://doi.org/10.1016/j.ecolind.2022.108676

McCauley K, Barlow K (2023) Regenerative agriculture: increasing plant diversity and soil carbon sequestration on agricultural landscapes. Surg J 15(1). https://doi.org/10.21083/surg.v15i1.7196

Meena VS, Meena SK, Verma JP, Kumrar A, Aeron A, Mishra PK, Bisht JK, Pattanayaka A, Naveed M, Dotaniya ML (2017) Plant beneficial rhizospheric microorganism (PBRM) strategies to improve nutrients use efficiency: a review. Ecol Eng 107:8–32. https://doi.org/10.1016/j.ecoleng.2017.06.058

Minasny B, McBratney AB (2018) Limited effect of organic matter on soil available water capacity. Eur J Soil Sci 69(1):39–47. https://doi.org/10.1111/ejss.12475

Mohamed I, El-Meihy R, Ali M, Chen F, Raleve D (2017) Interactive effects of biochar and micronutrients on faba bean growth, symbiotic performance, and soil properties. J Plant Nutr Soil Sci 180:729–738. https://doi.org/10.1002/jpln.201700293

Moreira MT, Noya I, Feijoo G (2017) The prospective use of biochar as adsorption matrix–a review from a lifecycle perspective. Bioresour Technol 246:135–141. https://doi.org/10.1016/j.biortech.2017.08.041

Morillo E, Villaverde J (2017) Advanced technologies for the remediation of pesticide contaminated soils. Sci Total Environ 586:576–597. https://doi.org/10.1016/j.scitotenv.2017.02.020

Mosa A, Mansour MM, Soliman E, El-Ghamry A, El Alfy M, El Kenawy AM (2023) Biochar as a soil amendment for restraining greenhouse gases emission and improving soil carbon sink: current situation and ways forward. Sustainability 15:1206. https://doi.org/10.3390/su15021206

Mosier S, Córdova SC, Robertson GP (2021) Restoring soil fertility on degraded lands to meet food, fuel, and climate security needs via perennialization. Front Sustain Food Syst 5:706142. https://doi.org/10.3389/fsufs.2021.706142

Mourya NR, Bargali K, Bargali SS (2019) Effect of Coriaria nepalensis Wall. colonization in a mixed conifer forest of Indian Central Himalaya. J For Res 30(1):305–317. https://doi.org/10.1007/s11676-018-0613-x

Muhammad N, Hussain M, Ullah W, Khan TA, Ali S, Akbar A, Aziz R, Rafiq MK, Bachmann RT, Al-Wabel MI, Rizwan M (2018) Biochar for sustainable soil and environment: a comprehensive review. Arab J Geosci 11(23):1–14. https://doi.org/10.1007/s12517-018-4074-5

Mukherjee S, Sarkar B, Aralappanavar VK, Mukhopadhyay R, Basak BB, Srivastava P, Marchut-Mikołajczyk O, Bhatnagar A, Semple KT, Bolan N (2022) Biochar–microorganism interactions for organic pollutant remediation: challenges and perspectives. Environ Pollut 308:119609. https://doi.org/10.1016/j.envpol.2022.119609

Munera-Echeverri JL, Martinsen V, Strand LT, Zivanovic V, Cornelissen G, Mulder J (2018) Cation exchange capacity of biochar: an urgent method modification. Sci Total Environ 642:190–197. https://doi.org/10.1016/j.scitotenv.2018.06.017

Murariu DT (2024) The ecological role of some functional groups of invertebrates—a review. Sci Pap Ser A Agron 67(1):775–782
Murtaza G, Ahmed Z, Usman M (2022) Feedstock type, pyrolysis temperature and acid modification effects on physiochemical attributes of biochar and soil quality. Arab J Geosci 15(3):305. https://doi.org/10.1007/s12517-022-09539-9
Nair VD, Nair PKR, Dari B, Freitas AM, Chatterjee N, Pinheiro FM (2017) Biochar in the agroecosystem–climate–change–sustainability nexus. Front Plant Sci 8:2051. https://doi.org/10.3389/fpls.2017.02051
Nandini R, Prakasha HC (2022) Use of mulberry stalk as biochar and its effect on growth, yield and quality of mulberry Morus alba L. Mysore J Agric Sci 56(4):296–303
Neher DA, Barbercheck ME (2019) Soil microarthropods and soil health: intersection of decomposition and pest suppression in agroecosystems. Insects 10(12):414. https://doi.org/10.3390/insects10120414
Neuberger P, Romero C, Kim K, Hao X, McAllister TA, Ngo S et al (2024) Biochar is colonized by select arbuscular mycorrhizal fungi in agricultural soils. Mycorrhiza 34(3):191–201. https://doi.org/10.1007/s00572-024-01149-5
Nguyen TTN, Xu CY, Tahmasbian I, Che R, Xu Z, Zhou X, Wallace HM, Bai SH (2017) Effects of biochar on soil available inorganic nitrogen: a review and meta-analysis. Geoderma 288:79–96. https://doi.org/10.1016/j.geoderma.2016.11.004
Nguyen TTN, Wallace HM, Xu CY, Van Zwieten L, Weng ZH, Xu Z, Che R, Tahmasbian I, Hu HW, Bai SH (2018) The effects of short term, long term and reapplication of biochar on soil bacteria. Sci Total Environ 636:142–151. https://doi.org/10.1016/j.scitotenv.2018.04.278
Nguyen MK, Lin C, Hoang HG, Sanderson P, Dang BT, Bui XT, Nguyen NSH, Vo DN, Tran HT (2022) Evaluate the role of biochar during the organic waste composting process: a critical review. Chemosphere 299:134488. https://doi.org/10.1016/j.chemosphere.2022.134488
Nguyen TB, Nguyen VT, Hoang HG, Cao NDT, Nguyen TT, Vo TDH, Nguyen NKQ, Pham MDT, Nghiem DL, Vo TKQ, Dong CD (2023) Recent development of algal biochar for contaminant remediation and energy application: a state-of-the art review. Curr Pollut Rep 9(1):73–89. https://doi.org/10.1007/s40726-022-00243-6
Nikitin DA, Semenov MV, Chernov TI, Ksenofontova NA, Zhelezova AD, Ivanova EA, Khitrov NB, Stepanov AL (2022) Microbiological indicators of soil ecological functions: a review. Eurasian Soil Sci 55(2):221–234. https://doi.org/10.1134/S1064229322020090
Nim YS, Wong KB (2019) The maturation pathway of nickel urease. Inorganics 7:85. https://doi.org/10.3390/inorganics7070085
Ning Y, Xiao Z, Weinmann M, Li Z (2019) Phosphate uptake is correlated with the root length of celery plants following the association between arbuscular mycorrhizal fungi, *Pseudomonas sp*. and biochar with different phosphate fertilization levels. Agronomy 9(12):1–12. https://doi.org/10.3390/agronomy9120824
Nocentini M, Panettieri M, de Castro Barragán JMG, Mastrolonardo G, Knicker H (2021) Recycling pyrolyzed organic waste from plant nurseries, rice production and shrimp industry as peat substitute in potting substrates. J Environ Manag 277:111436. https://doi.org/10.1016/j.jenvman.2020.111436
Ojha SK, Benjamin JC, Singh AK (2018) Effect of compost in combination with PGPR on growth of tomato (*Lycopersicon esculentum*) plant. Int J Agric Sci Res 6:63–72
Oladele SO (2019) Effect of biochar amendment on soil enzymatic activities, carboxylate secretions and upland rice performance in a sandy clay loam Alfisol of Southwest Nigeria. Sci Afr 4:e00107. https://doi.org/10.1016/j.sciaf.2019.e00107
Olagoke FK, Bettermann A, Nguyen PTB, Redmile-Gordon M, Babin D, Smalla K, Nesme J, Sørensen SJ, Kalbitz K, Vogel C (2022) Importance of substrate quality and clay content on microbial extracellular polymeric substances production and aggregate stability in soils. Biol Fertil Soils 58(4):435–457. https://doi.org/10.1007/s00374-022-01632-1
Oliveira FR, Patel AK, Jaisi DP, Adhikari S, Lu H, Khanal SK (2017) Environmental application of biochar: current status and perspectives. Bioresour Technol 246:110–122. https://doi.org/10.1016/j.biortech.2017.08.122

Ortiz A, Sansinenea E (2021) Recent advancements for microorganisms and their natural compounds useful in agriculture. Appl Microbiol Biotechnol 105:891–897. https://doi.org/10.1007/s00253-020-11030-y

Padalia K, Bargali SS, Bargali K, Manral V (2022) Soil microbial biomass phosphorus under different land use systems. Trop Ecol 63:30–48. https://doi.org/10.1007/s42965-021-00184-z

Paetsch L, Mueller CW, Kögel-Knabner I, Lutzow M, Girardin C, Rumpel C (2018) Effect of in-situ aged and fresh biochar on soil hydraulic conditions and microbial C use under drought conditions. Sci Rep 8:1–11. https://doi.org/10.3390/agronomy10060824

Palansooriya KN, Wong JTF, Hashimoto Y, Huang L, Rinklebe J, Chang SX, Bolan N, Wang H, Ok YS (2019) Response of microbial communities to biochar-amended soils: a critical review. Biochar 1(1):3–22. https://doi.org/10.1007/s42773-019-00009-2

Pallarés J, González-Cencerrado A, Arauzo I (2018) Production and characterization of activated carbon from barley straw by physical activation with carbon dioxide and steam. Biomass Bioenergy 115:64–73. https://doi.org/10.1016/j.biombioe.2018.04.015

Pan SY, Dong CD, Su JF, Wang PY, Chen CW, Chang JS, Kim H, Huang CP, Hung CM (2021) The role of biochar in regulating the carbon, phosphorus, and nitrogen cycles exemplified by soil systems. Sustainability 13(10):5612. https://doi.org/10.3390/su13105612

Pandey V, Patel A, Patra DD (2016) Biochar ameliorates crop productivity, soil fertility, essential oil yield and aroma profiling in basil (*Ocimum basilicum* L.). Ecol Eng 90:361–366. https://doi.org/10.1016/j.ecoleng.2016.01.020

Pandey H, Maaren I, Shah K, Maraseni T (2020) Response of topographic and biodiversity variables on biomass and carbon density in community forests of Himalayan foot-hills. J For Livelihood 19:51–65. https://doi.org/10.3126/jfl.v19i1.59627

Pandey R, Bargali SS, Bargali K, Pandey VC (2023) Temporal variability in fine root dynamics in relation to tree girth size in sub-tropical Shorea robusta forests. Land Degrad Dev 34(5):1522–1537. https://doi.org/10.1002/ldr.4550

Pandey R, Bargali SS, Bargali K, Karki H, Chaturvedi RK (2024) Dynamics of nitrogen mineralization and fine root decomposition in sub-tropical Shorea robusta Gaertner f. forests of Central Himalaya, India. Sci Total Environ:170896. https://doi.org/10.1016/j.scitotenv.2024.170896

Pandit NR, Mulder J, Hale SE, Zimmerman AR, Pandit BH, Cornelissen G (2018) Multi-year double cropping biochar field trials in Nepal: finding the optimal biochar dose through agronomic trials and cost-benefit analysis. Sci Total Environ 637:1333–1341. https://doi.org/10.1016/j.scitotenv.2018.05.107

Pant M, Negi GCS, Kumar P (2017) Macrofauna contributes to organic matter decomposition and soil quality in Himalayan agroecosystems, India. Appl Soil Ecol 120:20–29. https://doi.org/10.1016/j.apsoil.2017.07.019

Panth M, Hassler SC, Baysal-Gurel F (2020) Methods for management of soilborne diseases in crop production. Agriculture 10:16. https://doi.org/10.3390/agriculture10010016

Panwar NL, Pawar A (2022) Influence of activation conditions on the physicochemical properties of activated biochar: a review. Biomass Convers Biorefinery 12(3):925–947. https://doi.org/10.1007/s13399-020-00870-3

Park Y, Solhtalab M, Thongsomboon W, Aristilde L (2022) Strategies of organic phosphorus recycling by soil bacteria: acquisition, metabolism, and regulation. Environ Microbiol Rep 14:3–24. https://doi.org/10.1111/1758-2229.13040

Pattnaik S, Mohapatra B, Gupta A (2021) Plant growth-promoting microbe mediated uptake of essential nutrients (Fe, P, K) for crop stress management: microbe–soil–plant continuum. Front Agron 3:689972. https://doi.org/10.3389/fagro.2021.689972

Philippot L, Chenu C, Kappler A, Rillig MC, Fierer N (2024) The interplay between microbial communities and soil properties. Nat Rev Microbiol 22(4):226–239. https://doi.org/10.1038/s41579-023-00980-5

Pittol M, Scully E, Miller D, Durso L, Mariana Fiuza L, Valiati VH (2018) Bacterial community of the rice floodwater using cultivation-independent approaches. Int J Microbiol 2018:6280484. https://doi.org/10.1155/2018/6280484

Poeydebat C, Tixier P, Chabrier C, de Bellaire LD, Vargas R, Daribo MO, Carval D (2017) Does plant richness alter multitrophic soil food web and promote plant-parasitic nematode regulation in banana agroecosystems? Appl Soil Ecol 117:137–146. https://doi.org/10.1016/j.apsoil.2017.04.017

Pokharel P, Ma Z, Chang SX (2020) Biochar increases soil microbial biomass with changes in extra- and intracellular enzyme activities: a global meta-analysis. Biochar 2(1):65–79. https://doi.org/10.1007/s42773-020-00039-1

Polláková N, Šimanský V, Kravka M (2018) The influence of soil organic matter fractions on aggregates stabilization in agricultural and forest soils of selected Slovak and Czech hilly lands. J Soils Sediments 18(8):2790–2800. https://doi.org/10.1007/s11368-017-1842-x

Potapov AM (2022) Multifunctionality of belowground food webs: resource, size and spatial energy channels. Biol Rev 97:1691–1711. https://doi.org/10.1111/brv.12857

Prasad M, Chrysargyris A, McDaniel N, Kavanagh A, Gruda NS, Tzortzakis N (2020) Plant nutrient availability and pH of biochars and their fractions, with the possible use as a component in a growing media. Agronomy 10:10. https://doi.org/10.3390/agronomy10010010

Pukalchik M, Mercl F, Terekhova V, Tlustoš P (2018) Biochar, wood ash and humic substances mitigating trace elements stress in contaminated sandy loam soil: evidence from an integrative approach. Chemosphere 203:228–238. https://doi.org/10.1016/j.chemosphere.2018.03.181

Purakayastha TJ, Bera T, Bhaduri D, Sarkar B, Mandal S, Wade P, Kumari S, Biswas S, Menon M, Pathak H et al (2019) A review on biochar modulated soil condition improvements and nutrient dynamics concerning crop yields: pathways to climate change mitigation and global food security. Chemosphere 227:345–365. https://doi.org/10.1016/j.chemosphere.2019.03.170

Pylro VS, Fulthorpe R, Roesch LFW (2020) Editorial: Microbe-mediated processes in soils. Front Environ Sci 8:124

Qi L, Pokharel P, Chang SX, Zhou P, Niu H, He X, Wang Z, Gao M (2020) Biochar application increased methane emission, soil carbon storage and net ecosystem carbon budget in a 2-year vegetable–rice rotation. Agric Ecosyst Environ 292:106831. https://doi.org/10.1016/j.agee.2020.106831

Qin Y, Li G, Gao Y, Zhang L, Ok YS, An T (2018) Persistent free radicals in carbon-based materials on transformation of refractory organic contaminants (ROCs) in water: a critical review. Water Res. https://doi.org/10.1016/j.watres.2018.03.012

Qiu Z, Chen JH, Tang JW, Zhang QZ (2018) A study of cadmium remediation and mechanisms: improvements in the stability of walnut shell-derived biochar. Sci Total Environ 636:80–84. https://doi.org/10.1016/j.scitotenv.2018.04.215

Rafique M, Sultan T, Ortas I, Chaudhary HJ (2017) Enhancement of maize plant growth with inoculation of phosphate-solubilizing bacteria and biochar amendment in soil. Soil Sci Plant Nutr 63(5):460–469. https://doi.org/10.1080/00380768.2017.1373599

Rahman MT, Zhu QH, Zhang ZB, Zhou H, Peng X (2017) The roles of organic amendments and microbial community in the improvement of soil structure of a Vertisol. Appl Soil Ecol 111:84–93. https://doi.org/10.1016/j.apsoil.2016.11.018

Rahman M, Guo Z, Zhang Z, Zhou H, Peng X (2018) Wetting and drying cycles improving aggregation and associated C stabilization differently after straw or biochar incorporated into a Vertisol. Soil Tillage Res 175:28–36. https://doi.org/10.1016/j.still.2017.08.007

Rahmanian M, Khadem A (2024) The effects of biochar on soil extra and intracellular enzymes activity. Biomass Convers Biorefinery 14(18):21993–22005. https://doi.org/10.1007/s13399-023-04330-6

Rahul R, Sharma P, Singh A, Singh J, Kumar M (2022) Soil enzymes and their role in soil health improvement. In: Advances in agricultural and industrial microbiology: volume 1: microbial diversity and application in agroindustry. Springer Nature Singapore, Singapore, pp 39–61. https://doi.org/10.1007/978-981-16-8918-5_3

Ramezanzadeh H, Zarehaghi D, Baybordi A, Bouket AC, Oszako T, Alenezi FN, Belbahri L (2023) The impacts of biochar-assisted factors on the hydrophysical characteristics of amended soils: a review. Sustainability 15(11):8700. https://doi.org/10.3390/su15118700

Rana KL, Kour D, Kaur T, Sheikh I, Yadav AN, Kumar V et al (2020) Endophytic microbes from diverse wheat genotypes and their potential biotechnological applications in plant growth promotion and nutrient uptake. Proc Natl Acad Sci India Sect B Biol Sci 90:969–979. https://doi.org/10.1007/s40011-020-01168-0

Rasool M, Akhter A, Haider MS (2021) Molecular and biochemical insight into biochar and Bacillus subtilis induced defense in tomatoes against *Alternaria solani*. Sci Hortic 285:110203

Raya-Moreno I, Cañizares R, Domene X, Carabassa V, Alcañiz JM (2017) Comparing current chemical methods to assess biochar organic carbon in a Mediterranean agricultural soil amended with two different biochars. Sci Total Environ 598:604–618. https://doi.org/10.1016/j.scienta.2021.110203

Razzaghi F, Obour PB, Arthur E (2020) Does biochar improve soil water retention? A systematic review and meta-analysis. Geoderma 361:114055. https://doi.org/10.1016/j.geoderma.2019.114055

Ren YS, Khan H, Khan B, Ilyas M, Iqbal S (2023) Application of eucalyptus wood waste-derived biochar for adsorption of aluminum from aqueous solutions. Desalin Water Treat 287:116–123. https://doi.org/10.5004/dwt.2023.29347

Rittl TF, Oliveira DMS, Canisares LP, Sagrilo E, Butterbach-Bahl K, Dannenmann M, Cerri CEP (2021) High application rates of biochar to mitigate N2O emissions from a N-fertilized tropical soil under warming conditions. Front Environ Sci 8:1. https://doi.org/10.3389/fenvs.2020.611873

Rivka FB, Laird DA, Spokas KA (2018) Sorption of ammonium and nitrate to biochars is electrostatic and pH-dependent. Sci Rep 8:17627. https://doi.org/10.1038/s41598-018-35534-w

Rodrigues M, Withers PJA, Soltangheisi A, Vargas V, Holzschuh M, Pavinato PS (2021) Tillage systems and cover crops affecting soil phosphorus bioavailability in Brazilian Cerrado Oxisols. Soil Till Res 205:104770. https://doi.org/10.1016/j.still.2020.104770

Rosenthal J, Gunn KM (2025) Arbuscular mycorrhizal fungi in biochar-amended soils: a review. Soil Res 63(5). https://doi.org/10.1071/SR24123

Roy P, Dias G (2017) Prospects for pyrolysis technologies in the bioenergy sector: a review. Renew Sust Energ Rev 77:59–69. https://doi.org/10.1016/j.rser.2017.03.136

Saccá ML, Barra Caracciolo A, Di Lenola M, Grenni P (2017) Ecosystem services provided by soil microorganisms. In: Soil biological communities and ecosystem resilience. Springer International Publishing, Cham, pp 9–24. https://doi.org/10.1007/978-3-319-63336-7_2

Sansinenea E (2021) Application of biofertilizers: current worldwide status. In: Rakshit A, Meena VS, Parihar M, Singh HB, Singh AK (eds) Biofertilizers. Volume 1: Advances in bio-inoculants. Elsevier/Woodhead Publishing, Amsterdam/Sawston, pp 183–190. https://doi.org/10.1016/B978-0-12-821667-5.00004-X

Santorufo L, Memoli V, Santini G, Di Natale G, Trifuoggi M, Barile R, Maisto G (2023) Responses of soil microarthropods to soil metal fractions under different Mediterranean vegetation covers. Catena 232:107438. https://doi.org/10.1016/j.catena.2023.107438

Sarfraz R, Hussain A, Sabir A, Ben Fekih I, Ditta A, Xing S (2019) Role of biochar and plant growth promoting rhizobacteria to enhance soil carbon sequestration—a review. Environ Monit Assess. https://doi.org/10.1007/s10661-019-7400-9

Saxena A, Srivastava PC, Pathak A, Pachauri SP, Srivastava P (2024) Effect of fortified pine needle biochar on green fodder yield, nutrient uptake of oat and soil properties. Soil Adv 2:100019. https://doi.org/10.1016/j.soilad.2024.100019

Schlatter DC, Kahl K, Carlson B, Huggins DR, Paulitz T (2018) Fungal community composition and diversity vary with soil depth and landscape position in a no-till wheat-based cropping system. FEMS Microbiol Ecol 94(7):fiy098. https://doi.org/10.1093/femsec/fiy098

Schmidt HP, Kammann C, Hagemann N, Leifeld J, Bucheli TD, Sánchez Monedero MA, Cayuela ML (2021) Biochar in agriculture–a systematic review of 26 global meta-analyses. GCB Bioenergy 13(11):1708–1730. https://doi.org/10.1111/gcbb.12889

Schreefel L, Schulte RPO, De Boer IJM, Schrijver AP, Van Zanten HHE (2020) Regenerative agriculture – the soil is the base. Glob Food Sec 26:100404. https://doi.org/10.1016/j.gfs.2020.100404

Schreiter IJ, Schmidt W, Kumar A, Graber ER, Schüth C (2020) Effect of water leaching on biochar properties and its impact on organic contaminant sorption. Environ Sci Pollut Res 27:691–703. https://doi.org/10.1007/s11356-019-06904-2

Sechi V, De Goede RGM, Rutgers M, Brusaard L, Mulder CA (2017) Community trait-based approach to ecosystem functioning in soil. Agric Ecosyst Environ 239:265–273. https://doi.org/10.1016/j.agee.2017.01.036

Semida WM, Beheiry HR, Sétamou M, Simpson CR, Abd El-Mageed TA, Rady MM, Nelson SD (2019) Biochar implications for sustainable agriculture and environment: a review. South Afr J Bot 127:333–347. https://doi.org/10.1016/j.sajb.2019.11.015

Seneviratne M, Weerasundara L, Ok YS, Rinklebe J, Vithanage M (2017) Phytotoxicity attenuation in Vigna radiata under heavy metal stress at the presence of biochar and N fixing bacteria. J Environ Manag 186:293–300. https://doi.org/10.1016/j.jenvman.2016.07.024

Shah KK, Tripathi S, Tiwari I, Shrestha J, Modi B, Paudel N, Das BD (2021) Role of soil microbes in sustainable crop production and soil health: a review. Agric Sci Technol 13(2):109–118. https://doi.org/10.15547/ast.2021.02.019

Shahi C, Bargali SS, Bargali K, Vibhuti (2023) Dry matter dynamics and CO_2 mitigation in the herb layer of Central Himalayan agroecosystems along an altitudinal gradient, India. Trop Ecol 64(1):180–192. https://doi.org/10.1007/s42965-022-00258-6

Shaikh SD, Shelke US, Shaikh AS, Pawar NB (2025) Eco-friendly cultivation of onion (*Allium cepa* L.) using vesicular-arbuscular mycorrhizal (VAM) fungi. IJSAT Int J Sci Technol 16(2). https://doi.org/10.71097/IJSAT.v16.i2.5192

Sharma R, Maanik and Thakur N (2023) Biochar applications in fruit crop production: enhancing yield, quality and sustainability. In: Encyclopedia of agriculture and allied sciences (Royal Book Publishing) Jammu & Kashmir, India.

Shirzad M, Karimi M, Rodrigues AE, Silva JA (2023) Biochar in carbon sequestration. In: Biochar composites: and fundamentals its and applications. Springer Nature Singapore, Singapore, pp 73–105. https://doi.org/10.1007/978-981-99-5239-7_4

Siegelman RL, Milner PJ, Kim EJ, Weston SC, Long JR (2019) Challenges and opportunities for adsorption based CO_2 capture from natural gas combined cycle emissions. Energy Environ Sci 12:2161–2173. https://doi.org/10.1039/C9EE00505F

Silva JMD, Medeiros EV, Duda GP, Barros JAD, Santos UJD (2017) FAMES and microbial activities involved in the suppression of cassava root rot by organic matter. Rev Caatinga 30(3):708–717. https://doi.org/10.1590/1983-21252017v30n3

Silva LG, de Andrade CA, Bettiol W (2020) Biochar amendment increases soil microbial biomass and plant growth and suppresses Fusarium wilt in tomato. Trop Plant Pathol 45(1):73–83. https://doi.org/10.1007/s40858-020-00332-1

Šimanský V (2017) Is the period of 18 years sufficient for an evaluation of changes in soil organic carbon under a variety of different soil management practices? Commun Soil Sci Plant Anal 48:37–42. https://doi.org/10.1080/00103624.2016.1253717

Sokol NW, Slessarev E, Marschmann GL, Nicolas A, Blazewicz SJ, Brodie EL et al (2022) Life and death in the soil microbiome: how ecological processes influence biogeochemistry. Nat Rev Microbiol 20:415–430. https://doi.org/10.1038/s41579-022-00695-z

Song D, Tang J, Xi X, Zhang S, Liang G, Zhou W, Wang X (2018) Responses of soil nutrients and microbial activities to additions of maize straw biochar and chemical fertilization in a calcareous soil. Eur J Soil Biol 84:1–10. https://doi.org/10.1016/j.ejsobi.2017.11.003

Stirling G, Hayden H, Pattison T, Stirling M (2017) Soil health, soil biology, soil borne diseases and sustainable agriculture: a guide. Aust Plant Pathol 46(4):387

Sumbul A, Ansari R, Rizvi R, Mahmood I (2020) Azotobacter: a potential bio-fertilizer for soil and plant health management. Saudi J Biol Sci 27. https://doi.org/10.1016/j.sjbs.2020.08.004

Sun Q, Meng J, Lan Y, Shi G, Yang X, Cao D, Chen W, Han X (2021) Long-term effects of biochar amendment on soil aggregate stability and biological binding agents in brown earth. Catena 205:105460. https://doi.org/10.1016/j.catena.2021.105460

Sun J, Jia Q, Li Y, Zhang T, Chen J, Ren Y et al (2022) Effects of arbuscular mycorrhizal fungi and biochar on growth, nutrient absorption, and physiological properties of maize (Zea mays L.). J Fungi 8(12):1275. https://doi.org/10.3390/jof8121275

Swagathnath G, Rangabhashiyam S, Murugan S, Balasubramanian P (2019) Influence of biochar application on growth of *Oryza sativa* and its associated soil microbial ecology. Biomass Convers Biorefinery 9:341–352. https://doi.org/10.1007/s13399-018-0365-z

Taheri MAR, Astaraei AR, Lakzian A, Emami H (2024) The role of biochar and sulfur-modified biochar on soil water content, biochemical properties and millet crop under saline-sodic and calcareous soil. Plant Soil 499(1):221–236. https://doi.org/10.1007/s11104-023-05912-z

Tan Z, Lin CS, Ji X, Rainey TJ (2017) Returning biochar to fields: a review. Appl Soil Ecol 116:1–11. https://doi.org/10.1016/j.apsoil.2017.03.017

Tan XF, Zhu SS, Wang RP, Chen YD, Show PL, Zhang FF, Ho SH (2021) Role of biochar surface characteristics in the adsorption of aromatic compounds: pore structure and functional groups. Chin Chem Lett 32(10):2939–2946. https://doi.org/10.1016/j.cclet.2021.04.059

Tan S, Zhou G, Yang Q, Ge S, Liu J, Cheng YW, Yek PNY, Mahari WAW, Kong SH, Chang JS et al (2023) Utilization of current pyrolysis technology to convert biomass and manure waste into biochar for soil remediation: a review. Sci Total Environ 864:160990. https://doi.org/10.1016/j.scitotenv.2022.160990

Tang F, Xu Z, Gao M, Li L, Li H, Cheng H, Zhang C, Tian G (2019) The dissipation of cyazofamid and its main metabolite in soil response oppositely to biochar application. Chemosphere 218:26–35. https://doi.org/10.1016/j.chemosphere.2018.11.094

Tang J, Zhang L, Zhang J, Ren L, Zhou Y, Zheng Y, Luo L, Yang Y, Huang H, Chen A (2020) Physicochemical features, metal availability and enzyme activity in heavy metal-polluted soil remediated by biochar and compost. Sci Total Environ 701:134751. https://doi.org/10.1016/j.scitotenv.2019.134751

Tang Z, Liu X, Li G, Liu X (2022) Mechanism of biochar on nitrification and denitrification to N_2O emissions based on isotope characteristic values. Environ Res 212:113219. https://doi.org/10.1016/j.envres.2022.113219

Tao S, Wu Z, Wei M, Liu X, He Y, Ye BC (2019) *Bacillus subtilis* SL-13 biochar formulation promotes pepper plant growth and soil improvement. Can J Microbiol 65(5):333–342. https://doi.org/10.1139/cjm-2018-0333

Tibbett M, Fraser TD, Duddigan S (2020) Identifying potential threats to soil biodiversity. PeerJ 8:e9271. https://doi.org/10.7717/peerj.9271

Timofeeva A, Galyamova M, Sedykh S (2022) Prospects for using phosphate-solubilizing microorganisms as natural fertilizers in agriculture. Plants 11:2119. https://doi.org/10.3390/plants11162119

Tomczyk A, Sokołowska Z, Boguta P (2020) Biochar physicochemical properties: pyrolysis temperature and feedstock kind effects. Rev Environ Sci Biotechnol 19:191–215. https://doi.org/10.1007/s11157-020-09523-3

Torabian S, Farhangi-Abriz S, Denton MD (2019) Do tillage systems influence nitrogen fixation in legumes? A review. Soil Tillage Res 185:113–121. https://doi.org/10.1016/j.still.2018.09.006

Totsche KU, Ray N, Kögel-Knabner I (2024) Structure–function co-evolution during pedogenesis—microaggregate development and turnover in soils. J Plant Nutr Soil Sci 187(1). https://doi.org/10.1002/jpln.202400012

Trivedi P, Leach JE, Tringe SG, Sa T, Singh BK (2020) Plant-microbiome interactions: from community assembly to plant health. Nat Rev Microbiol 18:607–621. https://doi.org/10.1038/s41579-020-0412-1

Ujjainiya P, Choudhary M, Jatav HS, Tokala VY, Rajput VD, Minkina T (2022) Impact of weed management practices on soil microflora and dehydrogenase enzyme activity under varying levels of nitrogen in winter season onion (*Allium cepa* L.). Bull Environ Contam Toxicol 108:430–436. https://doi.org/10.1007/s00128-021-03265-w

Ullah S, Liang H, Ali I, Zhao Q, Iqbal A, Wei S, Shah T, Yan B, Jiang L (2020) Biochar coupled with contrasting nitrogen sources mediated changes in carbon and nitrogen pools, microbial and

enzymatic activity in paddy soil. J Saudi Chem Soc 24(11):835–849. https://doi.org/10.1016/j.jscs.2020.08.008
Vaskevicius L, Malunavicius V, Jankunec M, Lastauskiene E, Talaikis M, Mikoliunaite L, Maneikis A, Gudiukaite R (2023) Insights in MICP dynamics in urease-positive *Staphylococcus Sp.* H6 and Sporosarcinapasteurii bacterium. Environ Res 234:116588. https://doi.org/10.1016/j.envres.2023.116588
Veiga TRLA, Lima JT, Dessimoni ALA, Pego MFP, Soares JR, Trugilho PF (2017) Different plant biomass characterizations for biochar production. Cerne 23:529–536
Ventura M, Alberti G, Panzacchi P, Delle Vedove G, Miglietta F, Tonon G (2019) Biochar mineralization and priming effect in a poplar short rotation coppice from a 3-year field experiment. Biol Fertil Soils 55(1):67–78. https://doi.org/10.1007/s00374-018-1329-y
Verma P, Kumar S, Shakya M, Sandhu SS (2021) VAM: an alternate strategy. In: Microbial rejuvenation of polluted environment: volume 1, vol 25, p 153. https://doi.org/10.1007/978-981-15-7447-4_6
Voisin R, Horwitz P, Godrich S, Sambell R, Cullerton K, Devine A (2024) What goes in and what comes out: a scoping review of regenerative agricultural practices. Agroecol Sustain Food Syst 48(1):124–158. https://doi.org/10.1080/21683565.2023.2270441
Wang J, Wang S (2019) Preparation, modification and environmental application of biochar: a review. J Clean Prod 227:1002–1022. https://doi.org/10.1016/j.jclepro.2019.04.282
Wang J, Xiong Z, Kuzyakov Y (2016) Biochar stability in soil: meta-analysis of decomposition and priming effects. GCB Bioenergy 8:512–523. https://doi.org/10.1111/gcbb.12266
Wang D, Fonte SJ, Parikh SJ, Six J, Scow KM (2017) Biochar additions can enhance soil structure and the physical stabilization of C in aggregates. Geoderma 303:110–117. https://doi.org/10.1016/j.geoderma.2017.05.027
Wang H, Shi Y, Wang D, Yao Z, Wang Y, Liu J, Zhang S, Wang A (2018) A biocontrol strain of Bacillus subtilis WXCDD105 used to control tomato *Botrytis cinerea* and *Cladosporium fulvum* Cooke and promote the growth of seedlings. Int J Mol Sci 19(5):1371. https://doi.org/10.3390/ijms19051371
Wang M, Wang JJ, Tafti ND, Hollier CA, Myers G, Wang X (2019a) Effect of alkali-enhanced biochar on silicon uptake and suppression of gray leaf spot development in perennial ryegrass. Crop Protect 119:9–16. https://doi.org/10.1016/j.cropro.2019.01.013
Wang ZH, Tang CS, Wang HY, Zhao CJ, Yin DW, Yuan Y, Yang KJ, Li ZT (2019b) Effect of different amounts of biochar on meadow soil characteristics and maize yields over three years. Bioresources 14(2):4194–4209
Wang Q, Xu L, Guo D, Wang G, Song X, Ma Y (2021) The continuous application of biochar in field: effects on P fraction, P sorption and release. Chemosphere 263:128084. https://doi.org/10.1016/j.chemosphere.2020.128084
Wang L, Hamel C, Lu P, Wang J, Sun D, Wang Y, Lee SJ, Gan GY (2023) Using enzyme activities as an indicator of soil fertility in grassland—an academic dilemma. Front Plant Sci 14:1175946. https://doi.org/10.3389/fpls.2023.1175946
Wang X, Li Z, Li Q, Hu Z (2025) Alleviation of plant abiotic stress: mechanistic insights into emerging applications of phosphate-solubilizing microorganisms in agriculture. Plants 14(10):1558. https://doi.org/10.3390/plants141015588
Weber K, Quicker P (2018) Properties of biochar. Fuel 217:240–261. https://doi.org/10.1016/j.fuel.2017.12.054
Wei W, Zhang S, Wu L, Cui D, Ding X (2020) Biochar and phosphorus fertilization improved soil quality and inorganic phosphorus fractions in saline-alkaline soils. Arch Agron Soil Sci:1–14. https://doi.org/10.1080/03650340.2020.1784879
Wei X, Xie B, Wan C, Song R, Zhong W, Xin S, Song K (2024) Enhancing soil health and plant growth through microbial fertilizers: mechanisms, benefits, and sustainable agricultural practices. Agronomy 14(3):609. https://doi.org/10.3390/agronomy14030609
Weisser WW, Roscher C, Meyer ST, Ebeling A, Luo G, Allan E, Beßler H, Barnard RL, Buchmann N, Buscot F et al (2017) Biodiversity effects on ecosystem functioning in a 15-year grassland

experiment: patterns, mechanisms, and open questions. Basic Appl Ecol 23:1–73. https://doi.org/10.1016/j.baae.2017.06.002

Weng Z, Van Zwieten L, Singh BP, Tavakkoli E, Joseph S, Macdonald LM, Rose TJ, Rose MT, Kimber SWL, Morris S, Cozzolino D, Araujo JR, Archanjo BS, Cowie A (2017) Biochar built soil carbon over a decade by stabilizing rhizodeposits. Nat Clim Chang 7:371–376. https://doi.org/10.1038/nclimate3276

Weng ZH, Van Zwieten L, Singh BP, Tavakkoli E, Kimber S, Morris S, Macdonald LM, Cowie A (2018) The accumulation of rhizodeposits in organo-mineral fractions promoted biochar-induced negative priming of native soil organic carbon in Ferralsol. Soil Biol Biochem 118:91–96. https://doi.org/10.1016/j.soilbio.2017.12.008

Winck BR, Chauvat M, Coulibaly SFM, Santonja M, Saccol de Sá EL, Forey E (2020) Functional collembolan assemblages induce different plant responses in Lolium perenne. Plant Soil 452(1–2):347–358. https://doi.org/10.1007/s11104-020-04579-0

Wolna-Maruwka A, Piechota T, Niewiadomska A, Kamiński A, Kayzer D, Grzyb A, Pilarska AA (2021) The effect of biochar-based organic amendments on the structure of soil bacterial community and yield of maize (*Zea mays* L.). Agronomy 11(7):1286. https://doi.org/10.3390/agronomy11071286

Wong JTF, Chen Z, Chen X, Ng CWW, Wong MH (2017) Soil-water retention behavior of compacted biochar-amended clay: a novel landfill final cover material. J Soils Sediments 17:590–598. https://doi.org/10.1007/s11368-016-1401-x

Wong JTF, Chen X, Deng W, Chai Y, Ng CWW, Wong MH (2019) Effects of biochar on bacterial communities in a newly established landfill cover topsoil. J Environ Manag 236:667–673. https://doi.org/10.1016/j.jenvman.2019.02.010

Wood R, Mašek O, Erastova V (2024) Developing realistic molecular models of biochars. Cell Rep Phys Sci 5(7):102037. https://doi.org/10.1016/j.xcrp.2024.102037

Wu C, Zhi D, Yao B, Zhou Y, Yang Y, Zhou Y (2022) Immobilization of microbes on biochar for water and soil remediation: a review. Environ Res 212:113226. https://doi.org/10.1016/j.envres.2022.113226

Wu M, Wang M, Shi W, Zhang Q, Guo T, Li P, Han Y, Li H (2025) Biochar-mediated effects on changes in soil quality and microbial communities. Agronomy 15(8):1861. https://doi.org/10.3390/agronomy15081861

Wystalska K, Kwarciak-Kozłowska A, Włodarczyk R (2022) Influence of technical parameters of the pyrolysis process on the surface area, porosity, and hydrophobicity of biochar from sunflower husk pellet. Sustainability 15(1):394. https://doi.org/10.3390/su15010394

Xiong X, Yu IKM, Cao L, Tsang DCW, Zhang S, Ok YS (2017) A review of biochar based catalysts for chemical synthesis, biofuel production, and pollution control. Bioresour Technol 246:254–270. https://doi.org/10.1016/j.biortech.2017.06.163

Xu N, Tan G, Wang H, Gai X (2016) Effect of biochar additions to soil on nitrogen leaching, microbial biomass and bacterial community structure. Eur J Soil Biol 74:1–8. https://doi.org/10.1016/j.ejsobi.2016.02.004

Xu G, Zhang Z, Deng L (2018a) Adsorption behaviors and removal efficiencies of inorganic, polymeric and organic phosphates from aqueous solution on biochar derived from sewage sludge of chemically enhanced primary treatment process. Water 10:869. https://doi.org/10.3390/w10070869

Xu W, Wang G, Deng F, Zou X, Ruan H, Chen HYH (2018b) Responses of soil microbial biomass, diversity and metabolic activity to biochar applications in managed poplar plantations on reclaimed coastal saline soil. Soil Use Manag 34:597–605. https://doi.org/10.1111/sum.12460

Xu Y, Seshadri B, Sarkar B, Rumpel C, Sparks D, Bolan NS (2018c) Microbial control of soil carbon turnover, the future of soil carbon. Elsevier, Amsterdam, pp 165–194. https://doi.org/10.1016/B978-0-12-811687-6.00006-7

Xu M, Wu J, Yang G, Zhang XH, Peng H, Yu XY, Xiao YL, Qi H (2019) Biochar addition to soil highly increases P retention and decreases the risk of phosphate contamination of waters. Environ Chem Lett 17(1):533–541. https://doi.org/10.1007/s10311-018-0802-z

Xue Y, Wang C, Hu Z, Zhou Y, Xiao Y, Wang T (2019) Pyrolysis of sewage sludge by electromagnetic induction: biochar properties and application in adsorption removal of Pb(II), Cd(II) from aqueous solution. Waste Manag 89:48–56. https://doi.org/10.1016/j.wasman.2019.03.047

Yadav AN, Rastegari AA, Yadav N, Kour D (2020) Advances in plant microbiome and sustainable agriculture. Springer, Singapore. https://doi.org/10.1007/978-981-15-3208-5

Yan T, Xue J, Zhou Z, Wu Y (2021) Impacts of biochar-based fertilization on soil arbuscular mycorrhizal fungal community structure in a karst mountainous area. Environ Sci Pollut Res 28:66420–66434. https://doi.org/10.1007/s11356-021-15499-6

Yan S, Zhang S, Yan P, Aurangzeib M (2022) Effect of biochar application method and amount on the soil quality and maize yield in Mollisols of Northeast China. Biochar 4(1):56. https://doi.org/10.1007/s42773-022-00180-z

Yang CD, Lu SG (2021) Pyrolysis temperature affects phosphorus availability of rice straw and canola stalk biochars and biochar amended soils. J Soils Sediments 21(8):2817–2830. https://doi.org/10.1007/s11368-021-02993-0

Yao T, Zhang W, Gulaqa A, Cui Y, Zhou Y, Weng W, Wang X, Liu Q, Jin F (2021) Effects of peanut shell biochar on soil nutrients, soil enzyme activity, and rice yield in heavily saline-sodic paddy field. J Soil Sci Plant Nutr 21(1):655–664. https://doi.org/10.1007/s42729-020-00390-z

Yavari S, Malakahmad A, Sapari NB, Yavari S (2017) Sorption properties optimization of agricultural wastes-derived biochars using response surface methodology. Process Saf Environ Prot 109:509–519. https://doi.org/10.1016/j.psep.2017.05.002

Ye J, Joseph SD, Ji M, Nielsen S, Mitchell DR, Donne S, Horvat J, Wang J, Munroe P, Thomas T (2017) Chemolithotrophic processes in the bacterial communities on the surface of mineral enriched biochars. ISME J 11:1087. https://doi.org/10.1038/ismej.2016.187

Yin J, Zhao L, Xu X, Li D, Qiu H, Cao X (2022) Evaluation of long-term carbon sequestration of biochar in soil with biogeochemical field model. Sci Total Environ 822:153576. https://doi.org/10.1016/j.scitotenv.2022.153576

You J, Sun L, Liu X, Hu X, Xu Q (2019) Effects of sewage sludge biochar on soil characteristics and crop yield in loamy sand soil. Pol J Environ Stud 28:2973–2980. https://doi.org/10.15244/pjoes/93294

Yu KL, Lau BF, Show PL, Ong HC, Ling TC, Chen WH, Ng EP, Chang JS (2017) Recent developments on algal biochar production and characterization. Bioresour Technol 246:2–11. https://doi.org/10.1016/j.biortech.2017.08.009

Yu L, Yu M, Lu X, Tang C, Liu X, Brookes PC, Xu J (2018) Combined application of biochar and nitrogen fertilizer benefits nitrogen retention in the rhizosphere of soybean by increasing microbial biomass but not altering microbial community structure. Sci Total Environ 640:1221–1230. https://doi.org/10.1016/j.scitotenv.2018.06.018

Yudina A, Kuzyakov Y (2023) Dual nature of soil structure: the unity of aggregates and pores. Geoderma 434:116478. https://doi.org/10.1016/j.geoderma.2023.116478

Zaborowska M, Kucharski J, Wyszkowska J (2017) Brown algae and basalt meal in maintaining the activity of arylsulfatase of soil polluted with cadmium. Water Air Soil Pollut 228:267. https://doi.org/10.1007/s11270-017-3449-7

Zaheer MS, Ali HH, Soufan W, Iqbal R, Habib-ur-Rahman M, Iqbal J, Israr M, El Sabagh A (2021) Potential effects of biochar application for improving wheat (*Triticum aestivum* L.) growth and soil biochemical properties under drought stress conditions. Land 10(11):1125. https://doi.org/10.3390/land10111125

Zaid F, Al-Awwal N, Yang J, Anderson SH, Alsunuse BT (2024) Effects of biochar-amended composts on selected enzyme activities in soils. Processes 12(8):1678. https://doi.org/10.3390/pr12081678

Zhang H, Voroney RP, Price GW (2015) Effects of temperature and processing conditions on biochar chemical properties and their influence on soil C and N transformations. Soil Biol Biochem 83:19–28. https://doi.org/10.1016/j.soilbio.2015.01.006

Zhang X, Gao B, Zheng Y, Hu X, Creamer AE, Annable MD, Li Y (2017) Biochar for volatile organic compound (VOC) removal: sorption performance and governing mechanisms. Bioresour Technol 245:606–614. https://doi.org/10.1016/j.biortech.2017.09.025

Zhang G, Guo X, Zhu Y, Liu X, Han Z, Sun K, Ji L, He Q, Han L (2018a) The effects of different biochars on microbial quantity, microbial community shift, enzyme activity, and biodegradation of polycyclic aromatic hydrocarbons in soil. Geoderma 328:100–108. https://doi.org/10.1016/j.geoderma.2018.05.009

Zhang L, Jing Y, Xiang Y, Zhang R, Lu H (2018b) Responses of soil microbial community structure changes and activities to biochar addition: a meta-analysis. Sci Total Environ 643:926–935. https://doi.org/10.1016/j.scitotenv.2018.06.231

Zhang L, Xiang Y, Jing Y, Zhang R (2019) Biochar amendment effects on the activities of soil carbon, nitrogen, and phosphorus hydrolytic enzymes: a meta-analysis. Environ Sci Pollut Res 26(22):22990–23001. https://doi.org/10.1007/s11356-019-05604-1

Zhang L, Jing Y, Chen C, Xiang Y, Rezaei Rashti M, Li Y, Deng Q, Zhang R (2021a) Effects of biochar application on soil nitrogen transformation, microbial functional genes, enzyme activity, and plant nitrogen uptake: a meta-analysis of field studies. GCB Bioenergy 13(12):1859–1873

Zhang Y, Wang J, Feng Y (2021b) The effects of biochar addition on soil physicochemical properties: a review. Catena 202:105284. https://doi.org/10.1016/j.catena.2021.105284

Zhang J, Jin K, Luo Y, Du L, Tian R, Wang S, Shen Y, Zhang J, Li N, Shao W et al (2023) Responses of soil enzyme activity to long-term nitrogen enrichment and water addition in a typical steppe. Agronomy 13:1920. https://doi.org/10.3390/agronomy13071920

Zhao Y, Zhao L, Mei Y, Li F, Cao X (2018) Release of nutrients and heavy metals from biochar-amended soil under environmentally relevant conditions. Environ Sci Pollut Res Int 25:2517–2527. https://doi.org/10.1007/s11356-017-0668-9

Zheng H, Wang X, Chen L, Wang Z, Xia Y, Zhang Y, Wang H, Luo X, Xing B (2018) Enhanced growth of halophyte plants in biochar-amended coastal soil: roles of nutrient availability and rhizosphere microbial modulation. Plant Cell Environ 41:517–532. https://doi.org/10.1111/pce.12944

Zheng BX, Ding K, Yang XR, Wadaan MAM, Hozzein WN, Peñuelas J, Zhu YG (2019) Straw biochar increases the abundance of inorganic phosphate solubilizing bacterial community for better rape (*Brassica napus*) growth and phosphate uptake. Sci Total Environ 647:1113–1120. https://doi.org/10.1016/j.scitotenv.2018.07.454

Zheng X, Lu X, Zhou M, Huang W, Zhong Z, Wu X, Zhao B (2022) Experimental study on mechanical properties of root–soil composite reinforced by MICP. Materials 15:3586. https://doi.org/10.3390/ma15103586

Zhou H, Zhang D, Wang P, Liu X, Cheng K, Li L, Zheng J, Zhang X, Zheng J, Crowley D, Van Zwieten L, Pan G (2017) Changes in microbial biomass and the metabolic quotient with biochar addition to agricultural soils: a meta-analysis. Agric Ecosyst Environ 239:80–89. https://doi.org/10.1016/j.agee.2017.01.006

Zhou X, Chen Z, Li Z, Wu H (2020) Impacts of aeration and biochar addition on extracellular polymeric substances and microbial communities in constructed wetlands for low C/N wastewater treatment: implications for clogging. Chem Eng J 396:125349. https://doi.org/10.1016/j.cej.2020.125349

Zhu X, Chen B, Zhu L, Xing B (2017) Effects and mechanisms of biochar-microbe interactions in soil improvement and pollution remediation: a review. Environ Pollut 227:98–115. https://doi.org/10.1016/j.envpol.2017.04.032

Zhu J, Li M, Whelan M (2018) Phosphorus activators contribute to legacy phosphorus availability in agricultural soils: a review. Sci Total Environ 612:522–537. https://doi.org/10.1016/j.scitotenv.2017.08.095

Zhu X, Li X, Shen B, Zhang Z, Wang J, Shang X (2022) Bioremediation of lead-contaminated soil by inorganic phosphate-solubilizing bacteria immobilized on biochar. Ecotoxicol Environ Saf 237:113524. https://doi.org/10.1016/j.ecoenv.2022.113524

Zilberman D, Laird D, Rainey C, Song J, Kahn G (2023) Biochar supply-chain and challenges to commercialization. GCB Bioenergy 15:7–23. https://doi.org/10.1111/gcbb.12952

Biochar and Soil Hydrological Properties: Yield Improvement Through Drought Mitigation

Bhumika Sharma, Peeyush Sharma, Vikas Abrol, Altaf Hussain, Jyotsana Kalsi, and Owais Ali Wani

Abstract

Global crop productivity and food security are seriously threatened by abiotic stressors as heat, salinity, heavy metals and drought, with the latter becoming one of the most damaging and frequent natural calamities. Extended stretches of decreased precipitation cause acute water scarcity, impacting agricultural, socio-economic and ecological systems. Of the three types of droughts viz. agricultural, meteorological and hydrological, the latter is more closely associated with decreased crop output and soil moisture. Plant physiological, anatomical and morphological processes are disrupted by drought stress, which lowers chlorophyll synthesis, speeds up leaf senescence and reduces photosynthesis, all of which reduce crop productivity. The potential for biochar (BC), a carbon-rich soil additive, to boost crop yield as well as water use efficiency (WUE) during drought has garnered increased attention recently. In moisture stress conditions, BC enhances crop growth by enhancing soil quality, leaf conductance, carbon assimilation, antioxidant activity, nutrient absorption and chlorophyll synthesis. Additionally, the application of BC alters important physical characteristics of the soil, particularly its structure, soil particle composition, porosity, distribution of pore sizes, accessible water capacity and hydraulic conductivity. This improves the capacity of the soil to hold water, allow for infiltration as well as provide water to root zone. The extent to which mass density, pore size, and moisture

B. Sharma · P. Sharma · V. Abrol · A. Hussain
Division of Soil Science and Agricultural Chemistry, SKUAST Jammu, Jammu and Kashmir, India

J. Kalsi
Division of Plant Physiology, SKUAST Jammu, Jammu and Kashmir, India

O. A. Wani (✉)
Division of Agronomy, Faculty of Agriculture, SKUAST-Kashmir, Wadura, Sopore, Jammu and Kashmir, India

O. A. Wani et al. (eds.), *Biochar for Sustainable Soil Health and Crop Production*, https://doi.org/10.1007/978-981-92-2230-8_6

content have improved depends on the pyrolysis settings, texture of soil and biochar feedstock. Improved drought resistance is supported by increased moisture-holding ability and soil hydraulic conductivity, which keep water available to plants and regulate the circulation of water in the root zone, during dry spells. The extent and consistency of BC's effects vary depending on the context, despite encouraging results. This underscores the need for additional study to optimize feedstocks, application rates and integration tactics across a range of soil and climatic circumstances. To fulfil the potential of biochar's as a eco-friendly technique for improving crop resilience under drought stress, these gaps must be filled.

Keywords

Drought stress · Biochar · Soil hydraulic properties · Crop productivity · Water use efficiency

1 Introduction

Crop productivity and the stability of the worldwide supply of food are seriously threatened by abiotic stressors. Crop productivity is being adversely affected by the steadily rising severity of non-living factors (abiotic) stressors, i.e. temperature, salinity, heavy metals and moisture (Hassan et al. 2021). Among these abiotic stressors, moisture stress is a severe abiotic factor that significantly lower agricultural yields worldwide. Being one of the most destructive natural disaster, drought happens regularly and poses major risks to ecosystem services and human livelihoods (Alahacoon and Edirisinghe 2022; Jalayer et al. 2023). It mostly results from extended stretches of very low precipitation, which causes severe water shortages that affect ecological, socioeconomic and agricultural systems (Das et al. 2022a). However, based on how it manifests and the context in which it affects, drought is usually categorized into three types: hydrological, agricultural and meteorological (AghaKouchak et al. 2015; Gholinia and Abbaszadeh 2024). The most common of these is agricultural drought, which happens when a protracted lack of precipitation causes a deficiency in soil moisture, which lowers crop water availability and consequently lowers crop yields (Ding et al. 2024; Sun et al. 2023). Drought stress is a primary abiotic stressor that dramatically reduces agricultural output (Liang et al. 2020; Latif et al. 2022). Numerous processes are disrupted by drought stress, ranging from physiological and anatomical levels to morphological levels (Zhao et al. 2022). A significant drop in photosynthesis and crop productivity results from DS's rise in leaf senescence and decrease in chlorophyll synthesis (Sunaina et al. 2019; Ma et al. 2021).

Biochar (BC) is an excellent method for raising agricultural productivity and abiotic stress tolerance. Additionally, BC has been acknowledged as an essential tool for increasing crop productivity and water use efficiency (WUE) (Singh et al. 2019). Therefore, the use of BC-enhanced absorption of nutrients, carbon assimilation, and antioxidant activities greatly increases plant development under DS (Sorrenti et al. 2016). Furthermore, under DS, BC improves stomata conductance, WUE, and chlorophyll synthesis, all of which support improved plant growth (Ramzani et al. 2017; Langeroodi et al. 2019). Biochar has become more popular as a soil supplement due to its advantages for carbon absorption, reducing greenhouse gas emissions, boosting crop productivity, and enhancing soil quality (Wang et al. 2020). Soil's texture, structure, porosity, distribution of pore sizes, available water capacity, and drainage capacities have all been shown to be altered by the insertion of biochar (Singh et al. 2019). It has been acknowledged that porosity is a significant soil characteristic that influences root zone functions like water uptake and respiration. Biochar soil amendment has been demonstrated to enhance overall structure and porosity of soil (Moragues-Saitua et al. 2023).

However, the type of soil and charcoal feedstocks employed varied in how much these attributes were improved. For example, there were significant changes in the bulk density, porosity, and water content of a sandy clay loam soil (Abrol et al. 2016). Furthermore, the insertion of biochar raised the water content of the soil at each matric potential (Haider et al. 2020). The temperature at which biochar was pyrolyzed affected how much it increased the available water capacity (AWC) in soils with coarse textures (Paneque et al. 2016).

However, the texture of the soil affected this effect. Soil hydraulic conductivity is an important soil property that influences the likelihood of soil runoff after a period of heavy rainfall or irrigation and regulates water infiltration and movement within the soil profile. Soil saturation hydraulic conductivity (Ksat) may increase when biochar is used (Abrol et al. 2016; Sharma et al. 2025). The foundation of drought resistance is soil water retention and hydraulic stability, which regulate the duration of water availability for plants in the root zone and how consistently it can flow through the soil to the roots during hot, dry conditions. The critical soil-moisture thresholds, where ecosystems transition from energy-limited to water-limited and stomata start to shut thresholds that are strongly controlled by the soil hydraulic conductivity curve and vary with texture, are delayed by high retention (from finer textures, organic matter and a well-developed pore network), which prevents matric potential from dropping too quickly.

Furthermore, using BC improves the soil's biological and physiochemical properties, which benefits the growth of crops under DS (Agbna et al. 2017). As a result, we have included details on the different ways that BC helps plants cope with drought stress in this chapter. We also identified a number scientific gaps that must be addressed if biochar is to achieve its promised destiny as a soil addition.

2 Properties of Biochar Related to Hydrology

A carbonaceous porous material made from biomass is called biochar. The chemical makeup and abundance of the carbon component differ based on the parent material. Biochar has been made from a range of feedstocks, including wastewater, wood debris, dung from livestock, and agricultural leftovers (Bera et al. 2018). Biochar can be produced by a number of thermochemical processes, including gasification, pyrolysis, and hydrothermal carbonization (Tripathi et al. 2016). Because of its huge surface area, high porosity, and abundance of functional groups, biochar is a viable option for environmental cleanup. It can be utilized as a catalyst and adsorbent to eliminate several dangerous contaminants from both waste water and water (Sharma et al. 2018). It has long been recognized that biochar can improve soil fertility by acting as an adsorbent and a source of nitrogen (Vijayaraghavan 2019).

Biochar is created by highly aromatizing organic waste, such as urban sewage, chicken manure, and farming straw, under anaerobic conditions and high temperatures in order to reduce pollution and optimize resource utilization (Nguyen et al. 2019). Numerous studies have shown that biochar has high levels of organic carbon (OC), large charge densities (Rajapaksha et al. 2016), low bulk densities (Jain et al. 2017), high surface areas, and high porous structures (Villagra-Mendoza and Horn 2018). Biochar has been shown to improve the biological and chemical properties of soil, such as pH, EC, zeta potential, and cation exchange capacity (CEC) (Abujabhah et al. 2016).

Organic additives have been found to be a significant alternative for encouraging the development and stabilization of soil aggregates and microbial activity. Soil erosion and filtration may be affected by the mean weight diameter (MWD), an important measure of soil quality. Biochar improves soil structure by boosting aggregate stability and encouraging the creation of macroaggregates. Porosity has long been recognized as a crucial soil property that influences root zone processes like plant respiration and root water uptake. The use of biochar improved the porosity and structure of the soil (Saxena et al. 2017).

A crucial component of crop productivity is soil water. Crops that receive too little or too much water will not produce the desired yields, and both positive and negative effects on crop quality are possible. Gravity, capillarity (pressure), adsorption (electrostatic), and osmosis (solute) all affect soil water content (h), which can be expressed gravimetrically (hg; g water/g soil) or volumetrically (hv; cm^3 water/ cm^3 soil). However, the forces of adsorption and capillarity specifically affect the amount of water that plants can access (hpaw). The difference between water stored at field capacity (hfc) and that at the permanent wilting point (hpwp) determines plant accessible water (PAW) (Liu et al. 2017).

Only measurements of the soil's water potential (w) at hpwp (1000–1500 kPa) and after gravity drainage (~0 kPa) can accurately determine PAW. According to the soil matrix potential (wm), the permanent wilting point is the point at which plant water availability stops because of the large counter pressure needed by roots to extract it from the soil particle matrix. Total SWC and PAW can be impacted by soil type, maturation/development, and management, all of which change the overall characteristics of horizons.

Although increases in productivity and growth of crops (Liu et al. 2017) and soil organic carbon (SOC) mineralization have drawn a lot of attention, little study has looked at how biochar affects the hydrological and physical properties of soil. In the few studies that were conducted, the impact of biochar on the physical attributes of

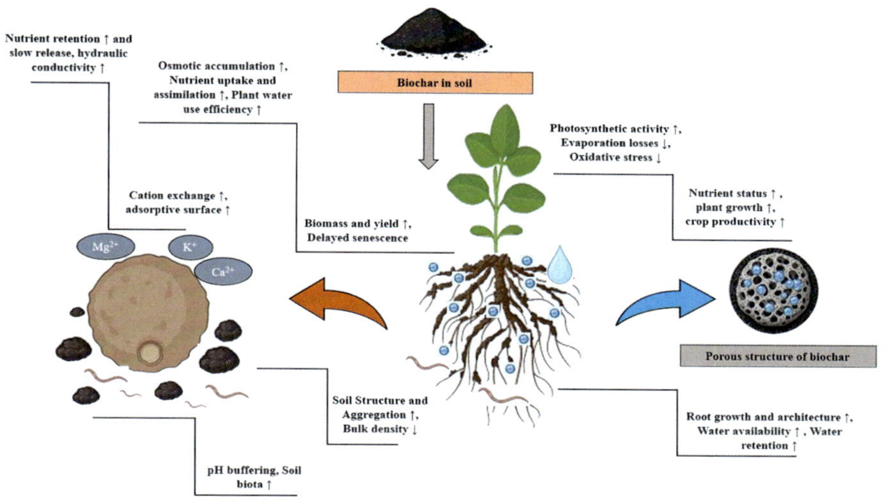

Fig. 1 Mechanisms by which biochar enhances soil moisture storage potential and mitigates drought stress in crops

soil were likewise non-uniform, and they varied based on a wide range of soil and biochar circumstances (Fig. 1).

3 Biochar's Influence on Soil Hydraulic Properties

(a) Pore space of soil: Apart from impacting the water transmission, conservation, and consumption, soil pores supply oxygen to plants and animals. The soil pore structure is significantly impacted by the pore distribution, connectivity and particle size of biochar. The pores of biochar are loose and porous, determining its surface area. As SA increases, pore space and porosity increase, and the pyrolysis temperature increases, the volatile matter decreases (Abrol et al. 2016). By boosting soil microorganisms, a large surface area promotes microbial activity and the growth of plant roots in soil pores. Conversely, crop roots loosen the soil and subtly change its porosity. Sandy soil is characterized by its large pores and poor capacity to hold water. Soil porosity, permeability and saturated hydraulic conductivity increase dramatically after adding biochar (Liu et al. 2018a). Crop development is aided by biochar's ability to cause the soil's pore distribution (PD) to change to a narrower pore size distribution (Dokoohaki et al. 2017). Following the application of biochar, an increase in soil porosity was noted in the 5–10 and 25 m ranges (Rasa et al. 2018). According to Wang and Zhang (2019), biochar gives frozen soil its maximum structural stability index by increasing overall porosity and the number of pores larger than 0.25 mm in diameter. Straw biochar has been shown to increase macropores and mesopores in clay soils, and it can be combined with soil particles to form large, stable agglomerates (Saxena et al. 2017). The soil becomes more permeable when biochar is added. It permits soil pore reorganization, which modifies soil pore distribution.

(b) Water-Holding Capacity of Soil (SWHC): According to Fu et al. (2019), soil texture and precipitation levels impact the composition of moisture, which is a crucial component of soil health. Understanding soil water flow and solute transfer is based on the hydraulic characteristics of the soil (Palviainen et al. 2020). The high porosity and specific SA of biochar change the water residence duration and flow direction in soil, increase water-holding capacity and decrease soil water permeability resistance (Abrol et al. 2016). An important factor in improving soil water-holding capacity (SWHC) is biochar's microporous structure. According to studies, using biochar greatly increases SWHC, especially in sandy soils, which have a lesser capacity to retain water than clay soils and so provide more advantages in these textures (Wang et al. 2018). According to Fu et al. (2019), soil texture and precipitation levels impact the composition of moisture, which is a crucial component of soil health. According to Verheijen et al. (2019), adding biochar resulted in a significant increase in SWHC and a decrease in soil bulk density. Compared to loamy soils, this effect was more noticeable in sandy soils. For example, compared to control treatments, adding 25% and 5% biochar to sandy soils increased water retention by 260% and 370%, respectively (Butnan et al. 2018). Increased agricultural yield may follow from such increases in soil water content (SWC). Long-term studies showed that, after three consecutive years, soil water content improved consistently and maize yields increased by 28%, 30%, and 140%, respectively, compared to untreated controls, despite short-term field trials with biochar application at 20 t ha^{-1} showing no discernible impact on yield (Vitkova et al. 2017). The association with biochar dosage and the ability to hold water is yet unknown, though, because incremental additions do not necessarily result in commensurate improvements.

(c) Soil Water Retention Curve or Soil Water Characteristic Curve (SWCC): One of most important hydraulic characteristics of soils is the soil water characteristic curve (SWCC), which shows the connection between soil water suction (also known as matric potential) and soil water content (Bayat et al. 2019). The SWCC is mostly determined by the content and form of the soil (Hussain et al. 2020). The curve is heavily impacted by variables that alter pore size distribution, porosity and pore connectivity because it represents the interplay between soil pore architecture and water content. By changing the bulk density, porosity, pore continuity and pore dimensions of the soil, the biochar amendment modifies these pore-related characteristics, which impacts mechanical stability and water retention. Interestingly, soils modified with biochar have been shown to conserve more water, as evidenced by smaller average pore diameters and lower total pore volumes. Therefore, adding biochar tends to improve water availability and soil hydraulic conductivity, especially in coarse-textured soils like sand. Overall, the soil water retention curve can be shifted and reshaped by biochar, highlighting its significance in soil hydraulic management (Ahmad Bhat et al. 2022).

(d) Soil Hydraulic Conductivity (K): Based on the kind of soil, the biochar's properties and the rate of application, the impact of biochar on soil saturation hydraulic conductivity (Ksat) can vary greatly, exhibiting increases, decreases, or barely noticeable variations. When compared to untreated soils, the addition of biochar increased Ksat by 28–176% in a number of experiments (Lim et al. 2016). On the other hand, significant decreases were noted in coarse sandy soils, with the most noticeable drop being 270% after 5% biochar was applied (Lim et al. 2016). Igalavithana et al. (2017a) found that pyrolyzing maize residue biochar at 500 °C gradually reduced Ksat at increasing application rates: reductions of 46.6%, 63.4%, 76.7%, and 83.5% with 2.5%, 5%, 7.5%, and 10% biochar, respectively. However, in many instances, even comparatively high application rates (such as 4%), did not significantly impact K (Glab et al. 2016). All things considered; these findings imply that soil texture has a significant impact on how biochar affects soil hydraulic conductivity (K). Applying biochar typically results in a considerable decrease in K in coarse-textured soils like sands, whereas it typically increases K in fine-textured soils like clays. However, the effect is usually small or irregular for medium-textured soils, with K increases being insignificant or fluctuating erratically. Rearranged soil particles, enhanced microporosity and improved pore connectivity, all of which improve drainage are responsible for the favourable reaction in clay soils. Tomczyk et al. (2020) proposed that high rates of biochar application can occasionally generate a modest increase in water flow, which might be related to the reorganization of delicate soil pore structures brought on by sample preparation. The observed decrease in Ksat in coarse-textured soils after applying biochar is frequently ascribed to the fine biochar particles the majority of which are usually less than 2 mm in size clogging macropores. The partial blockage of these pores lowers total porosity and restricts water transmission because sandy soils mostly depend on macropore movement. In addition, adding more biochar tends to increase the fraction of tiny and medium pores, which further reduces hydraulic conductivity by filling pores. On the other hand, a compensating equilibrium between the quantities of micropores and macropores may explain the insignificant impact of biochar on medium-textured soils. Even while biochar particles may fill some of the macropores in these types of soils, new pores may be created by the simultaneous reorganization of the soil structure, resulting in a comparatively steady water flow (Fig. 2).

(e) Plant-Available Water (PAW): Research shows that biochar often improves plant-available water (PAW). Biochar's porous nature allows it to hold water in both pore and intra-particle spaces. Micropores increase the reservoir of water that plants may access by retaining water by capillary and adhesion forces, in contrast to macropores and mesopores that drain easily. As a result, the application of biochar alters the bulk density, water flow, soil porosity and overall water retention (WR). Biochar can lower irrigation frequency by increasing PAW, which is especially beneficial in semi-arid and water-limited areas. It is

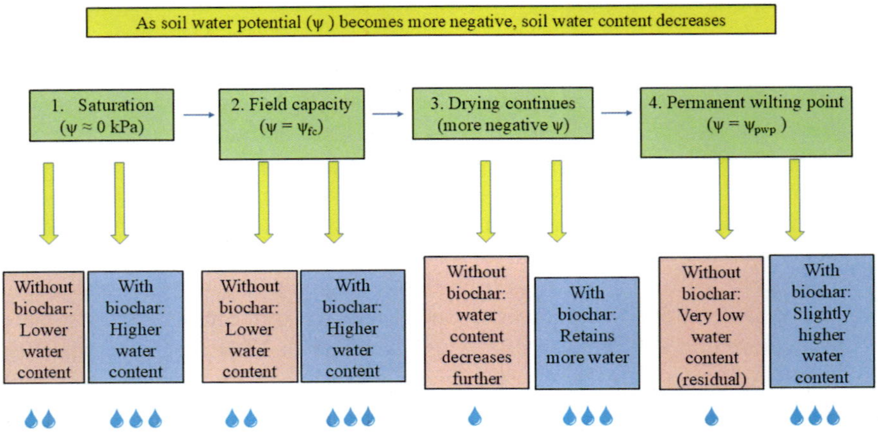

Fig. 2 Soil water retention curves with and without biochar amendment

anticipated that the beneficial effects of biochar on water retention will be more noticeable in sandy soils, where microporosity and SSA are naturally low.

The reason for this discrepancy is that direct comparisons are challenging due to variations in biochar application rates, feedstock types, pyrolysis temperatures, trial length and soil texture classes among research. Furthermore, low application rates frequently have little impacts; for example, multiple studies have found that 10 Mg ha^{-1} had no discernible impact on available water (Xiao et al. 2016), while larger dosages of biochar consistently produced notable increases in AW. Most investigations were carried out in controlled laboratory settings, and findings about the impact of biochar on available water (AW) were often made soon after it was incorporated into soils.

To adequately capture the dynamic interactions between soil and biochar over time, however, long-term field trials are required. Butnan et al. (2018) found that whereas woodchip biochar applied at the same concentration had no discernible effect, straw-derived biochar improved AW. Additionally, seven trials had no discernible effect, highlighting the fact that biochar does not always raise AW and that its efficacy is highly reliant on the type of soil and rate of application, indicating a site-specific response. In contrast to coarser-textured soils, which often exhibit more consistent improvements, clayey soils may see an increase, decrease, or no effect from the application of biochar. It is crucial to remember that increased soil water retention does not always result in more water being available to plants. Water retention may occasionally be limited by the hydrophobicity of biochar, particularly in fresh, low-temperature biochar with organic residues on their surface. Nevertheless, when biochar matures and experiences surface oxidation, its hydrophobicity decreases, eventually improving plants' access to water.

(f) Infiltration: One of the most important hydrological processes that controls surface runoff and soil erosion while also recharging soil moisture stores is infiltration, or the downward flow of water through the soil surface. Adequate water storage for plant growth is ensured by proper infiltration, which promotes biomass production and aids in the build-up of soil organic matter (SOM). On the other hand, inadequate management techniques can lower infiltration rates, resulting in ponding or surface runoff, which increases evaporation losses and reduces the amount of soil water available for crops. This has a negative impact on soil composition and structure in addition to reducing plant productivity. Table 1 provides a summary of how biochar affects soil infiltration rates. Based on the information now available, the impacts of biochar may vary based on the kind of soil, the characteristics of the biochar and the circumstances surrounding its application. For instance, a number of research found that biochar improved infiltration (Novak et al. 2016), whereas other studies found decreases or insignificant effects. These divergent results are in line with findings for other physical and hydrological aspects of the soil, where the combination of soil texture, pore structure and biochar qualities may cause biochar to enhance, degrade, or have no discernible impact. After applying biochar, the infiltration rate may decrease as a result of either the physical breakdown of the biochar particles or the filling of the pores with water. According to Ding et al. (2024), soil compaction may also result from the structural degradation of biochar brought on by tillage activities, heavy machinery movement during application or water flushing. Particles that have been dislodged may then plug soil pores, decreasing water flow. For example, when agitated in water, biochar made from pelletized lignocellulosic wastes and manure physically broke down into flake-like particles. These micrometer- to nanometer-sized particles frequently featured sharp edges, which allowed them to move downhill through the soil profile and stay suspended in percolating water. They can reduce infiltration by blocking soil micropores due to their uneven sizes and forms. Because clay soils are dominated by micropores and biochar is very porous, its interaction with fine-textured soils may boost infiltration and pore formation. To completely understand the intricate relationships between biochar and soil structure that control infiltration behaviour, however, more investigation is necessary.

(g) Water Repellence: Soil water repellence (SWR) is greatly influenced by a number of hydrological processes, including runoff, bypass flow, infiltration, water retention and structural integrity. The addition of biochar influences a number of processes that influence the physical behaviour of soil, including microbial activity, organic matter (OM) turnover, soil aggregation and water input and dispersion. According to Omondi et al. (2016), extreme SWR can significantly limit water infiltration and interfere with associated hydrological processes. For instance, soils in forests damaged by fire frequently show high levels of hydrophobicity. Whether biochar itself affects SWR is a crucial query. When applied to soils, biochar's hydrophobic qualities may cause water to be repelled due to its microstructure.

Biochar decreased repellence in various studies: Liu et al. (2017) found that 1.5%, 2.5%, and 5.0% biochar applications decreased SWR, with the 5.0% treatment reducing repellence five times more than the control. Similarly, Glab

Table 1 Effect of biochar on edaphic hydrological properties across different edaphic types

Edaphic texture	Biochar feedstock	Soil hydraulic properties		References
Sloping black soil	Corn straw	Improved infiltration and decreased runoff; increased field capacity and saturated water content		Yu et al. (2021)
Sand	Pinewood, pine bark	Increased water retention		Suliman et al. (2017)
Clay loam	Wood biochar	Increased moisture holding capacity; decreased bulk density		Libutti et al. (2021)
Sandy clay loam	Vinasse derived biochar	Increased infiltration; reduced run off		Nicosia et al. (2021)
Silt enriched sand	Groundnut-shell	Increased soil water retention		Ni et al. (2018)
Clayey sand	*Eichhornia crassipes*	Increased soil moisture holding capacity		Bordoloi et al. (2018)
Sandy loam/silt loam	Almond shell and softwood biochar	Reduced hydraulic conductivity		Gelardi et al. (2021)
Sandy clay loam	*Miscanthus* sp.	Increased soil water retention		Moragues-Saitua et al. (2017)
Silt	Wood biochar	Decreased water repellency; increased infiltration; increased retention		Yang et al. (2020)
Clay loam	Rice husk	Increased hydraulic conductivity		Abdillah et al. (2024)
Sandy loam	Maize cob, paddy husk @ 0–40 g kg^{-1}	Improved plant available water		Obia et al. (2016)
Sand	Mixed feedstock	Increased infiltration; reduced evaporation		Wang et al. (2023)
Coarse sandy subsoil	Wood biochar	Increased plant available water		Bruun et al. (2023)
Silty loam	Various biochars	Increased plant available water		Tokova et al. (2023)
Silt	Wood biochar	Reduced infiltration time		Yang et al. (2020)
Sandy loam	Poultry manure (80:20) and pine chips (50:50) 100% poultry litter 100% pine chips	Biochar recommended dose: 0% 2% 2% 2%	Water infiltration (mL min^{-1}) 0.086[b] 0.0168[a] 0.110[b] 0.047[b] 0.119[b]	Novak et al. (2016)

[a]DMRT Post hoc

et al. (2016) found that 5% application resulted in the largest decrease of repellence, with slight increases observed at 4% compared to 0.5% and no effect at 1–2%. Pyrolysis temperature, feedstock type, placement technique, soil texture, and application duration all affect how hydrophobic the biochar is. Since surface oxidation and repellent layer erosion reduce repellence over time, fresh biochar typically has more hydrophobicity than old biochar. The temperature of pyrolysis is especially crucial because, although the degree of this impact varies depending on the feedstock, low-temperature biochars are generally more water-repellent than high-temperature ones. Furthermore, it has been demonstrated that adding biochar to soil decreases repellence in comparison to applying biochar on the surface, particularly when soil moisture levels are intermediate (~25%) (Hussain et al. 2020). Overall, most research indicates that biochar has a negligible impact on SWR, with soil–biochar interactions tending to decrease hydrophobicity and moderate repellence over time, even though it can affect SWR under specific circumstances.

(h) Soil Temperature: Through its inherent thermal and electrical characteristics as well as by changing the physical features of the soil, the application of biochar can affect the temperature of the soil. The addition of biochar has been shown in numerous studies to alter the dynamics of soil temperature under various circumstances. Corn cob biochar applied at 4.5 and 9.0 Mg ha^{-1} in a winter wheat–corn rotation, for example, decreased daily soil temperature swings in the North China Plain, dropping surface soil temperature by 0.8 °C in extremely hot conditions and elevating it by 0.6 °C in cooler soils (Igalavithana et al. 2017a, b). In Poland, biochar made from wood decreased the average soil temperature amplitude in grasslands while increasing it in fallow areas. However, the mean temperature at a depth of 0–15 cm was unaffected (Usowicz et al. 2016). Overall, there is evidence that biochar can buffer soil by lowering daytime peaks and raising night time temperatures, which can moderate excessive thermal swings, even if it does not seem to significantly change mean soil temperature.

4 Drought Mitigation Strategies via Biochar

According to Siebielec et al. (2020), biochar has become a highly effective strategy for decreasing the adverse impacts of moisture stress. Biochar, black gold of agriculture,, has drawn a lot of attention lately as a way to counteract the harmful impacts of DS (Zhang et al. 2019). According to Odugbenro et al. (2020), the addition of BC increases the stability of aggregates, CEC, intake of nutrients and water, soil moisture levels, and SOC to aid in drought resilience. Soils treated with biochar have shown promise in reducing the adverse effects of drought stress. By enhancing the physical properties of the soil, including water absorption and infiltration, biochar contributes to increased water accessibility in the soil for plants (Li et al. 2021). Additionally, by encouraging good microbial communities, that can positively affect crop growth and development in water-limited conditions, biochar may indirectly increase drought tolerance (Bolan et al. 2023; Wu et al. 2023). It's use improves plant shape by increasing nutrient availability (Lupo and Moshelion 2024). Additionally, it improves soil water availability, which helps plants that are experiencing osmotic stress. While biochar primarily increases the water holding capacity of soil, it also encourages water infiltration and drainage in fine-textured soils by improving saturated hydraulic conductivity. By strengthening fine roots and root systems, its application helps plants better bind soil particles. In dry and semi-arid areas in particular, biochar is essential for conserving soil and water. Biochar's porous structure, which frequently has micro- and mesopores (less than 10 μm), is important for retaining soil water. In accordance with Wei et al. (2023), these pores serve as micro-reservoirs that collect and hold water at matric potentials that are reachable by plant roots. By enhancing soil pore dispersion and lowering quick drainage losses, biochar improves plant-available water and field capacity when added to sandy soils (Li et al. 2021). Biochar frequently increases aggregation and changes pore connectivity in fine-textured soils, improving the rate of infiltration and moisture-retention ability without causing undue waterlogging (Gaurav et al. 2025). These increases the available water in the rhizosphere immediately improves soil's ability that protect plants from sporadic drought.

Maintaining soil moisture during protracted dry spells is one of biochar's most important drought-mitigation functions. By enhancing the soil's architecture and decreasing bulk density, biochar promotes deeper root penetration and soil volume exploration (Zhang et al. 2025). Plants may access water stored at deeper depths and maintain transpiration during extended dry spells thanks to this root–soil interaction. Furthermore, as top layers dry up, biochar promotes unsaturated hydraulic conductivity, which makes it easier for water to be redistributed to plant roots (Hussain et al. 2020). These impacts eventually result in increased crop survival, more consistent yields during periods of recurrent drought and increased water-use efficiency. A promising method for boosting root characteristics in crops like maize is to incorporate biochar up to 20 cm into the soil. This improves soil structure and nutrient availability, which is especially crucial during drought. Improved soil aeration, increased porosity and the immobilization of nitrogen and phosphorus due to biochar's potent adsorption capacity are all associated with the growth in root length and volume with biochar application (Gaurav et al. 2025). These changes allow roots to reach deeper into the edaphic horizons when surface soils dry out, giving them access to nutrients as well as moisture from layers that remain moist even during periods of minimal rainfall. This promotes plant growth. Biochar significantly delays the onset of water stress and allows plants to maintain metabolic activity during the early and mid-stages of drought by promoting deeper rooting and expanding the volume of soil investigated by roots. In contrast to untreated soils, biochar tends to decrease root diameter and angle, promoting the growth of finer roots with a larger surface area. When water availability becomes limited, this finer root system, along with improved root metabolism and nitrogen uptake, increases the effectiveness of water absorption and nutrient acquisition (Wan et al. 2023).

Furthermore, biochar encourages the development of fine roots and stimulates root–microbe interactions, resulting in rhizosphere surroundings which encourages the development of beneficial microorganisms. These beneficial organisms enable osmotic adjustment in plants, enhance soil aggregation and aid in nutrient cycling, all of which are essential adaptive responses during drought (Das et al. 2022a, b). For example, in wheat, biochar strengthens capacity of the plant to withstand low moisture circumstances by enhancing rhizosphere architecture and encouraging effective nutrient uptake by the roots (Olmo et al. 2016). In leguminous crops like red clover, biochar enhances biological nitrogen fixation and potassium availability. These processes not only provide vital nutrients but also boost osmotic control and stress tolerance mechanisms, which helps plants remain robust in water-deficient environments. By offering carbon substrates and mineral nutrients that support microbial activity during drought stress, biochar enhances the soil microbial population in addition to its direct benefits for roots. Its ash concentration provides vital inorganic nutrients and stabilizes microbial metabolites through organic mineral interactions, guaranteeing the durability of beneficial microbial populations that improve root functioning in the face of reduced water availability (Zhu et al. 2020). Applying biochar as a whole creates a soil root microbial system that is better able to absorb, hold, and distribute water and nutrients when rainfall is scarce. Furthermore, by fortifying the architecture of rhizosphere system, this synergy guarantees longer plant growth, postpones the onset of water stress and enhances plant survival during protracted drought.

Because soils typically absorb and retain more energy than they release, the soil ecosystem is significantly impacted by rising temperatures. This results in changes to thermal properties like conductivity, diffusivity, reflectance and overall soil temperature. Biochemical activities, microclimate regulation and soil–plant interactions are all directly impacted by these changes in soil thermal characteristics. Applying biochar turns out to be a useful tactic for controlling these thermal dynamics. By influencing the flow and accumulation of heat and water in the soil, biochar modifies surface-energy partitioning and encourages more uniform heat distribution throughout the soil profile, stabilizing the soil microclimate (Usowicz et al. 2020).

These impacts are demonstrated by field evidence. Zhao et al. (2016) reported that applying biochar to sandy loams reduced their thermal diffusivity (0.6–21.5%) and thermal conductivity (0.3–32.2%). Significant decreases in bulk density (24.7–34.6%), increases in soil moisture content and reductions in thermal conductivity of up to 59.8% were also linked to these modifications. When taken as a whole, these changes produce more consistent soil temperatures and ideal circumstances for microbial activity and plant growth. Additionally supporting this finding, Xiong et al. (2020), the addition of biochar significantly affects the soil's thermal properties, which are negatively correlated with bulk density and positively correlated with soil moisture. According to their research, depending on the application rate and moisture regime, biochar not only lessens thermal conductivity and diffusivity but also lessens temperature variations across soil depths. According to these findings, biochar improves soil resilience against extreme temperature variability in agricultural settings by moderating soil temperature dynamics.

When it comes to seed germination, seedling establishment, organic matter decomposition, CO_2 emissions, mineralization of nutrients, agrochemical degradation, soil water interactions, microbial activity and crop growth, soil temperature is a crucial regulator of microclimatic conditions (Liu et al. 2018a, b). A rise in soil temperature usually results in a reduction in water viscosity and a rise in the pace at which soil moisture evaporates, two processes that have conflicting effects on interactions between soil and plants. On the one hand, plants experience water scarcity as a result of faster evaporation, which prevents water from migrating downhill into the profile of soil (Onwuka and Mang 2018). Conversely, lower water viscosity improves photosynthetic efficiency and nutrient transport by increasing infiltration and root absorption (Olmo et al. 2016).

Additionally, high temperatures promote photosynthesis, food uptake and root cell metabolism, which promotes root elongation and the growth of lateral roots. Raising the temperature of the soil also boosts microbial activity, which enhances the cycling of nutrients and increases the concentrations of nutrients in plant tissues. Increased temperatures also speed up the flow of water-soluble nutrients like silicon, calcium, magnesium, phosphate, sulphate and nitrate in soils by intensifying diffusive processes. In light of these dynamics, biochar is being investigated more and more as an amendment to enhance soil moisture retention in hot climates. Results, however, are not always consistent; some studies found reductions in water retention (Carvalho et al. 2016), while others reported improvements (Głąb et al. 2018). For instance, Villagra-Mendoza and Horn (2018) showed that applying 5% biochar made from *Mangifera indica* wood improved water absorption by 11% in sandy loams and 137% in sandy soils.

These contradictory results highlight how complex the interactions between biochar and soil are. Similarly, Baiamonte et al. (2019) found that applying biochar made from forest waste increased plant-available water in arid soils by 226%. Climate, pyrolysis conditions, texture, and biochar feedstock all have a significant impact on results. Madari et al. (2017) found that no discernible impact on water availability was seen in the last year following 5 years of applying savannah wood biochar on sandy clay loam Haplic Ferralsols at rates ranging from 0 to 16 Mg ha^{-1}. Notwithstanding these variances, biochar continuously shows promise in enhancing water-use efficiency (WUE) and mitigating adverse impact of elevated soil temperatures on plants (Sharma et al. 2025). By reducing day-night temperature swings, biochar moderates' soil thermal regimes beyond hydraulic alterations. In paddy soils, decreases of roughly 0.66–1.39 °C have been noted (Liu et al. 2018a, b). Additionally, adding biochar improves the thermal characteristics of the soil and stabilizes the soil environment by increasing the amount of organic matter in the soil and decreasing bulk density (Obia et al. 2016). In addition to protecting soils from heat-induced stress, these combined effects also foster the growth of plants in a sustainable manner in rising climates.

By changing clay–mineral interactions, aggregate stability and cation exchange capacity (CEC), high soil temperatures have a negative impact on the structural integrity and functionality of soils. Heat stress-induced dehydration of soil clay minerals increases inter-particle interactions, frequently resulting in clay loss and silt fraction enrichment. This ultimately lowers CEC and weakens the stability of soil aggregates. Long-term exposure to high temperatures can cause cracking and disintegration in sandy soils, which can alter the soil's texture and cause the loss of sand-sized particles. Furthermore, by changing their binding properties, the heat transformation of iron and aluminium oxides further compromises the stability of soil aggregates.

One possible amendment method to combat these heat-induced degradation processes is the addition of biochar. Organic and inorganic components including silica, metal oxides and phenolic compounds, which are present in biochar, serve as signalling agents for microbial reactions to stressors like drought, pH changes and root desiccation. These substances improve microbial colonization and soil microenvironment stability by encouraging the development of exopolysaccharide-rich biofilms in bacteria (Costa et al. 2018). By secreting binding metabolites like lipids, glomalin, and polysaccharides, microbes linked to macro-aggregates help stabilize the aggregate and gain from habitat enhancement mediated by biochar.

Additionally, adding biochar to soils greatly increases enzymatic activity, which directly improves nutrient cycling and aggregate formation. While enzymes like urease, arylsulfatase and acid phosphatase facilitate the mobilization of phosphorus, nitrogen and sulphur, increasing nutrient availability for plants, elevated activities of β-glucosidase catalyse cellulose hydrolysis, providing glucose and energy for microbial proliferation (Lopes et al. 2021). Numerous studies verify that biochar increases laccase and manganese peroxidase activity, which promote fungal growth and increase fungal biomass in soils (Taskin et al. 2019a), despite some worries that recalcitrant aromatic compounds in biochar may inhibit fungal proliferation at higher application rates (Taskin et al. 2019b). By strengthening aggregate stability, this enhanced fungal colonization further fortifies the soil structure.

In addition to microbial mediation, biochar itself stabilizes soil aggregates by acting as a physical binding agent, promoting organo-mineral complexes and stimulating beneficial interactions with mycorrhizal fungus. The surface polar and non-polar interactions between soil clay minerals and biochar particles have a substantial impact on aggregation formation and stability (Du et al. 2017). Additionally, improvements in electrical conductivity (EC), pH, moisture retention, surface area, porosity, and soil organic matter (OM) brought about by biochar encourage microbial growth, which accelerates aggregation processes (Burrell et al. 2016).

5 Increase in Crop Yield Under Water Stress Conditions

To assess how biochar affects the 1000-grain weight (TGW), biological yield (BY), and grain yield (GY) of four significant crops: viz. *Triticum aestivum*, corn, paddy and *Glycine max*, Farhangi-Abriz et al. (2021) carried out a thorough meta-analysis. According to the results, biochar considerably increased GY in maize (28%) and wheat (13%), but neither rice nor soybeans showed any discernible advantages. The benefits were especially noticeable for TGW in wheat and GY in maize and wheat. The capacity of biochar to improve nitrogen (N) use efficiency, by lowering N losses (Dong et al. 2020), which may help to explain these gains. Additionally, by mineralizing native organic matter, biochar may add more nitrogen to the soil (Kätterer et al. 2019). Remarkably, maize responded to biochar more strongly than wheat. The carboxylation pathway in maize, which imparts greater water and N usage efficiency than in other cereals, may be connected to this (Omara et al. 2019). Ye et al. (2020) reported that, in general, maize reacts better to biochar than other cereals, with these variances being significantly influenced by the climate. Actually, the maximum dataset's maize research were carried out in subtropical climates, which demonstrated the greatest yield increases. Although additional examination of other yield parameters, such as spikelet or grain number, was not possible due to data restrictions, the 13% increase in GY for wheat was closely linked to a 4.3% increase in TGW. In contrast, rice and soybeans were not significantly impacted by biochar. This may be related to soybeans' capacity to fix nitrogen from the atmosphere, which lessens the crop's reliance on outside N inputs (Adams et al. 2018). The benefits of biochar for legumes are unclear because some reports even claim that it can inhibit biological N fixing (Horel et al. 2018). Likewise, biochar frequently has little to no effect on rice output, especially in flooded paddy circumstances where gains in soil water retention are negligible (Liu et al. 2019; Ye et al. 2020). The advantages of biochar for soil chemical characteristics were sometimes outweighed by yield decreases, which were ascribed to microbial N immobilization (Carvalho et al. 2016).

All of these results demonstrate that biochar's efficacy is very crop- and environment-dependent, with cereals particularly maize and wheat exhibiting the highest potential advantages.

In the face of increasingly severe drought events, the ability of crop to survive, thrive as well as reproduce depends on its roots. Several studies have examined how the root properties of various plant species react to water stress; these studies have investigated the results at a specific scale and within specific plant species groups. A meta-analysis by Kou et al. (2022) found that water stress raised the fibrous root system's root-to-shoot ratio, root hair density, and root hair length by 21.6%, 35.8%, and 49.4%, respectively. Water stress resulted in a reduction of 19.8% length and 21.9% dry weight of root. Plants increase the root-to-shoot ratio in response to drought for a number of reasons. By lowering the buildup of dry matter in both the root and the shoot, water stress frequently improved the root-to-shoot ratio, particularly when the root system was more severely reduced. The root-to-shoot ratio rose as a result of both higher root dry weight and lower shoot dry weight. Water stress increased the root-to-shoot ratio, root growth angle, and root hair density in the tap root system by 42.6%, 25.3%, and 27.3%, respectively. Similar to the fibrous root system, water stress generally improved the root-to-shoot ratio by causing a maximum decrease in the root system and a lower dry weight in the root and the shoot. The variable drought responses of each attribute revealed the variety of drought-coping strategies used by plants.

The depth of a plant's roots determines its ability to absorb water in the deep soil layers (Nardini et al. 2016). Even with extremely low water availability, persistent and ongoing root system growth may increase the soil's mechanical strength and the roots' penetration capacity, leading to a deeper root system during drought. Root depth has been observed to influence seasonal variations in gas exchange, water status and vascular integrity maintenance (Nardini et al. 2016). Crops under various habitats have significantly difference in their root depths (Olmo et al. 2016), with drought-tolerant species often having deep roots and drought-sensitive species having shallow roots (Zhou et al. 2020).

Nonetheless, latest reports on Arabidopsis have also found that a shallow root system develops to take up water from the soil's surface under drought (Ogura et al. 2019). This new finding raises the question of whether plants with shallow or deep roots are more resilient to drought and questions accepted knowledge about deep roots and drought tolerance. An environment with frequent rainfall may give rise to this shallow rooting habit.

The length of the roots is a good indicator of the root system's overall growth. Not surprisingly, there have been reports of inconsistent and variable root length responses to drought. According to Sharma et al. (2018), the total measurable length of a root does not always indicate its ability to absorb water because only a section of it is actively involved in water and nutrient uptake. Even though they might not help with water intake right away, the long and deep roots are essential for transporting water and minerals from root regions to different plant parts (Sharma et al. 2018). Longer roots may have a significantly higher root metabolic cost during drought conditions. Therefore, under drought conditions, higher productivity may be possible for genotypes that can increase root length and obtain more water at lower metabolic costs. Therefore, while selecting crop genotypes for increased length of root, the metabolic cost to the plant must also be considered; otherwise, a higher metabolic cost could lead to a lower yield.

It has been established that adding biochar to the soil can enhance its capacity to hold onto nutrients and water. In order to evaluate the effects of applying biochar on the physical and morphological characteristics and yield performance of tomato plants under simultaneous salt and drought stress, Obadi et al. (2023) carried out a greenhouse experiment. The study included 16 treatment combinations with two irrigation water qualities (freshwater at 0.90 dS m^{-1} and saline water at 2.30 dS m^{-1}), four irrigation regimes (100%, 80%, 60%, and 40% of crop evapotranspiration, ETc), and two levels of biochar (5% by weight (BC5%) and no biochar (BC0%)). In comparison to the control plants (BC0%), the yield of *Solanum lycopersicum* crops under different irrigation treatments improved by 4.60%, 16.74%, 8.67%, and 2.97% for 100%, 80%, 60%, and 40% ETc, respectively, when 5% biochar was added to freshwater. For *Solanum lycopersicum* plants treated with biochar and irrigated with freshwater under a 40% ETc deficit irrigation, the WUE improved by 97.02% as compared to full irrigation. The capacity of the biochar to hold water, encourage porosity, and supply nutrients to the plant under water-stressed conditions may account for the improvement in yield and WUE. Stomatal closure and a decrease in transpiration rate (TR) in response to salt and water stress may be the cause of the increase in WUE with deficit watering. WUE and tomato plant yield increased by 2.7% and 1.11%, respectively, when biochar was added at the suggested rate (BC5%). This boost in yield and WUE is the result of biochar's activity in the soil, which promotes root growth (Table 2).

Table 2 Biochar feedstocks, application rates, and effects on crop yield & WUE under drought/water deficit conditions

Feedstock and pyrolysis temp (°C)	Crop	Biochar application rate	Effects on plant parameters	References
Agricultural residue	Maize (*Zea mays* L.)	@ 20 t ha^{-1}	Increased maize grain, improved WUE	Faloye et al. (2019)
Rice hull (500 °C)	Maize (*Zea mays* L.)	0%, 1%, 2%, 5%	Increased biomass, enhanced drought tolerance	Kim et al. (2016)
Conocarpus wood waste (400 °C)	Tomato (*Solanum lycopersicum* L.)	0%, 4%, 8%	Improved plant height, biomass, fruit yield	Usman et al. (2016)
Rice straw (450–550 °C)	Wheat (*Triticum aestivum* L.)	3%, 5% w/w	Improved growth, yield, photosynthesis, WUE	Abbas et al. (2018)
Rice straw (450 °C)	Rice (*Oryza sativa* L.)	3%, 5% w/w	Increased biomass, photosynthesis, WUE, reduced ROS	Rizwan et al. (2018)
Wheat straw (500 °C)	Wheat (*Triticum aestivum* L.)	7.18 g/kg	Increased tillers, spike length, grain yield, chlorophyll, stomatal conductance, RWC	Haider et al. (2020)
Date palm/ pistachio residues (560 °C)	Eggplant (*Solanum melongena* L.)	–	Enhanced growth, yield, chlorophyll, antioxidative activity	Ebrahimi et al. (2021)
Maize straw	Pumpkin	@ 10–15 t ha^{-1}	Enhanced WUE	Langeroodi et al. (2019)
Wheat straw	Wheat (*Triticum aestivum* L.)	3%, 5% w/w	@ 5%- improved WUE, grain yield and growth traits	Zulfiqar et al. (2022)
Hardwood-oak 350 °C & softwood -pine biochar, 500 °C	Sweet corn (*Zea mays* L. var. saccharata)	13 t ha^{-1}	Reduced bulk density, slightly improved water content, increased chlorophyll, plant height, and biomass	Singh et al. (2022)
Plant based	Chickpea (*Cicer arietinum* L.)	–	Reduced drought effects, high dry matter yield, increased WUE	Hashem et al. (2019)
–	Wheat (*Triticum aestivum* L.)	3.7% w/w	Increased WUE, drought resilience; improved grain weight	Wu et al. (2023)
Sunflower straw (400 °C)	Sunflower (*Helianthus annuus* L.)	0, 15, 30 t ha^{-1}	At 30 t ha^{-1}, biochar under deficit irrigation improved sunflower kernel quality, biomass (+1.3–28.3%), and yield (+1.1–33.5%)	Zhang et al. (2025)

(continued)

Table 2 (continued)

Feedstock and pyrolysis temp (°C)	Crop	Biochar application rate	Effects on plant parameters	References
Mixed/ woody biochar	Tomato (*Solanum lycopersicum* L.)	@ 5% w/w	Increased yield, improved water use efficiency	Obadi et al. (2023)
Wheat straw	Tomato (*Solanum lycopersicum* L.)	@ 25 t ha^{-1}	Increased WUE	Agbna et al. (2017)
–	Cucumber (*Cucumis sativus*)	@ 20 t ha^{-1}	Increased WUE	Ali et al. (2019)

6 Interaction with Management Practices

An innovative technique to enhance plant growth, nutrient absorption ae well as drought resilience in agroecosystems is the co-application of biochar with advantageous microbes. In addition to improving the physical characteristics of soil, biochar's special porous structure offers the perfect environment for microbial colonization, growth and survival. By serving as micro-niches, these holes shield microorganisms from environmental stressors such desiccation, salt, pH changes and the presence of harmful substances, enabling them to proliferate and thrive in the rhizosphere (Egamberdieva et al. 2017). Because of its potential as a microbial carrier, biochar is particularly helpful in soils that are prone to drying up, as these conditions frequently impair microbial activity and survival. Research has demonstrated that when combined with microbial inoculants, biochar greatly improves plant performance in water stress situations. For instance, Egamberdieva et al. (2017) demonstrated that combined application of biochar and Bradyrhizobium decreased water stress in moisture-stressed blue lupin seedlings by improving biomass formation, nutrient uptake, nodulation, and overall development. The synergy results from the fact that soils altered with biochar hold onto more water and nutrients, which promote plant metabolism and microbial activity. In addition, the injected bacteria improve biological nitrogen fixation and nutrient solubilization, giving plants vital resources when water supplies are scarce. This combination impact delays the start of drought-induced senescence, maintains turgor and sustains photosynthesis. By enhancing aeration, holding onto soil moisture and supplementing the soil with labile organic molecules that act as microbial substrates, biochar mechanistically encourages microbial activity. Once established, beneficial bacteria help solubilize important elements like magnesium, potassium, phosphorus and nitrogen, guaranteeing that, despite reduced water availability, plants acquire a consistent source of necessary minerals. Through the maintenance of metabolic systems that are otherwise compromised under stress, these mechanisms lessen the adverse effects of drought. Long-term soil resilience is aided by these biochar–microbe interactions, which also increase soil biological fertility and nutrient use efficiency. However, depending on the type of biochar, the conditions under which it is produced, and the microbial species involved, the results of interactions between biochar and microbes are not always favourable. When birchwood-derived biochar (made at 500 °C) and *Rhizophagus irregularis* were applied together to drought-stressed potato plants, Liu et al. (2017) found that the plants' nutrient uptake, water usage efficiency (WUE) and leaf area decreased. A few key elements were mostly responsible for the negative outcome. First, biochar's low capacity to hold onto nitrogen and phosphate decreased the amount of these vital elements that plants could access. Second, by inhibiting microbial activity and limiting root growth, its breakdown may release chemicals like ethylene or phenolic compounds that could have phytotoxic effects. Lastly, The advantages of biochar for soil chemistry were limited when it was applied at low rates since it was unable to significantly raise soil pH. These elements work together to explain why biochar's anticipated advantages were not entirely achieved.

In agricultural areas with limited water resources, deficit irrigation (DI) is a commonly used water management technique that aims to maximize and preserve finite water supplies. This method can increase water usage efficiency (WUE), which is measured as yield per unit of water applied, by providing crops with water below their full evapotranspiration (ET) demand. DI is highly advised for semi-dry and dryland settings where the primary aim is to maximize water productivity rather than land productivity. DI is especially successful in irrigated systems where water conservation is crucial. However, crop type, growth stage and varietal traits all affect how effective DI is. Stage-specific DI control has shown promise in a number of crops since crop water requirements vary among phenological stages, including vegetative and reproductive phases (Seidel et al. 2017). However, when irrigation scheduling is not well understood, poor water allocation can lead to decreased crop performance and, eventually, water productivity. Under water-limited or drought-prone conditions, deficit irrigation (DI) can be used in conjunction with water-saving agricultural techniques, such as the use of soil amendments, to increase productivity (Agbna et al. 2017). Soil additives are inputs that are added to the soil to improve its chemical, biological and physical characteristics, which in turn promotes improved crop performance. The highly porous and stable carbon-rich structure of biochar, which is made by pyrolyzing plant residues at high temperatures with little oxygen, has made it stand out among these as a promising amendment.

Biochar is becoming more and more acknowledged as a useful instrument for enhancing soil moisture availability in water-scarce areas because of its exceptional water retention capability. Research found that capacity of various soil to retain water has increased in around 90% of experiments by applying biochar. The ideal application rate, however, is still up for debate because it depends so heavily on factors including crop type, feedstock origin, pH of biochar, nutritional composition and soil properties (Yang et al. 2019). On regard to this, an experiment by Kafle et al. (2025) assessed the impact of both DI and charcoal amendment on hydraulic characteristics of soil, cucumber (*Cucumis sativus*) growth, physiology and yield.

An experiment was set up as a split-plot design, with three biochar rates (0, 15, and 20 t ha^{-1}) as subplots, each duplicated four times and four irrigation regimes applied as main plots viz. 100% crop evapotranspiration (ETc)-100% ETc, 80% ETc-60% ETc, 60% ETc-80% ETc and 40% ETc-40% ETc. The findings demonstrated that, in 2022, under hot and dry conditions, severe deficit irrigation (40–40% ETc) significantly decreased transpiration, photosynthesis, gas exchange rate, chlorophyll content as well as intrinsic water-use efficiency (iWUE). In comparison to the full irrigation treatment (100% ETc-100% ETc) in both years, the leaf area index (LAI) under 40% ETc-40% ETc decreased by 52% and the ultimate dry biomass by 47%, however in 2022, biomass was reduced by just 17% under 60% ETc-80% ETc. Over the course of two seasons, yield penalties under 80% ETc-60% ETc, 60% ETc-80% ETc and 40% ETc-40% ETc were 14%, 16%, and 34% lower than full irrigation, respectively. The use of biochar at 15 and 20 t ha^{-1} significantly raised soil saturation hydraulic conductivity by 1.6 and 2.2 times in 2022 as compared to the unaltered control; nevertheless, its effects on cucumber growth, physiological traits, and yield were insignificant in both years. In light of these results, Kafle et al. (2025) proposed that 80% ETc–60% ETc might be a viable substitute for full irrigation, lowering water consumption by 25% while only lowering yield by 14%. They also underlined the necessity of doing long-term research on biochar to ascertain its overall effects on crop productivity in areas with restricted water supplies.

The comparison that biochar is a "sponge-like" material helps to understand why its hydrological benefits might persist for years. Plant waste is pyrolyzed at high temperatures with minimal oxygen to create biochar, a highly porous and chemically stable substance. These holes serve as micro-reservoirs, storing water that would otherwise be lost by deep percolation or rapid evaporation. Oxidation increases the surface of biochar as it ages in soil, adding functional groups like hydroxyls and carboxyls that make it more attractive to water molecules and cations. As a result, biochar lessens the effects of recurrent droughts by temporarily enhancing soil water holding capacity and buffering soil moisture availability over time. Evidence from field and meta-analytical studies supports these theoretical conclusions. A worldwide systematic review found that biochar considerably raises the soil's capacity for holding water and available water content, especially in coarse-textured soils (Edeh et al. 2020). Extensive field research has demonstrated that these benefits are long-lasting; soils treated with biochar still show lower bulk density and higher porosity years later (Liu et al. 2017). Residual benefits can occasionally become more apparent as biochar interacts with organic matter and minerals in soil, stabilizing soil aggregates and improving saturated hydraulic conductivity. These findings demonstrate that biochar is more than just a one-season addition; it is a persistent soil conditioner with enduring hydrological benefits. The stability of biochar itself supports these long-term benefits. Unlike most organic inputs that mineralize in a few months or years, biochar has mean residence lengths ranging from several centuries to millennia, based on the pyrolytic and biomass parameters (Wang et al. 2016). Because of its exceptional durability, biochar will continue to perform hydrological and agronomic tasks for a very long time after application. In practice, these advantages are often apparent to farmers as decreased nutrient leaching, more drought-tolerant crops and soils that stay wet longer after irrigation. Both perspectives concur that biochar offers long-term carbon sinking and residual hydrological benefits, but researchers quantify this using soil-water retention curves and carbon stability models. Because of its dual purpose, it is a powerful tool for sustainable agriculture in the face of climate variability. Extreme weather events may have a long-term effect on the longevity of biochar, even though it could be used for improving the soils and plants impacted. Determining its stability is therefore essential for the long-term mitigation of negative effects. The biochar's lifespan and resistance in environments at which it decomposes gradually and resists both living and non-living deterioration are referred to as its stability. Biochar remains stable in the soil over time means that carbon will be sequestered for a long period, increasing its effectiveness in removing greenhouse gases from the air thus reducing the negative impact of climate change (Tisserant and Cherubini 2019). According to Wang et al. (2016), About 3% of biochar was bioavailable, indicating that a sizable amount of biochar remains persistent in soil for a considerable amount of time. Furthermore, Wang et al. (2016) state that the remaining 97% of biochar contributes directly to the long-term retention of carbon in soils.

Thermogravimetric analysis could also be used to evaluate the biochar's durability (Igalavithana et al. 2017a, b). According to some sources, the Hydrogen to Carbon and Oxygen to Carbon ratios are reliable measures of the stability of biochar. According to Klasson (2017), biochars with an Oxygen to Carbon ratio (O/C) below 0.4 and a Hydrogen to Carbon ratio (H/C) below 0.7 are regarded as extremely preferred ratios. When assessing the stability of biochar, the H/C ratio is preferred over the O/C ratio due to the possibility of errors in oxygen estimation. Adding biochar to soil enhances carbon sequestration and stabilizes organic matter by forming organo-mineral complexes, which in turn strengthens the stability of the soil biochar complex (Tsai and Chang 2020). Furthermore, adding biochar to soil may result in the release of biochar colloids into the soil solution; this could be influenced by the soil's pH, chemical composition, and surface functioning, all of which affect the stability and destiny of biochar. Additionally, adding natural organic ingredients like humic acid makes the biochar colloidal solution more stable (Yang et al. 2019). Many additional things also effect the stability of biochar. Highly aromatic carbonaceous structures that are resistant to degradation are primarily responsible for the high stability. It's interesting to note that high lignin content boosts aromaticity, which improves the stability of biochar (Leng et al. 2019). But according to Tomczyk et al. (2020), the complete conversion of lignin into hydrophobic polycyclic aromatic hydrocarbons is inhibited by biochar produced below 500 °C, enhancing its ability to repel water and decreasing its stability, which may be lessened by increasing hydrophobicity. Accordingly, biochars made at temperatures higher than 650 °C are regarded as being extremely thermostable. Another element that contributes to the stability of biochar is aggregation. Fungal hyphae can grow on biochar, which actively encourages aggregate formation and strengthens the stability of biochar. Biochar functions as a cementing substance for the formation of aggregates and is associated with soil micro-aggregates that prevent degradation.

7 Case Studies and Meta-Analytical Insights

Han et al. (2023) conducted a global synthesis to maximize the important role of biochar in agricultural output and to enhance its application strategies using extraction analysis, random forest simulation, and win-win analysis across 381 datasets drawn from 63 separate research globally. According to their research, the application of biochar considerably reported maximum production and productivity by 14.45%, water use efficiency (WUE) by 14.28% and nitrogen use efficiency (NUE) by 13.97%. These findings demonstrate that biochar can enhance resource performance and productivity in a variety of agroecosystems. The analysis also showed that efficiency of adding biochar was directly swayed by a number of soil and biochar characteristics. Soil bulk density, total soil nitrogen (Soil-TN), soil organic carbon (SOC), biochar carbon content (Biochar-TC) and the carbon-to-nitrogen ratio of biochar (Biochar-C:N) were important variables. Particularly, when amended with biochar of relatively low carbon content (TC $\leq$ 200 g kg^{-1}) and intermediate C:N ratios (50–100), soils with poor fertility and high bulk density showed the most noticeable gains.

Biochar-based fertilizers (BBFs) have garnered a lot of interest recently as a promising nutrient management innovation; nevertheless, until Melo et al.'s work in 2022, their effects on crop yield had not been thoroughly examined. In order to fill this gap, the authors synthesized 148 pairwise comparisons of BBF treatments, conventional fertilizer controls and non-fertilized controls by conducting a meta-analysis of published papers from 2011 to 2021. According to their research, using BBF at comparatively low rates (mean 0.9 t ha^{-1}) enhanced crop output by an astounding 186% when compared to non-fertilized controls and by 10% when compared to fertilized controls. Given that comparable yield increases are usually only documented when biochar is treated as a soil conditioner at significantly larger rates (15–30 t ha^{-1}), this performance is very noteworthy. According to these findings, biochar can function as a matrix in fertilizer formulations, significantly increasing the efficiency of nutrient usage above and beyond what is possible with traditional fertilizers alone. Important parameters affecting the size of BBF benefits were also identified by the study. According to cluster study, BBFs might increase yields by up to 15% and were particularly successful in soils that were not responding well to traditional fertilizers. Equally important were the production conditions: BBFs produced at higher heating temperatures (>400 °C) increased productivity by an average of 12%, while those produced at lower temperatures ($\leq$400 °C) had no discernible impact. Additionally, the final BBF mixture's carbon content was a significant factor; formulations with >30% C produced the biggest productivity gains (17%), whereas combinations with $\leq$30% C had no effect.

Prior to the global synthesis by Ye et al. (2020), the added value of biochar when applied in conjunction with fertilizers, beyond the effect of fertilizers alone, had not been precisely measured. Biochar applications $\leq$20 t ha^{-1} in conjunction with inorganic fertilizers (IF) and/or organic amendments (OA) were the focus of the authors' meta-analysis, which assessed short-term (1-year) field responses in crop yield across a wide range of climates, soils, biochar types and management practices.

Their results demonstrated that IF alone enhanced yields by 26% (CI: 15–40%) in comparison to a non-fertilized control, however IF and biochar together generated a significantly larger increase of 48% (CI: 30–70%).

Crucially, adding biochar with fertilizers increased yields by an extra 15% (CI: 11–19%) when compared directly with IF-only treatments, demonstrating that when combined with fertilizers, biochar was equally as effective for boosting productivity.

The limited short-term impacts of biochar when not paired with fertilizer inputs are highlighted by the fact that it did not significantly increase yields when compared to either fertilized or unfertilized controls. The authors pointed out that the short-term liming effects of plant-derived biochars (>90% of those evaluated) might have contributed to the improvements that were seen when IF was added. Their independent meta-analysis of multi-year research, however, offered additional information: three or more years following a single application of biochar, yields rose by 31% (CI: 17–49%). This suggests that rather than only liming, longer-term benefits were associated with more durable changes including better soil structure and higher cation exchange capacity (CEC).

Bai et al. (2022) carried out a global meta-analysis that offered more quantitative proof of the contribution of biochar and its interactions with fertilizers to increased agricultural output. The application of biochar alone was found to boost crop yield by an average of 25.3% across the synthesized research. The combined application of biochar and nutrient inputs was highlighted by the additional 10% improvement in yields when paired with inorganic fertilizers as opposed to fertilizer use alone. The investigation also showed that a number of traits, such as crop type, application rate and the type of fertilizer treatment used in conjunction with biochar, had a significant effect of biochar on the yield of crop. The highest yield increases among the evaluated soil conditions were observed in extremely acidic soils ($pH \leq 5$), where the liming impact and ability of biochar to enhance nutrient retention seemed to be especially advantageous. Various studies demonstrate the capability of biochar to improve problematic soils that might otherwise restrict crop output. Furthermore, because it not only increases yields but also improves nutrient usage efficiency, the study indicated that combining biochar with both organic and inorganic fertilizers offers a particularly promising approach for enhancing agricultural performance.

Conditions for pyrolysis also had an impact. In contrast to those Increased pyrolysis temperatures and more alkaline pH levels resulted in a beneficial relationship between biochar and nutrient inputs; biochar's produced at ≤550 °C with a final pH of ≤8 greatly enhanced crop growth and yield. This is likely due to their higher nutrient content and more labile organic fractions. This suggests that the benefits of biochar for crop productivity are highly context-dependent, and that crop species, climate, soil characteristics and the type of biochar must be considered.

The hydraulic properties of the soil were influenced by how long the biochar was left to age in the field. Based on an experiment that lasted 1–10 years, Jiang et al. (2025) investigated the long-term impacts of biochar on the hydraulic characteristics of soil. The soil's Total Porosity (TP), Field Capacity (FC) and Available Water Capacity (AWC) significantly changed with the addition of biochar. Interestingly, when field time increased in the second year, the treatment group showed the biggest increase in soil TP, FC, and AWC. In comparison to the control group, the soil TP, FC, and AWC in the treatment group increased by 9.83%, 13.57%, and 14.11%, accordingly, in the first year; 15.20%, 16.37%, and 16.59%, respectively, in the second year; 8.90%, 10.44%, and 12.26%, accordingly, in the third year; and 7.90%, 4.26%, and 7.63%, respectively, after 3 years. According to other studies that demonstrate how aging impacts biochar's hydrophilicity (Ivanova et al. 2023), the older layer of biochar in soil from farms is more hydrophilic than that of control biochar samples (Aller et al. 2017).

Over time, biochar particles fracture due to physical strain from freeze-thaw cycles in field settings, which reduces their ability to retain water and increases their hydrophobicity (Hussain et al. 2020). Although there was no mechanical damage in the first year, the overall positive effect of biochar on soil hydraulic properties peaked in the second. This could help to explain why, following the second year of the trial, the use of biochar reduced soil TP, FC, and AWC as the field tests continued. Additionally, as time went on, fine-particle soils clogged the biochar pores by moving from the outside to the inside (Ren et al. 2018). This reduction in total porosity causes the soil's FC and AWC to decrease.

8 Challenges and Research Gaps

A list of some of the most important research requirements and gaps identified during this evaluation is provided below. The application technique of biochar, particularly depth of soil is crucial for growers maintaining farms with limited and no tillage. Additionally, most farm research often involves only single application of biochar. The research does not adequately document how frequently biochar is applied, even though one study found that split application had a far higher positive impact than one-time treatment. To evaluate the financial and environmental advantages of applying biochar at varying depths and frequencies (once versus split), more field research is therefore needed (in a variety of soils and climate conditions).

Most of the published literature on biochar and emissions of greenhouse gases comes from studies conducted in labs, greenhouses, and mesocosms. To evaluate the economic and environmental advantages of biochar and to better understand how it affects soil processes, field studies under a variety of soil types, climates, and management techniques are required. Biochar can have a variety of impacts on greenhouse gas emissions, including positive, negative, and no effect, in addition to variations in soil, climate, biochar feedstock, pyrolysis temperature, and biochar properties. Research on how to modify biochar manufacturing procedures for certain variables (soil, crop, climate, etc.) is necessary to further investigate the issue of biochar's varying influence (Palansooriya et al. 2019).

Of the field studies analysed, 89% are short-term (1–3 years), whereas just 11% are medium-term (5–7 years). Some studies have shown that the reduction in N_2O emissions brought about by biochar can continue for up to 7 years after it is applied.

However, this has to be further explored through long-term field research and on-farm evaluations that combine GHG emission reduction with C sequestration in response to biochar treatments (Edeh et al. 2020).

Lack of financing may be one of the factors contributing to the dearth of long-term studies. For these kinds of projects to be developed and managed, it is therefore essential to promote long-term financial commitment. These long-term studies should yield the data required to further comprehend, assess, and validate the impact of biochar on soil C residence time, greenhouse gas mitigation, and the sustainability of net-zero farming systems.

One reason for the dearth of long-term research could be a lack of financing. It is therefore essential to promote ongoing funding for the creation and administration of these kinds of studies. Such long-term studies should yield the data required to further comprehend, assess, and validate the impacts of biochar on soil C residence time, greenhouse gas mitigation, and the sustainability of net-zero agricultural systems.

9 Conclusion

Drought and salt stress significantly reduce crop growth and production. According to the studies mentioned above, using biochar enhanced plant biomass and growth under salinity and moisture stress. Utilizing biochar in the soil improved the photosynthetic capacity, plant absorption of nutrients and modified gas exchange characteristics. Drought significantly alters plant growth and development by interfering with several physiological, biochemical, and molecular processes. Applying BC promotes nutrient absorption, membrane integrity, and nutrient homeostasis, all of which boost plant performance in these circumstances. Furthermore, the BC amendment boosts photosynthetic efficiency and antioxidant activity, protecting the plant from oxidative and osmotic stressors caused by dryness and maintaining hormonal balance. There is currently little proof that applying field biochar increases crop yields and water availability, especially in temperate climates. The relationship between biochar and crop water use has not yet been thoroughly investigated. Understanding biochar's longer-term capacity to sustain claimed effects like better crop yields may be impacted by the fact that the majority of biochar trials are carried out over brief time periods. This is particularly relevant in light of biochar's refractory and hydrophobic properties, as well as the potential structural, physical, chemical, and hydrological effects on crop productivity when combined with soil. Additionally, biochar can act as a aggregating agent to promote the formation of macro-aggregates and prevent their disintegration into smaller aggregates. Increased mesopore production may have resulted from improved macro-aggregates in soil supplemented with biochar. These pores are in charge of retaining water for plant absorption during drought stress, which raises the biomass, water use efficiency and chlorophyll content of plants in soils supplemented with biochar. In conclusion, by raising macro-aggregates, biochar expanded the volume of meso-pores containing water that is available to plants. Our study's findings may help us better understand the effects of biochar by linking changes in pore size distribution to modifications in soil aggregate structure caused by biochar. Additionally, we show that biochar is a potentially helpful management tactic for reducing drought conditions.

References

Abbas T, Rizwan M, Ali S, Adrees M, Mahmood A, Zia-ur-Rehman M, Ibrahim M, Arshad M, Qayyum MF (2018) Biochar application increased the growth and yield and reduced cadmium in drought stressed wheat grown in an aged contaminated soil. Ecotoxicol Environ Saf 148:825–833. https://doi.org/10.1016/j.ecoenv.2017.11.063

Abdillah AN, Shivanand P, Cheok Q, Bakar SA, Taha H, Metali F (2024) Impact of rice husk biochar on soil properties and microbial diversity for paddy cultivation. J Soil Sci Plant Nutr 24(4):7507–7524. https://doi.org/10.1007/s42729-024-02055-7

Abrol V, Ben-Hur M, Verheijen FG, Keizer JJ, Martins MAS, Tenaw H, Tchehansky L, Graber ER (2016) Biochar effects on soil water infiltration and erosion under seal formation conditions: rainfall simulation experiment. J Soils Sediments 16:2709–2719. https://doi.org/10.1007/s11368-016-1448-8

Abujabhah IS, Bound SA, Doyle R, Bowman JP (2016) Effects of biochar and compost amendments on soil physico-chemical properties and the total community within a temperate agricultural soil. Appl Soil Ecol 98:243–253. https://doi.org/10.1016/j.apsoil.2015.10.021

Adams MA, Buckley TN, Salter WT, Buchman N, Blessing CH, Turnbull TL (2018) Contrasting responses of crop legumes and cereals to nitrogen availability. New Phytol 217(4):1475–1483. https://doi.org/10.1111/nph.14918

Agbna GH, Dongli S, Zhipeng L, Elshaikh NA, Guangcheng S, Timm LC (2017) Effects of deficit irrigation and biochar addition on the growth, yield, and quality of tomato. Sci Hortic 222:90–101. https://doi.org/10.1016/j.scienta.2017.05.004

AghaKouchak A, Farahmand A, Melton FS, Teixeira J, Anderson MC, Wardlow BD, Hain CR (2015) Remote sensing of drought: progress, challenges and opportunities. Rev Geophys 53(2):452–480. https://doi.org/10.1002/2014RG000456

Ahmad Bhat S, Kuriqi A, Dar MUD, Bhat O, Sammen SS, Towfiqul Islam ARM, Elbeltagi A, Shah O, Ai-Ansari N, Ali R, Heddam S (2022) Application of biochar for improving physical, chemical, and hydrological soil properties: a systematic review. Sustainability 14(17):11104. https://doi.org/10.3390/su141711104

Alahacoon N, Edirisinghe M (2022) A comprehensive assessment of remote sensing and traditional based drought monitoring indices at global and regional scale. Geomat Nat Haz Risk 13(1):762–799. https://doi.org/10.1080/19475705.2022.2044394

Ali AB, Haofang Y, Hong L, Yan YW, Elshaikh NA, Hussein G, Pandab S, Hassan S (2019) Enhancement of depleted loam soil as well as cucumber productivity utilizing biochar under water stress. Commun Soil Sci Plant Anal 50(1):49–64. https://doi.org/10.1080/00103624.2018.1547389

Aller D, Rathke S, Laird D, Cruse R, Hatfield J (2017) Impacts of fresh and aged biochars on plant available water and water use efficiency. Geoderma 307:114–121. https://doi.org/10.1016/j.geoderma.2017.08.007

Bai SH, Omidvar N, Gallart M, Kämper W, Tahmasbian I, Farrar MB, Singh K, Zhou G, Muqadass B, Xu CY, Koech R (2022) Combined effects of biochar and fertilizer applications on yield: a review and meta-analysis. Sci Total Environ 808:152073. https://doi.org/10.1016/j.scitotenv.2021.152073

Baiamonte G, Crescimanno G, Parrino F, De Pasquale C (2019) Effect of biochar on the physical and structural properties of a desert sandy soil. Catena 175:294–303. https://doi.org/10.1016/j.catena.2018.12.019

Bayat H, Mazaheri B, Mohanty BP (2019) Estimating soil water characteristic curve using landscape features and soil thermal properties. Soil Tillage Res 189:1–14. https://doi.org/10.1016/j.still.2018.12.018

Bera T, Purakayastha TJ, Patra AK, Datta SC (2018) Comparative analysis of physicochemical, nutrient, and spectral properties of agricultural residue biochars as influenced by pyrolysis temperatures. J Mater Cycles Waste Manag 20(2):1115–1127. https://doi.org/10.1007/s10163-017-0675-4

Bolan S, Hou D, Wang L, Hale L, Egamberdieva D, Tammeorg P, Li R, Wang B, Xu J, Wang T (2023) The potential of biochar as a microbial carrier for agricultural and environmental applications. Sci Total Environ 886:163968. https://doi.org/10.1016/j.scitotenv.2023.163968

Bordoloi S, Garg A, Sreedeep S, Lin P, Mei G (2018) Investigation of cracking and water availability of soil-biochar composite synthesized from invasive weed water hyacinth. Bioresour Technol 263:665–677. https://doi.org/10.1016/j.biortech.2018.05.011

Bruun EW, Ravenni G, Müller-Stöver D, Petersen CT (2023) Small biochar particles added to coarse sandy subsoil greatly increase water retention and affect hydraulic conductivity. Eur J Soil Sci 74(6):e13442. https://doi.org/10.1111/ejss.13442

Burrell LD, Zehetner F, Rampazzo N, Wimmer B, Soja G (2016) Long-term effects of biochar on soil physical properties. Geoderma 282:96–102. https://doi.org/10.1016/j.geoderma.2016.07.019

Butnan S, Deenik JL, Toomsan B, Vityakon P (2018) Biochar properties affecting carbon stability in soils contrasting in texture and mineralogy. Agric Nat Resour 51(6):492–498. https://doi.org/10.1016/j.anres.2018.03.002

Carvalho MTM, Madari BE, Bastiaans L, Van Oort PAJ, Leal WGO, Heinemann AB, Da Silva MAS, Maia AHN, Parsons D, Meinke H (2016) Properties of a clay soil from 1.5 to 3.5 years after biochar application and the impact on rice yield. Geoderma 276:7–18. https://doi.org/10.1016/j.geoderma.2016.04.013

Costa OY, Raaijmakers JM, Kuramae EE (2018) Microbial extracellular polymeric substances: ecological function and impact on soil aggregation. Front Microbiol 23(9):1636. https://doi.org/10.3389/fmicb.2018.01636

Das S, Das J, Umamahesh NV (2022a) Investigating the propagation of droughts under the influence of large-scale climate indices in India. J Hydrol 610:127900. https://doi.org/10.1016/j.jhydrol.2022.127900

Das SK, Ghosh GK, Avasthe R (2022b) Valorising biomass to engineered biochar and its impact on soil, plant, water and microbial dynamics: a review. Biomass Convers Biorefinery 12(9):4183–4199. https://doi.org/10.1007/s13399-020-00836-5

Ding Y, Zhang L, He Y, Cao S, Wei X, Guo Y, Ran L, Filonchyk M (2024) Spatiotemporal evolution of agricultural drought and its attribution under different climate zones and vegetation types in the Yellow River Basin of China. Sci Total Environ 914:169687. https://doi.org/10.1016/j.scitotenv.2023.169687

Dokoohaki H, Miguez FE, Laird D, Horton R, Basso AS (2017) Assessing the biochar effects on selected physical properties of a sandy soil: an analytical approach. Commun Soil Sci Plant Anal 48(12):1387–1398. https://doi.org/10.1080/00103624.2017.1358742

Dong D, Wang C, Van Zwieten L, Wang H, Jiang P, Zhou M, Wu W (2020) An effective biochar-based slow-release fertilizer for reducing nitrogen loss in paddy fields. J Soils Sediments 20(8):3027–3040. https://doi.org/10.1007/s11368-019-02401-8

Du ZL, Zhao JK, Wang YD, Zhang QZ (2017) Biochar addition drives soil aggregation and carbon sequestration in aggregate fractions from an intensive agricultural system. J Soils Sediments 17(3):581–589. https://doi.org/10.1007/s11368-015-1349-2

Ebrahimi M, Souri MK, Mousavi A, Sahebani N (2021) Biochar and vermicompost improve growth and physiological traits of eggplant (*Solanum melongena* L.) under deficit irrigation. Chem Biol Technol Agric 8(1):19. https://doi.org/10.1186/s40538-021-00216-9

Edeh IG, Mašek O, Buss W (2020) A meta-analysis on biochar's effects on soil water properties–new insights and future research challenges. Sci Total Environ 714:136857. https://doi.org/10.1016/j.scitotenv.2020.136857

Egamberdieva D, Reckling M, Wirth S (2017) Biochar-based Bradyrhizobium inoculum improves growth of lupin (*Lupinus angustifolius* L.) under drought stress. Eur J Soil Biol 78:38–42. https://doi.org/10.1016/j.ejsobi.2016.11.007

Faloye OT, Alatise MO, Ajayi AE, Ewulo BS (2019) Effects of biochar and inorganic fertiliser applications on growth, yield and water use efficiency of maize under deficit irrigation. Agric Water Manag 217:165–178. https://doi.org/10.1016/j.agwat.2019.02.044

Farhangi-Abriz S, Torabian S, Qin R, Noulas C, Lu Y, Gao S (2021) Biochar effects on yield of cereal and legume crops using meta-analysis. Sci Total Environ 775:145869. https://doi.org/10.1016/j.scitotenv.2021.145869

Fu Q, Zhao H, Li T, Hou R, Liu D, Ji Y, Zhou Z, Yang L (2019) Effects of biochar addition on soil hydraulic properties before and after freezing-thawing. Catena 176:112–124. https://doi.org/10.1016/j.catena.2019.01.008

Gaurav S, Diptanu B, Mehta CM, Prasann K, Nishihara E, Inubushi K, Sudo S, Hayashida S, Patra PK, Minkina T, Rajput VD (2025) Effects of biochar amendment at various soil depths on maize roots and growth indices. Sci Rep 15(1):26310. https://doi.org/10.1038/s41598-025-09218-1

Gelardi DL, Ainuddin IH, Rippner DA, Patiño JE, Abou Najm M, Parikh SJ (2021) Biochar alters hydraulic conductivity and impacts nutrient leaching in two agricultural soils. Soil 7(2):811–825. https://doi.org/10.5194/soil-7-811-2021

Gholinia A, Abbaszadeh P (2024) Agricultural drought monitoring: a comparative review of conventional and satellite-based indices. Atmosphere 15(9):1129. https://doi.org/10.3390/atmos15091129

Glab T, Palmowska J, Zaleski T, Gondek K (2016) Effect of biochar application on soil hydrological properties and physical quality of sandy soil. Geoderma 281:11–20. https://doi.org/10.1016/j.geoderma.2016.06.028

Głąb T, Żabiński A, Sadowska U, Gondek K, Kopeć M, Mierzwa–Hersztek M, Tabor S (2018) Effects of co-composted maize, sewage sludge, and biochar mixtures on hydrological and physical qualities of sandy soil. Geoderma 315:27–35. https://doi.org/10.1016/j.geoderma.2017.11.034

Haider I, Raza MA, Iqbal R, Aslam MU, Habib-ur-Rahman M, Raja S, Khan MT, Aslam MM, Waqas M, Ahmad S (2020) Potential effects of biochar application on mitigating the drought stress implications on wheat (*Triticum aestivum* L.) under various growth stages. J Saudi Chem Soc 24(12):974–981. https://doi.org/10.1016/j.jscs.2020.10.005

Han M, Zhang J, Zhang L, Wang Z (2023) Effect of biochar addition on crop yield, water and nitrogen use efficiency: a meta-analysis. J Clean Prod 420:138425. https://doi.org/10.1016/j.jclepro.2023.138425

Hashem A, Kumar A, Al-Dbass AM, Alqarawi AA, Al-Arjani AB, Singh G, Farooq M, Abd_Allah EF (2019) Arbuscular mycorrhizal fungi and biochar improves drought tolerance in chickpea. Saudi J Biol Sci 26(3):614–624. https://doi.org/10.1016/j.sjbs.2018.11.005

Hassan MU, Chattha MU, Khan I, Chattha MB, Barbanti L, Aamer M, Iqbal MM, Nawaz M, Mahmood A, Ali A, Aslam MT (2021) Heat stress in cultivated plants: nature, impact, mechanisms, and mitigation strategies—a review. Plant Biosyst Int J Dealing All Aspects Plant Biol 155(2):211–234. https://doi.org/10.1080/11263504.2020.1727987

Horel A, Potyó I, Szili-Kovács T, Molnár S (2018) Potential nitrogen fixation changes under different land uses as influenced by seasons and biochar amendments. Arab J Geosci 11(18):1–10. https://doi.org/10.1007/s12517-018-3916-5

Hussain R, Ravi K, Garg A (2020) Influence of biochar on the soil water retention characteristics (SWRC): potential application in geotechnical engineering structures. Soil Tillage Res 204:104713. https://doi.org/10.1016/j.still.2020.104713

Igalavithana AD, Mandal S, Niazi NK, Vithanage M, Parikh SJ, Mukome FN, Rizwan M, Oleszczuk P, Al-Wabel M, Bolan N, Tsang DC (2017a) Advances and future directions of biochar characterization methods and applications. Crit Rev Environ Sci Technol 47(23):2275–2330. https://doi.org/10.1080/10643389.2017.1421844

Igalavithana AD, Ok YS, Niazi NK, Rizwan M, Al-Wabel MI, Usman AR, Moon DH, Lee SS (2017b) Effect of corn residue biochar on the hydraulic properties of sandy loam soil. Sustainability 9(2):266. https://doi.org/10.3390/su9020266

Ivanova N, Obaeed GL, Sulkarnaev F, Buchkina N, Gubin A, Yurtaev A (2023) Effect of biochar aging in agricultural soil on its wetting properties and surface structure. Biochar 5(1):75. https://doi.org/10.1007/s42773-023-00272-4

Jain S, Singh A, Khare P, Chanda D, Mishra D, Shanker K, Karak T (2017) Toxicity assessment of *Bacopa monnieri* L. grown in biochar amended extremely acidic coal mine spoils. Ecol Eng 108:211–219. https://doi.org/10.1016/j.ecoleng.2017.08.039

Jalayer S, Sharifi A, Abbasi-Moghadam D, Tariq A, Qin S (2023) Assessment of spatiotemporal characteristic of droughts using in situ and remote sensing-based drought indices. IEEE J Sel Top Appl Earth Obs Remote Sens 16:1483–1502. https://doi.org/10.1109/JSTARS.2023.3237380

Jiang Z, Huang S, Meng Z (2025) Long-term effects of biochar on the hydraulic properties of soil: a meta-analysis based on 1–10 years field experiments. Geoderma 458:117318. https://doi.org/10.1016/j.geoderma.2025.117318

Kafle A, Singh S, Singh M, Bajwa P, Deb S, Simpson C, Ritchie G (2025) Influence of deficit irrigation and biochar amendment on growth, physiology, and yield of cucumber in West Texas. Sci Rep 15(1):9615. https://doi.org/10.1038/s41598-025-94113-y

Kätterer T, Roobroeck D, Andrén O, Kimutai G, Karltun E, Kirchmann H, Nyberg G, Vanlauwe B, de Nowina KR (2019) Biochar addition persistently increased soil fertility and yields in maize-soybean rotations over 10 years in sub-humid regions of Kenya. Field Crop Res 235:18–26. https://doi.org/10.1016/j.fcr.2019.02.015

Kim HS, Kim KR, Yang JE, Ok YS, Owens G, Nehls T, Wessolek G, Kim KH (2016) Effect of biochar on reclaimed tidal land soil properties and maize (*Zea mays* L.) response. Chemosphere 142:153–159. https://doi.org/10.1016/j.chemosphere.2015.06.041

Klasson KT (2017) Biochar characterization and a method for estimating biochar quality from proximate analysis results. Biomass Bioenergy 96:50–58. https://doi.org/10.1016/j.biombioe.2016.10.011

Kou X, Han W, Kang J (2022) Responses of root system architecture to water stress at multiple levels: a meta-analysis of trials under controlled conditions. Front Plant Sci 13:1085409. https://doi.org/10.3389/fpls.2022.1085409

Langeroodi AR, Campiglia E, Mancinelli R, Radicetti E (2019) Can biochar improve pumpkin productivity and its physiological characteristics under reduced irrigation regimes? Sci Hortic 247:195–204. https://doi.org/10.1016/j.scienta.2018.11.059

Latif M, Bukhari SA, Alrajhi AA, Alotaibi FS, Ahmad M, Shahzad AN, Dewidar AZ, Mattar MA (2022) Inducing drought tolerance in wheat through exopolysaccharide-producing rhizobacteria. Agronomy 12(5):1140. https://doi.org/10.3390/agronomy12051140

Leng L, Huang H, Li H, Li J, Zhou W (2019) Biochar stability assessment methods: a review. Sci Total Environ 647:210–222. https://doi.org/10.1016/j.scitotenv.2018.07.402

Li L, Zhang YJ, Novak A, Yang Y, Wang J (2021) Role of biochar in improving sandy soil water retention and resilience to drought. Water 13(4):407. https://doi.org/10.3390/w13040407

Liang G, Liu J, Zhang J, Guo J (2020) Effects of drought stress on photosynthetic and physiological parameters of tomato. J Am Soc Hortic Sci 145(1):12–17. https://doi.org/10.21273/JASHS04725-19

Libutti A, Francavilla M, Monteleone M (2021) Hydrological properties of a clay loam soil as affected by biochar application in a pot experiment. Agronomy 11(3):489. https://doi.org/10.3390/agronomy11030489

Lim TJ, Spokas KA, Feyereisen G, Novak JM (2016) Predicting the impact of biochar additions on soil hydraulic properties. Chemosphere 142:136–144. https://doi.org/10.1016/j.chemosphere.2015.06.069

Liu Z, Dugan B, Masiello CA, Gonnermann HM (2017) Biochar particle size, shape, and porosity act together to influence soil water properties. PLoS One 12(6):e0179079. https://doi.org/10.1371/journal.pone.0179079

Liu Y, Yang S, Lu H, Wang Y (2018a) Effects of biochar on spatial and temporal changes in soil temperature in cold waterlogged rice paddies. Soil Tillage Res 181:102–109. https://doi.org/10.1016/j.still.2018.04.008

Liu Y, Zhu J, Ye C, Zhu P, Ba Q, Pang J, Shu L (2018b) Effects of biochar application on the abundance and community composition of denitrifying bacteria in a reclaimed soil from coal mining subsidence area. Sci Total Environ 625:1218–1224. https://doi.org/10.1016/j.scitotenv.2018.01.003

Liu X, Zhou J, Chi Z, Zheng J, Li L, Zhang X, Zheng J, Cheng K, Bian R, Pan G (2019) Biochar provided limited benefits for rice yield and greenhouse gas mitigation six years following an amendment in a fertile rice paddy. Catena 179:20–28. https://doi.org/10.1016/j.catena.2019.03.033

Lopes ÉM, Reis MM, Frazão LA, da Mata Terra LE, Lopes EF, Dos Santos MM, Fernandes LA (2021) Biochar increases enzyme activity and total microbial quality of soil grown with sugarcane. Environ Technol Innovation 21:101270. https://doi.org/10.1016/j.eti.2020.101270

Lupo Y, Moshelion M (2024) The balance of survival: comparative drought response in wild and domesticated tomatoes. Plant Sci 339:111928. https://doi.org/10.1016/j.plantsci.2023.111928

Ma S, Wang Z, Guo X, Wang F, Huang J, Sun B, Wang X (2021) Sourdough improves the quality of whole-wheat flour products: mechanisms and challenges—a review. Food Chem 360:130038. https://doi.org/10.1016/j.foodchem.2021.130038

Madari BE, Silva MA, Carvalho MT, Maia AH, Petter FA, Santos JL, Tsai SM, Leal WG, Zeviani WM (2017) Properties of a sandy clay loam Haplic Ferralsol and soybean grain yield in a five-year field trial as affected by biochar amendment. Geoderma 305:100–112. https://doi.org/10.1016/j.geoderma.2017.05.029

Moragues-Saitua L, Arias-González A, Gartzia-Bengoetxea N (2017) Effects of biochar and wood ash on soil hydraulic properties: a field experiment involving contrasting temperate soils. Geoderma 305:144–152. https://doi.org/10.1016/j.geoderma.2017.05.041

Moragues-Saitua L, Arias-González A, Blanco F, Benito-Carnero G, Gartzia-Bengoetxea N (2023) Effects of biochar and wood ash amendments in the soil-water-plant environment of two temperate forest plantations. Front For Glob Change 5:878217. https://doi.org/10.3389/ffgc.2022.878217

Nardini A, Casolo V, Dal Borgo A, Savi T, Stenni B, Bertoncin P, Zini L, McDowell NG (2016) Rooting depth, water relations and non-structural carbohydrate dynamics in three woody angiosperms differentially affected by an extreme summer drought. Plant Cell Environ 39(3):618–627. https://doi.org/10.1111/pce.12646

Nguyen VT, Nguyen TB, Chen CW, Hung CM, Vo TD, Chang JH, Dong CD (2019) Influence of pyrolysis temperature on polycyclic aromatic hydrocarbons production and tetracycline adsorption behavior of biochar derived from spent coffee ground. Bioresour Technol 284:197–203. https://doi.org/10.1016/j.biortech.2019.03.096

Ni JJ, Chen XW, Ng CW, Guo HW (2018) Effects of biochar on water retention and matric suction of vegetated soil. Géotech Lett 8(2):124–129. https://doi.org/10.1680/jgele.17.00180

Nicosia A, Pampalone V, Ferro V (2021) Effects of biochar addition on rill flow resistance. Water 13(21):3036. https://doi.org/10.3390/w13213036

Novak J, Sigua G, Watts D, Cantrell K, Shumaker P, Szogi A, Johnson MG, Spokas K (2016) Biochars impact on water infiltration and water quality through a compacted subsoil layer. Chemosphere 142:160–167. https://doi.org/10.1016/j.chemosphere.2015.06.038

Obadi A, Alharbi A, Alomran A, Alghamdi AG, Louki I, Alkhasha A (2023) Effect of biochar application on morpho-physiological traits, yield, and water use efficiency of tomato crop under water quality and drought stress. Plants 12(12):2355. https://doi.org/10.3390/plants12122355

Obia A, Mulder J, Martinsen V, Cornelissen G, Børresen T (2016) In situ effects of biochar on aggregation, water retention and porosity in light-textured tropical soils. Soil Tillage Res 155:35–44. https://doi.org/10.1016/j.still.2015.08.002

Odugbenro GO, Liu Z, Sun Y (2020) Soil aggregate size distribution and total organic carbon in intra-aggregate fractions as affected by addition of biochar and organic amendments. Polish J Soil Sci 53(1):41. https://doi.org/10.17951/pjss.2020.53.1.41

Ogura T, Goeschl C, Filiault D, Mirea M, Slovak R, Wolhrab B, Satbhai SB, Busch W (2019) Root system depth in Arabidopsis is shaped by EXOCYST70A3 via the dynamic modulation of auxin transport. Cell 178(2):400–412

Olmo M, Villar R, Salazar P, Alburquerque JA (2016) Changes in soil nutrient availability explain biochar's impact on wheat root development. Plant Soil 399(1):333–343. https://doi.org/10.1007/s11104-015-2700-5

Omara P, Aula L, Oyebiyi F, Raun WR (2019) World cereal nitrogen use efficiency trends: review and current knowledge. Agrosyst Geosci Environ 2(1):1–8. https://doi.org/10.2134/age2018.10.0045

Omondi MO, Xia X, Nahayo A, Liu X, Korai PK, Pan G (2016) Quantification of biochar effects on soil hydrological properties using meta-analysis of literature data. Geoderma 274:28–34. https://doi.org/10.1016/j.geoderma.2016.03.029

Onwuka B, Mang BJ (2018) Effects of soil temperature on some soil properties and plant growth. Adv Plants Agric Res 8(1):34–37. https://doi.org/10.15406/apar.2018.08.00288

Palansooriya KN, Wong JT, Hashimoto Y, Huang L, Rinklebe J, Chang SX, Bolan N, Wang H, Ok YS (2019) Response of microbial communities to biochar-amended soils: a critical review. Biochar 1(1):3–22. https://doi.org/10.1007/s42773-019-00009-2

Palviainen M, Aaltonen H, Laurén A, Köster K, Berninger F, Ojala A, Pumpanen J (2020) Biochar amendment increases tree growth in nutrient-poor, young Scots pine stands in Finland. For Ecol Manag 474:118362. https://doi.org/10.1016/j.foreco.2020.118362

Paneque M, José M, Franco-Navarro JD, Colmenero-Flores JM, Knicker H (2016) Effect of biochar amendment on morphology, productivity and water relations of sunflower plants under non-irrigation conditions. Catena 147:280–287. https://doi.org/10.1016/j.catena.2016.07.037

Rajapaksha AU, Chen SS, Tsang DC, Zhang M, Vithanage M, Mandal S, Gao B, Bolan NS, Ok YS (2016) Engineered/designer biochar for contaminant removal/immobilization from soil and water: potential and implication of biochar modification. Chemosphere 148:276–291. https://doi.org/10.1016/j.chemosphere.2016.01.043

Ramzani PM, Shan L, Anjum S, Khan WU, Ronggui H, Iqbal M, Virk ZA, Kausar S (2017) Improved quinoa growth, physiological response, and seed nutritional quality in three soils having different stresses by the application of acidified biochar and compost. Plant Physiol Biochem 116:127–138. https://doi.org/10.1016/j.plaphy.2017.05.003

Rasa K, Heikkinen J, Hannula M, Arstila K, Kulju S, Hyväluoma J (2018) How and why does willow biochar increase a clay soil water retention capacity? Biomass Bioenergy 119:346–353. https://doi.org/10.1016/j.biombioe.2018.10.004

Ren X, Sun H, Wang F, Zhang P, Zhu H (2018) Effect of aging in field soil on biochar's properties and its sorption capacity. Environ Pollut 242:1880–1886. https://doi.org/10.1016/j.envpol.2018.07.078

Rizwan M, Ali S, Abbas T, Adrees M, Zia-ur-Rehman M, Ibrahim M, Abbas F, Qayyum MF, Nawaz R (2018) Residual effects of biochar on growth, photosynthesis and cadmium uptake in rice (*Oryza sativa* L.) under Cd stress with different water conditions. J Environ Manag 206:676–683. https://doi.org/10.1016/j.jenvman.2017.10.035

Saxena J, Rawat J, Kumar R (2017) Conversion of biomass waste into biochar and the effect on mung bean crop production. CLEAN–Soil Air Water 45(7):1501020. https://doi.org/10.1002/clen.201501020

Seidel SJ, Werisch S, Schütze N, Laber H (2017) Impact of irrigation on plant growth and development of white cabbage. Agric Water Manag 187:99–111. https://doi.org/10.1016/j.agwat.2017.03.011

Sharma M, Singh J, Baskar C, Kumar A (2018) A comprehensive review on biochar formation and its utilization for wastewater treatment. Pollut Res 37:S1–S8

Sharma P, Abrol V, Nazir J, Samnotra RK, Gupta SK, Anand S, Biswas JK, Shukla S, Kumar M (2025) Optimizing soil properties, water use efficiency, and crop yield through biochar and organic manure integration in organic soil. J Environ Manag 373:123673. https://doi.org/10.1016/j.jenvman.2024.123673

Siebielec S, Siebielec G, Klimkowicz-Pawlas A, Gałązka A, Grządziel J, Stuczyński T (2020) Impact of water stress on microbial community and activity in sandy and loamy soils. Agronomy 10(9):1429. https://doi.org/10.3390/agronomy10091429

Singh M, Saini RK, Singh S, Sharma SP (2019) Potential of integrating biochar and deficit irrigation strategies for sustaining vegetable production in water-limited regions: a review. HortScience 54(11):1872–1878. https://doi.org/10.21273/HORTSCI14271-19

Singh M, Singh S, Parkash V, Ritchie G, Wallace RW, Deb SK (2022) Biochar implications under limited irrigation for sweet corn production in a semi-arid environment. Front Plant Sci 13:853746. https://doi.org/10.3389/fpls.2022.853746

Sorrenti G, Masiello CA, Toselli M (2016) Biochar interferes with kiwifruit Fe-nutrition in calcareous soil. Geoderma 272:10–19. https://doi.org/10.1016/j.geoderma.2016.02.017

Suliman W, Harsh JB, Abu-Lail NI, Fortuna AM, Dallmeyer I, Garcia-Pérez M (2017) The role of biochar porosity and surface functionality in augmenting hydrologic properties of a sandy soil. Sci Total Environ 574:139–147. https://doi.org/10.1016/j.scitotenv.2016.09.025

Sun H, Xu Q, Wang Y, Zhao Z, Zhang X, Liu H, Gao J (2023) Agricultural drought dynamics in China during 1982–2020: a depiction with satellite remotely sensed soil moisture. GI Sci Remote Sens 60(1):2257469. https://doi.org/10.1080/15481603.2023.2257469

Sunaina B, Kumar JR, Rupak K, Mahesh R (2019) A case study on soil fertility status and maize productivity in Dang District, Nepal. Malays J Sustain Agric (MJSA) 3(2):56–59

Taskin E, Branà MT, Altomare C, Loffredo E (2019a) Biochar and hydrochar from waste biomass promote the growth and enzyme activity of soil-resident ligninolytic fungi. Heliyon 5(7):e02051

Taskin E, de Castro Bueno C, Allegretta I, Terzano R, Rosa AH, Loffredo E (2019b) Multianalytical characterization of biochar and hydrochar produced from waste biomasses for environmental and agricultural applications. Chemosphere 233:422–430. https://doi.org/10.1016/j.chemosphere.2019.05.204

Tisserant A, Cherubini F (2019) Potentials, limitations, co-benefits, and trade-offs of biochar applications to soils for climate change mitigation. Land 8(12):179. https://doi.org/10.3390/land8120179

Toková L, Igaz D, Horák J, Aydın E (2023) Can application of biochar improve the soil water characteristics of silty loam soil? J Soils Sediments 23(7):2832–2847. https://doi.org/10.1007/s11368-023-03505-y

Tomczyk A, Sokołowska Z, Boguta P (2020) Biochar physicochemical properties: pyrolysis temperature and feedstock kind effects. Rev Environ Sci Biotechnol 19(1):191–215. https://doi.org/10.1007/s11157-020-09523-3

Tripathi M, Sahu JN, Ganesan P (2016) Effect of process parameters on production of biochar from biomass waste through pyrolysis: a review. Renew Sust Energ Rev 55:467–481. https://doi.org/10.1016/j.rser.2015.10.122

Tsai CC, Chang YF (2020) Kinetics of C mineralization of biochars in three excessive compost-fertilized soils: effects of feedstocks and soil properties. Agronomy 10(11):1749. https://doi.org/10.3390/agronomy10111749

Usman AR, Al-Wabel MI, Abdulaziz AH, Mahmoud WA, EL-Naggar AH, Ahmad M, Abdulelah AF, Abdulrasoul AO (2016) Conocarpus biochar induces changes in soil nutrient availability and tomato growth under saline irrigation. Pedosphere 26(1):27–38. https://doi.org/10.1016/S1002-0160(15)60019-4

Usowicz B, Lipiec J, Łukowski M, Marczewski W, Usowicz J (2016) The effect of biochar application on thermal properties and albedo of loess soil under grassland and fallow. Soil Tillage Res 164:45–51. https://doi.org/10.1016/j.still.2016.03.009

Usowicz B, Lipiec J, Łukowski M, Bis Z, Usowicz J, Latawiec AE (2020) Impact of biochar addition on soil thermal properties: modelling approach. Geoderma 376:114574. https://doi.org/10.1016/j.geoderma.2020.114574

Verheijen FG, Zhuravel A, Silva FC, Amaro A, Ben-Hur M, Keizer JJ (2019) The influence of biochar particle size and concentration on bulk density and maximum water holding capacity of sandy vs sandy loam soil in a column experiment. Geoderma 347:194–202. https://doi.org/10.1016/j.geoderma.2019.03.044

Vijayaraghavan K (2019) Recent advancements in biochar preparation, feedstocks, modification, characterization and future applications. Environ Technol Rev 8(1):47–64. https://doi.org/10.1080/21622515.2019.1631393

Villagra-Mendoza K, Horn R (2018) Effect of biochar addition on hydraulic functions of two textural soils. Geoderma 326:88–95. https://doi.org/10.1016/j.geoderma.2018.03.021

Vitkova J, Kondrlova E, Rodny M, Surda P, Horak J (2017) Analysis of soil water content and crop yield after biochar application in field conditions. Plant Soil Environ 63(12):569–573

Wan H, Liu X, Shi Q, Chen Y, Jiang M, Zhang J, Cui B, Hou J, Wei Z, Hossain MA, Liu F (2023) Biochar amendment alters root morphology of maize plant: its implications in enhancing nutrient uptake and shoot growth under reduced irrigation regimes. Front Plant Sci 14:1122742. https://doi.org/10.3389/fpls.2023.1122742

Wang YZ, Zhang K (2019) Heavy metal pollution in soil and agricultural products on roadside of highway in seasonally frozen soil area. Rev Fac Agron 36(3):754–763

Wang J, Xiong Z, Kuzyakov Y (2016) Biochar stability in soil: meta-analysis of decomposition and priming effects. GCB Bioenergy 8(3):512–523. https://doi.org/10.1111/gcbb.12266

Wang Y, Xia H, Yao Z, Jiang C (2018) Effects of biochar and other amendments on the cabbage growth and soil fertility in yellow-brown soil and red soil. Sci Agric Sinica 51:4306–4315

Wang H, Ren T, Feng Y, Liu K, Feng H, Liu G, Shi H (2020) RETRACTED: Wang et al. Effects of the application of biochar in four typical agricultural soils in China. Agronomy 10:351. https://doi.org/10.3390/agronomy10111649

Wang J, Shi D, Li Y, Chen A, Feng S, Liu C (2023) A study on infiltration characteristics and one-dimensional algebraic model simulation in reclaimed soil with biochar. Water 15(16):2985. https://doi.org/10.3390/w15162985

Wei B, Peng Y, Lin L, Zhang D, Ma L, Jiang L, Li Y, He T, Wang Z (2023) Drivers of biochar-mediated improvement of soil water retention capacity based on soil texture: a meta-analysis. Geoderma 437:116591. https://doi.org/10.1016/j.geoderma.2023.116591

Wu Y, Wang X, Zhang L, Zheng Y, Liu X, Zhang Y (2023) The critical role of biochar to mitigate the adverse impacts of drought and salinity stress in plants. Front Plant Sci 14:1163451. https://doi.org/10.3389/fpls.2023.1163451

Xiao Q, Zhu LX, Zhang HP, Li XY, Shen YF, Li SQ (2016) Soil amendment with biochar increases maize yields in a semi-arid region by improving soil quality and root growth. Crop Pasture Sci 67(5):495–507

Xiong J, Yu R, Islam E, Zhu F, Zha J, Sohail MI (2020) Effect of biochar on soil temperature under high soil surface temperature in coal mined arid and semiarid regions. Sustainability 12(19):8238. https://doi.org/10.3390/su12198238

Yang W, Shang J, Sharma P, Li B, Liu K, Flury M (2019) Colloidal stability and aggregation kinetics of biochar colloids: effects of pyrolysis temperature, cation type, and humic acid concentrations. Sci Total Environ 658:1306–1315. https://doi.org/10.1016/j.scitotenv.2018.12.269

Yang X, Fan H, Du Y, Zhang L (2020) Effect of biochar application on physicochemical and hydraulic properties of loess. Adv Civil Eng 2020(1):4592092. https://doi.org/10.1155/2020/4592092

Ye L, Camps-Arbestain M, Shen Q, Lehmann J, Singh B, Sabir M (2020) Biochar effects on crop yields with and without fertilizer: a meta-analysis of field studies using separate controls. Soil Use Manag 36(1):2–18. https://doi.org/10.1111/sum.12546

Yu P, Li T, Fu Q, Liu D, Hou R, Zhao H (2021) Effect of biochar on soil and water loss on sloping farmland in the black soil region of Northeast China during the spring thawing period. Sustainability 13(3):1460. https://doi.org/10.3390/su13031460

Zhang M, Riaz M, Zhang L, El-Desouki Z, Jiang C (2019) Biochar induces changes to basic soil properties and bacterial communities of different soils to varying degrees at 25 mm rainfall: more effective on acidic soils. Front Microbiol 10:1321. https://doi.org/10.3389/fmicb.2019.01321

Zhang R, Qu Z, Wang L, Yang W, Zhang D, Zhao Y, Li J (2025) Enhancing soil microbial diversity, quality, and crop growth through deficit irrigation combined with biochar application. Front Sustain Food Syst 9:1528044. https://doi.org/10.3389/fsufs.2025.1528044

Zhao J, Ren T, Zhang Q, Du Z, Wang Y (2016) Effects of biochar amendment on soil thermal properties in the North China Plain. Soil Sci Soc Am J 80(5):1157–1166. https://doi.org/10.2136/sssaj2016.01.0020

Zhao ZY, Wang PY, Xiong XB, Wang YB, Zhou R, Tao HY, Grace UA, Wang N, Xiong YC (2022) Environmental risk of multi-year polythene film mulching and its green solution in arid irrigation region. J Hazard Mater 435:128981. https://doi.org/10.1016/j.jhazmat.2022.128981

Zhou Y, Wigley BJ, Case MF, Coetsee C, Staver AC (2020) Rooting depth as a key woody functional trait in savannas. New Phytol 227(5):1350–1361. https://doi.org/10.1111/nph.16613

Zhu P, Zhu J, Pang J, Xu W, Shu L, Hu H, Wu Y, Tang C (2020) Biochar improves the growth performance of maize seedling in response to antimony stress. Water Air Soil Pollut 231(4):154. https://doi.org/10.1007/s11270-020-04521-1

Zulfiqar B, Raza MA, Saleem MF, Aslam MU, Iqbal R, Muhammad F, Amin J, Ibrahim MA, Khan IH (2022) Biochar enhances wheat crop productivity by mitigating the effects of drought: insights into physiological and antioxidant defense mechanisms. PLoS One 17(4):e0267819. https://doi.org/10.1371/journal.pone.0267819

Biochar for Climate Change Mitigation and Sustainable Agriculture

Rumisa Ashraf, Sumyrah Mukhtar, Javid Ahmad Bhat, Najmu Saqib, Mehraj-u-Din Lone, Tajamul Mansoor, Owais Ali Wani, Tsering Dolker, Sabreena Ashraf, and Abid Shabir Bali

Abstract

High rates of population growth and land degradation make it difficult to sustain agricultural productivity, of which soil organic matter preservation is an important consideration. A new technology identified for agronomic use is biochar, which is a product of the pyrolysis of organic waste and is rich in carbon. The use of the biochar enhances soil texture, microbe habitat and diversity and promotes the steady release of nutrients. The effect of this innovation promotes sustainable plant cultivation through the intensification of seed germination and development and it also improves the yield and nutrient contents of plants. In addition to this, it reduces global warming by providing carbon sinks. On the economic front, while the viability depends on the region, there are successful models in the voluntary carbon markets and national schemes such as the Emissions Reduction Fund in Australia that encourage the use of the system in agricultural management. To encourage the adoption of the system on a broader scale, it is

R. Ashraf
Division of Entomology, FoH, Shalimar, SKUAST-Kashmir, Srinagar, Jammu and Kashmir, India

S. Mukhtar (✉) · J. A. Bhat · N. Saqib · Mehraj-u-Din Lone · T. Mansoor · T. Dolker
Division of Soil Science and Agricultural Chemistry, FoA, Wadura, SKUAST-Kashmir, Srinagar, Jammu and Kashmir, India

O. A. Wani
Division of Agronomy, Faculty of Agriculture, SKUAST-Kashmir, Wadura, Sopore, Jammu and Kashmir, India

S. Ashraf
Punjab Agricultural University, Ludhiana, Punjab, India

A. S. Bali
Division of Agronomy, Sher-e-Kashmir University of Agricultural Sciences and Technology of Kashmir, Srinagar, Jammu and Kashmir, India

O. A. Wani et al. (eds.), *Biochar for Sustainable Soil Health and Crop Production*, https://doi.org/10.1007/978-981-92-2230-8_7

essential that a universal sustainable approach and sustainability guidelines are developed. The current assessment of biochar explores recent developments in crop productivity enhancement and commercialization. In analyzing the socio-economic aspects of biochar use at local and global perspectives, biochar can be situated at the critical point of sustainability and climate change. Nevertheless, biochar application in agricultural practices can be considered an important stride in balancing agricultural and eco-conservational requirements.

Keywords

Biochar · Sustainable agriculture · Soil health · Crop productivity · Carbon sequestration · Climate change mitigation

1 Introduction

1.1 Climate Change, Agriculture, and Soil Systems: A Global Context of Urgency

Climate change, with its far-reaching consequences on the physical, ecological, and socio-economic environment of our planet, has turned out to be one of the most important and pressing concerns of the twenty-first century. The unprecedented level of global warming and climate variability is a consequence of the rising atmospheric concentration of green house gases (GHGs), mainly carbon dioxide (CO_2), methane (CH_2), and nitrous oxide (N_2O), which have altered the energy balance of our planet. The rise in GHGs emission, mainly by human-induced activities, has contributed significantly to glacier melting, ocean acidification, and biodiversity extinction, besides increasing the occurrences and severity of extreme weather phenomena such as heatwaves, floods, droughts, and tropical cyclones (Lehmann and Joseph 2015). The Paris Climate Accord, signed in 2015, emphasizes the seriousness of limiting global warming to well below 2 °C above the level of pre-industrial era and pursues an intention of 1.5 °C above the increasing level referred to above. This calls for a drastic reduction in global GHGs emission and also emphasizes positive CO_2 absorption through technological and terrestrial sinks (IPCC 2022).

The terrestrial system, and particularly the massive capacity of soils for organic carbon sequestration, represents one of the highly promising biotic partners in combatting climate change. Nevertheless, this can only be achieved through an overall paradigm shift in agricultural use and management of soils. Agriculture is an important sector playing a critical role in maintaining life on earth by sustaining humanity; however, at the same time, it is a major source of GHGs and has strong GHG mitigation potentials. The agriculture sector accounts for a big chunk of overall GHG emissions, specifically through CO_2 emissions because of forest clearing and biomass combustion in agriculture itself, N_2O emissions from fertilized soils, and CH_4 emissions associated with agriculture's domestic and export-related rice and cattle industries (Smith et al. 2021). Also, unsustainable agricultural practices like intensive agriculture involving intensive use of heavy agriculture equipment and practices like burning of agricultural residues, overuse of chemical fertilizers, and clearing of natural forests for agricultural expansion contribute substantially to already existing GHG emissions. These unsustainable agriculture practices lead to reduction in soil structure, enhanced rates of decomposition of soil organic materials, and subsequent carbon and nutrient losses, thus affecting soils' quality and health (Lal 2015).

This leads to degraded soils being net sources of carbon instead of being sinks and thereby contributing to the climate issue. However, the nexus between soils, agriculture, and climate change also holds a great challenge in the form of a great opportunity. This is based on the fact that agricultural practices occupy more than a third of the Earth's surface; thereby, a simple change in soils can make a great difference in the climate state of the Earth. Various methods of agricultural practices like conservation tillage, integral nutrient management, organic inputs, crop diversification, cover crops, and agroforestry practices have shown great potential to enhance the organic carbon content of soils and reduce the GHG emissions emanating from agricultural activities by improving soil-plant interactions and thereby improving soil microbial processes and carbon sequestration.

Soil, also referred to as the "living skin of the Earth," is a complex interface connecting the atmosphere, biosphere, and lithosphere and plays a vital role in carbon cycling in the Earth's system. The addition of healthy soils and their high organic carbon stocks not only enhances agricultural productivity but also affects water infiltration, nutrient cycling, and soil biota. The role of increasing carbon stocks in soil through sustainable land management therefore has dual benefits: it not only helps to reduce carbon content in the atmosphere but also improves soil and food security. Recent breakthroughs in soil science research using a variety of molecular and microbial methods and techniques have improved understanding of the role of soil biodiversity for carbon sequestration and nutrient cycling processes and patterns in diversified land use patterns and change processes. In this regard, it is also important that the carbon-prone agricultural sector converts to a carbon-aware and climate-resililient agricultural system. In this regard, it is therefore important that research and development activities and policy interventions should be aimed at reaping synergies in improving climate mitigation and agricultural productivity patterns and processes and linkages of improved soil management and monitoring through the use of modern and cutting-edge technology such as artificial intelligence and remote sensing and precision agriculture techniques and methods to monitor and address soil and crop conditions. Strengthening synergies between climate change mitigation and agricultural productivity is essential for achieving the Sustainable Development Goals (SDGs), especially SDG 2 (Zero Hunger), SDG 13 (Climate Action), and SDG 15 (Life on Land), through the promotion of sustainable practices, improved resource management, enhanced ecosystem resilience, and informed decision-making based on scientific knowledge and evidence.

In conclusion, agriculture and soil constitute a key focus of the climate agenda, which is both part of the problem and also part of the solution. There is potential in tapping into agriculture and soil using innovative solutions, which will form a basis of a positive future.

2 Biochar as a Sustainable Solution

Biochar, a carbon-rich, porous material produced through the pyrolysis of organic biomass under limited oxygen, offers a sustainable and multifaceted solution to climate change and agricultural challenges. Pyrolysis converts the labile carbon in biomass into a stable, aromatic structure resistant to microbial decomposition, allowing it to persist in soils for centuries to millennia (Kuzyakov et al. 2009). This transformation enables biochar to act as a long-term carbon sink, effectively removing atmospheric CO_2 and reducing net greenhouse gas emissions.

Beyond carbon sequestration, biochar significantly enhances soil physicochemical and biological properties. Its high surface area and cation exchange capacity improve soil aeration, water-holding capacity, and nutrient retention, while also creating favorable habitats for beneficial soil microbes. These improvements lead to better root development, increased fertilizer use efficiency, and higher crop yields. Furthermore, biochar application can substantially reduce emissions of non-CO_2 GHGs like methane (CH_2) and nitrous oxide (N_2O) by altering soil redox conditions and influencing microbial processes involved in nitrification and denitrification (Jeffery et al. 2011).

Biochar also contributes to environmental restoration and soil resilience. It is particularly beneficial in degraded, sandy, or acidic soils, where it can increase aggregate stability, reduce erosion, and buffer soil acidity. Produced from forestry residues, agricultural waste, or other organic by-products, biochar promotes a circular bioeconomy by transforming waste into a valuable soil amendment. This dual role in waste management and soil enhancement underscores biochar's significance as a climate-smart technology with wide-ranging environmental and agricultural benefits. In essence, biochar represents an innovative approach that integrates soil health improvement, carbon sequestration, and sustainable agriculture, making it a crucial component of future strategies aimed at achieving carbon neutrality and long-term ecosystem resilience.

2.1 Biochar and Carbon Sequestration: Labile and Non-labile Pools

A key mechanism through which biochar helps to counter climate change is by its actions on soil carbon. Soil organic carbon can take many different forms with varying turnover rates. These can be grouped into labile and non-labile compartments. The labile pool consists of organic matter with a particle size smaller than sand, as well as sugar and protein molecules. These materials can easily be degraded by microorganisms. On the other hand, the non-labile pool usually consists of humic matter and black carbon. These materials have a very low rate of turnover and can remain in soil for thousands of years. Biochar directly adds to the non-labile carbon pool. This is because the pyrolysis mechanism used in making biochar results in the formation of aromatic rings that are highly unstable. Additionally, biochar can indirectly increase the stability of the indigenous soil organic carbon. This is due to the fact that biochar has a porous structure that can bind and protect the labile soil organic matter from microbial degradation. This not only adds to the non-labile carbon pool but also makes the soil carbon more stable. Consequently, biochar is a dual-purpose tool that can be used for directly adding carbon to the non-labile carbon pool while helping in making the existing soil carbon more stable. In this matter, the role played by biochar is crucial. It can change the soil within agricultural land from a source of carbon emissions into a stable soil carbon sink.

3 Objectives and Scope of the Chapter

This chapter aims to provide a comprehensive overview of biochar's role in sustainable agriculture and climate change mitigation. It examines the properties and production methods of biochar, elucidates its mechanisms for carbon sequestration and soil health improvement, and presents global case studies demonstrating its practical applications. The chapter also discusses current challenges and limitations in biochar adoption, explores its integration into climate policies and carbon trading frameworks, and outlines future prospects for its deployment within the circular bioeconomy. By bridging scientific knowledge with practical implementation, this chapter highlights how biochar can contribute to climate resilience and sustainable land management.

4 Biochar: Properties and Production

The effectiveness of biochar in agricultural and environmental applications is intrinsically linked to its production processes and resulting physicochemical properties (International Biochar Initiative 2023). Key factors such as feedstock selection, pyrolysis conditions, and post-production treatment determine its functionality for specific uses. Understanding these variables is essential for optimizing biochar performance across diverse environments.

4.1 Feedstock Options

The choice of feedstock is a primary determinant of biochar characteristics and functional qualities. Commonly used feedstocks include various organic wastes and residues.

- **Agricultural Residues**: Biochar derived from materials such as rice husks, straw, cassava peels, and coconut shells can enhance soil structure, porosity, and water-holding capacity (Liu et al. 2019).
- **Forestry Waste**: Feedstocks like wood chips, sawdust, and other lignocellulosic materials typically yield biochar with high carbon content (60–85%) and

enhanced structural stability, making it ideal for long-term carbon sequestration (Singh et al. 2010).

- **Manure**: Biochar produced from animal dung is rich in nutrients such as nitrogen and phosphorus, offering immediate fertility benefits, although it generally exhibits lower carbon stability compared to woody biochars (Głąb et al. 2016).

4.2 Pyrolysis Conditions and Their Influence

Pyrolysis, the thermal decomposition of biomass at 300–700 °C under limited oxygen, is the primary method for biochar production (Bridgwater 2012). Specific pyrolysis conditions significantly influence the final product's properties (Tomczyk et al. 2020):

- **Temperature**: Higher pyrolysis temperatures (>500 °C) typically yield biochar with greater surface area, porosity, and carbon stability, but lower nutrient content. Lower temperatures (300–400 °C) retain more volatile nutrients and functional groups, enhancing short-term soil fertility but reducing long-term carbon stability.
- **Residence Time and Heating Rate**: These parameters affect pore size distribution, surface chemistry, and structural integrity, thereby influencing biochar's interaction with soil components and its nutrient retention capacity (Manyà 2012).

4.3 Key Physical and Chemical Characteristics

- **Physical Properties**: High porosity and large surface area improve soil water retention and provide habitat for beneficial soil microorganisms (Atkinson et al. 2010).
- **Chemical Properties**: Biochar's cation exchange capacity (CEC) tends to increase as it weathers in soil, enhancing its ability to retain nutrients like ammonium (NH_4^+) and potassium (K^+) (Laird et al. 2010). Fresh biochar often has an alkaline pH, which can help ameliorate acidic soils (Fig. 1).

Pyrolysis converts biomass into three main products: biochar, bio-oil, and syngas. While biochar serves as a stable carbon sink for long-term sequestration, bio-oil and syngas can be utilized as renewable energy sources, displacing fossil fuels and contributing to a carbon-negative cycle (Mohan et al. 2006).

5 Mechanisms of Climate Change Mitigation

Biochar contributes to climate change mitigation through direct carbon sequestration, reduction of non-CO_2 GHG emissions, and indirect pathways such as fossil fuel substitution and enhanced soil fertility (Smith 2016).

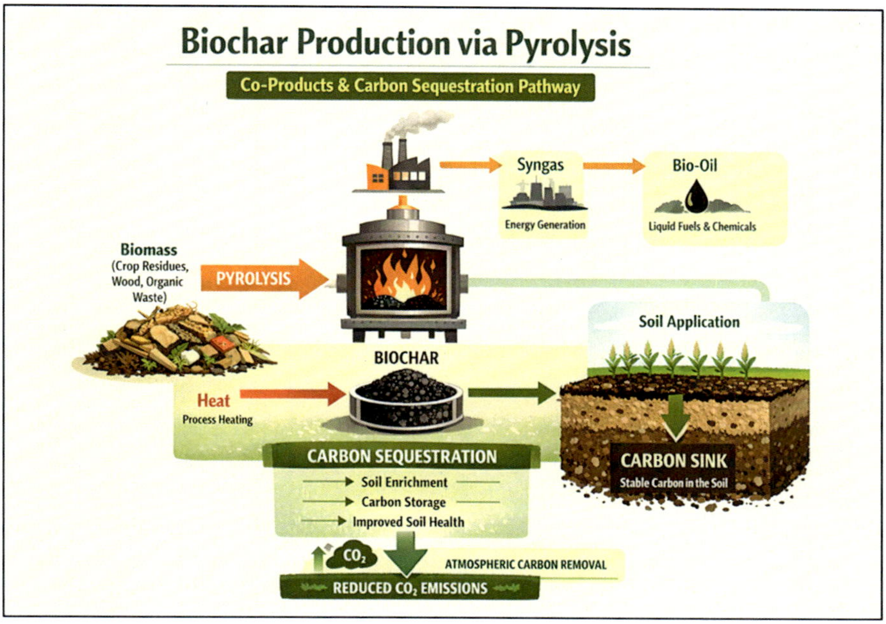

Fig. 1 Schematic diagram of biochar production through pyrolysis, highlighting co-products and carbon sequestration pathway

5.1 Carbon Sequestration

Approximately 50% of the carbon in biomass is stabilized into an aromatic, recalcitrant structure during pyrolysis (Kuzyakov et al. 2009). This carbon, which would otherwise be released through decomposition or burning, can be stored in soils for centuries to millennia, making the process inherently carbon-negative.

5.2 Reduction of GHG Emissions

- **Nitrous Oxide (N_2O)**: Biochar can reduce N_2O emissions by up to 50% through mechanisms such as improved soil aeration, adsorption of nitrogen substrates, and promotion of complete denitrification to N_2 gas (Cayuela et al. 2014).
- **Methane (CH_4)**: Biochar is particularly effective in reducing CH_4 emissions from flooded soils like rice paddies, with reduction rates ranging from 10% to 50%. This is achieved by altering redox conditions and providing a habitat for methane-oxidizing bacteria (Brown et al. 2011).
- **Energy Substitution**: The syngas and bio-oil co-produced during pyrolysis can be captured for heat and power generation, offsetting fossil fuel use and improving the system's overall energy balance and emission profile (Roberts et al. 2010).

5.3 Indirect Mitigation via Soil Fertility Enhancement

By improving soil health, water retention, and nutrient use efficiency, biochar can increase plant biomass production (Biederman and Harpole 2013). This enhanced growth leads to greater photosynthetic carbon uptake, creating a positive feedback loop that further augments soil carbon inputs (Major et al. 2010) (Table 1).

5.4 Biochar and GHG Mitigation Mechanisms

The mitigation of non-CO_2 GHGs by biochar involves complex biogeochemical interactions. For N_2O, biochar affects the microbial processes of nitrification (aerobic conversion of NH_4^+ to NO_3^-) and denitrification (anaerobic reduction of NO_3^- to N_2O or N_2). Its porous structure can improve soil oxygen diffusion, suppressing denitrification. Additionally, biochar's adsorption capacity can retain NH_4^+ and NO_3^-, making them less available for N-cycling microbes that produce N_2O. Some biochars also harbor specific microbial communities that favor the complete reduction of N_2O to benign N_2 gas.

For CH_4, biochar application in anaerobic environments like wetlands or rice fields can shift the balance between methanogenesis (CH_4 production) and methanotrophy (CH_4 consumption). The increased surface area and altered pH can enhance the activity of methane-oxidizing bacteria. Furthermore, biochar can influence the electron flow in soil, potentially diverting electrons away from methanogenic pathways. These multifaceted mechanisms underscore biochar's potential as a tool for integrated GHG management in agricultural systems.

Table 1 Mechanisms of biochar's role in climate change mitigation

Mechanism	Observed Impact/Potential	Key Processes Involved
Fossil Fuel Substitution	Significant net negative CO_2 emissions when co-products are utilized (Roberts et al. 2010).	Syngas and bio-oil from pyrolysis are used for energy generation.
N_2O Emission Reduction	Reduction potentials of up to 50% (Cayuela et al. 2014).	Alters soil nitrification/denitrification; improves aeration; adsorbs NH_4^+/NO_3^-.
Direct Carbon Sequestration	Mean residence time of 100s to 1000s of years. Wood-based biochar can sequester 60–85% of feedstock carbon (Singh et al. 2010).	Pyrolysis converts labile biomass carbon into stable aromatic carbon structures.
Indirect Sequestration	Crop yield increases up to 30% in degraded soils, enhancing soil organic carbon (Biederman and Harpole 2013).	Improved plant growth leads to higher biomass production and subsequent carbon input to soil.
CH_4 Emission Reduction	Reduction rates vary widely (10–50%) in paddy soils (Brown et al. 2011).	Provides habitat for methanotrophic bacteria; alters redox conditions in saturated soils.

6 Modulation of Soil Biota for Climate Resilience

Soil microorganisms are fundamental drivers of nutrient cycling, organic matter decomposition, and greenhouse gas fluxes. Biochar application can significantly modulate soil microbial community structure and function, thereby influencing climate feedbacks.

The porous architecture of biochar provides refuge for microbes, protecting them from predators and desiccation. This can increase overall microbial biomass and diversity. Specific changes include:

- **Promotion of Beneficial Microbes**: Biochar can enhance populations of arbuscular mycorrhizal fungi, which improve plant nutrient and water uptake, and plant growth-promoting rhizobacteria.
- **Influence on GHG-related Microbes**: As discussed, biochar can suppress nitrifying and denitrifying bacteria responsible for N_2O production while favoring methanotrophs that consume CH_4.
- **Stimulation of Carbon-Sequestering Microbes**: By providing a stable carbon substrate, biochar can promote microbial communities that produce persistent, sticky compounds (e.g., glomalin) that help bind soil particles into stable aggregates, physically protecting organic matter from decomposition and enhancing carbon storage.

Thus, biochar acts as a microbial habitat engineer, steering the soil biome towards functions that support both agricultural productivity and climate mitigation a process crucial for building long-term soil health and ecosystem resilience under changing climatic conditions.

7 Global Research and Applications

Biochar is being researched and applied across diverse agro-ecological zones, highlighting its global relevance (Liu et al. 2013).

7.1 Examples from Various Continents

- **Asia**: In China and India, rice husk biochar is widely used to improve the fertility and water retention of paddy soils, increasing yields while reducing methane emissions.
- **Africa**: Projects in Sub-Saharan Africa utilize biochar made from cassava peels and other local waste to restore degraded, nutrient-poor soils, thereby enhancing food security (Adekiya et al. 2019).
- **North America and Europe**: Research focuses on integrating biochar into organic farming and precision agriculture systems, as well as its role in carbon removal markets (Schmidt and Wilson 2014).

7.2 Integration with Carbon Markets and Policies

Biochar is increasingly recognized as a durable carbon removal solution in voluntary carbon markets (Gold Standard 2022). Verified biochar projects generate carbon credits, providing a financial incentive for adoption by farmers, entrepreneurs, and industries (Mathews 2008). Supportive policy frameworks including carbon pricing, tax incentives, and subsidies for sustainable land management are vital for scaling biochar implementation. Incorporating biochar into national carbon policies and international mitigation pledges, such as Nationally Determined Contributions (NDCs) under the Paris Agreement, could significantly accelerate progress toward net-zero emission targets (Galinato et al. 2011).

7.3 Synergies with Sustainable Agriculture and Circular Economy

Biochar exemplifies circular bioeconomy principles by converting low-value biomass (agricultural residues, forestry waste, manure) into a high-value soil amendment. This transformation minimizes waste disposal issues, recycles nutrients within the agricultural system, and reduces reliance on synthetic fertilizers (O'Connor et al. 2021). Integrating biochar into farming systems promotes a closed-loop agricultural paradigm that enhances soil fertility, microbial health, and resource efficiency, supporting long-term environmental and economic sustainability (Brassard et al. 2016) (Table 2).

8 Challenges and Limitations

Despite its significant potential, several challenges hinder the widespread adoption of biochar as a climate-smart solution (Petter et al. 2012).

- **Feedstock Availability and Competition**: Sustainable biochar production requires biomass sources that do not compete with food production or induce land-use change (Berndes et al. 2011). While agricultural and forestry residues are ideal, their collection, transport, and processing often entail high logistical and economic costs, limiting scalability (Shackley et al. 2011).
- **Variability in Biochar Properties and Soil Interactions**: Biochar performance is highly context-specific, depending on complex interactions among biochar type, soil properties, climate, and crop species (Glaser and Lehr 2019). For instance, a biochar that benefits sandy, acidic soils may have negligible or even negative effects in fertile, neutral-pH soils (Ippolito et al. 2012). This necessitates localized testing and tailored application strategies.
- **Socioeconomic Barriers**: The initial investment for pyrolysis technology, along with ongoing production and transportation costs, remains prohibitive for many smallholder farmers (McCarl et al. 2009). Yield increases alone often do not jus-

Table 2 Summary of field trials and case studies on biochar for mitigation

Location (Context)	Application Rate (t/ha)	Feedstock Type	Key Outcomes
USA (Corn Belt)	1–10	Corn Stover	N_2O emissions reduced by 20–50%; variable impact on yield in fertile soils (Spokas et al. 2012).
China (Paddy Fields)	10–20	Rice Husk	~15% yield increase; ~30% reduction in CH_4 emissions; improved N use efficiency.
Brazil (Acidic Soil)	5–15	Wood Chip	Soil pH increased; Al toxicity reduced; ~25% increase in maize yield (O'Connor et al. 2021).
Ghana (Degraded Soil)	5–10	Cassava Peel	Up to 30% yield increase for legumes; significant improvement in water retention (Adekiya et al. 2019).

tify the economic investment unless coupled with financial incentives such as carbon credit revenue (Hammond et al. 2011).
- **Monitoring, Reporting, and Verification (MRV) Challenges**: Integrating biochar into carbon trading systems requires standardized and transparent MRV protocols to ensure the credibility of carbon sequestration claims (IPCC 2019). Developing consistent MRV methods across diverse production systems, feedstocks, and environmental conditions remains a significant technical and institutional hurdle.

9 Future Prospects and Pathways for Biochar Deployment

The future of biochar hinges on continued technological innovation, strong policy support, and its strategic integration into climate action frameworks at national and international levels.

- **Contribution to Net-Zero Goals**: Biochar is recognized as one of the most readily deployable Negative Emission Technologies (NETs). Its scalability, coupled with agronomic and environmental co-benefits, makes it a vital component of global efforts to achieve net-zero GHG emissions (Fuss et al. 2018).
- **Integration into Climate Policy**: Key pathways include incorporating biochar projects into Nationally Determined Contributions (NDCs) under the Paris Agreement, developing subsidies or tax rebates for production and application, and creating transparent carbon markets that recognize biochar as a durable carbon sink.

Technological Innovations:
- **Engineered Biochars**: Advanced production and modification techniques allow for the design of biochars with specific properties for targeted uses, such as

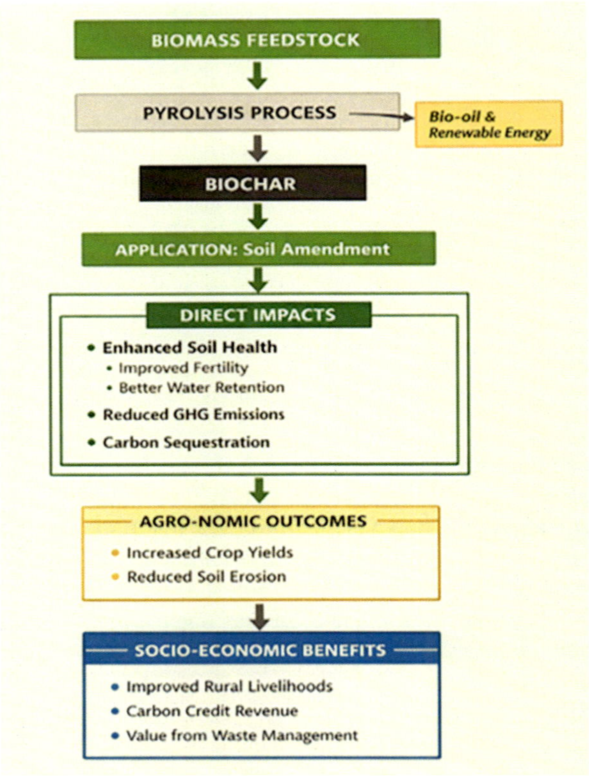

Fig. 2 Schematic representation of biochar production from biomass feedstock via pyrolysis and its environmental, agronomic, and socio-economic benefits (Chia et al. 2020)

enhanced nutrient retention, heavy metal immobilization, or pollutant remediation (Ghorbani et al. 2024).

- **Biochar-Based Fertilizers**: Incorporating nutrients into biochar matrices creates slow-release fertilizers that improve nutrient use efficiency and reduce environmental losses (Joseph et al. 2021).
- **Integration with BECCS**: Coupling biochar production with Bioenergy with Carbon Capture and Storage (BECCS) could establish advanced biorefinery systems that deliver renewable energy while ensuring permanent carbon sequestration (Fajardy and Mac Dowell 2017) (Fig. 2).

10 Conclusion

The growing body of scientific evidence positions biochar as a transformative, multi-beneficial solution to the intertwined challenges of agricultural sustainability and climate change. Its proven capacity to sequester carbon over centennial timescales and mitigate non-CO_2 GHG emissions marks it as a key technology in the transition to net-zero emissions. Beyond climate benefits, biochar improves soil structure, nutrient retention, and biological activity, restoring degraded lands and bolstering global food security. While challenges related to MRV standardization, context-dependent performance, and economic viability persist, they can be overcome through targeted innovation and concerted policy action. Ongoing research to optimize production parameters, tailor applications to local conditions, and improve verification protocols will further unlock biochar's potential. More than just a mitigation tool, biochar acts as a catalyst for systemic environmental regeneration, aligning seamlessly with circular bioeconomy principles and sustainable farming practices. Realizing its full potential will require robust carbon incentive mechanisms, sustained investment in research and development, and cross-sectoral collaboration among scientists, policymakers, and industry stakeholders. The strategic integration of biochar into global climate frameworks can significantly advance the goal of a carbon-neutral, resilient, and sustainable future for both humanity and the planet.

References

Adekiya AO, Agbede TM, Aboyeji CM, Dunsin O, Simeon VT (2019) Effects of biochar and poultry manure on soil properties, growth, yield and quality of cocoyam. Sci Rep 9(1):1–9

Atkinson CJ, Fitzgerald JD, Hipps NA (2010) Potential mechanisms for achieving agricultural benefits from biochar application to temperate soils: a review. Plant Soil 337(1):1–18

Berndes G, Bird N, Cowie A (2011) Bioenergy, land use change and climate change mitigation. IEA Bioenergy

Biederman LA, Harpole WS (2013) Biochar and its effects on plant productivity and nutrient cycling: a meta-analysis. GCB Bioenergy 5(2):202–214

Brassard P, Godbout S, Raghavan V (2016) Soil biochar amendment as a climate change mitigation tool: key parameters and mechanisms involved. J Environ Manag 181:484–497

Bridgwater AV (2012) Review of fast pyrolysis of biomass and product upgrading. Biomass Bioenergy 38:68–94

Brown TR, Wright MM, Brown RC (2011) Estimating profitability of two biochar production scenarios: slow pyrolysis vs fast pyrolysis. Biofuels Bioprod Biorefin 5(1):54–68

Cayuela ML, van Zwieten L, Singh BP, Jeffery S, Roig A, Sánchez-Monedero MA (2014) Biochar's role in mitigating soil nitrous oxide emissions: a review and meta-analysis. Agric Ecosyst Environ 191:5–16

Chia WY, Ying Tang DY, Khoo KS, Kay Lup AN, Chew KW (2020) Nature's fight against plastic pollution: algae for plastic biodegradation and bioplastics production. Environ Sci Ecotechnol 4:100065

Fajardy M, Mac Dowell N (2017) Can BECCS deliver sustainable and resource efficient negative emissions? Energy Environ Sci 10(6):1389–1426

Fuss S, Lamb WF, Callaghan MW, Hilaire J, Creutzig F, Amann T, Minx JC (2018) Negative emissions—part 2: costs, potentials and side effects. Environ Res Lett 13(6):063002

Galinato SP, Yoder JK, Granatstein D (2011) The economic value of biochar in crop production and carbon sequestration. Energy Policy 39(10):6344–6350

Ghorbani M, Amirahmadi E, Neugschwandtner RW (2024) How do different feedstocks and pyrolysis conditions effectively change biochar modification scenarios? A critical analysis of engineered biochars under H2O2 oxidation. Energy Convers Manag 301:117995

Głąb T, Palmowska J, Zaleski T, Gondek K (2016) Effect of biochar application on soil hydrological properties and physical quality of sandy soil. Geoderma 281:11–20

Glaser B, Lehr VI (2019) Biochar effects on phosphorus availability in agricultural soils: a meta-analysis. Sci Rep 9(1):1–9

Gold Standard (2022) Biochar methodology: for biochar production and application to soil. Gold Standard Foundation

Hammond J, Shackley S, Sohi S, Brownsort P (2011) Prospective life cycle carbon abatement for pyrolysis biochar systems in the UK. Energy Policy 39(5):2646–2655

International Biochar Initiative (2023) Standardized product definition and product testing guidelines for biochar that is used in soil. IBI

IPCC (2019) 2019 Refinement to the 2006 IPCC guidelines for national greenhouse gas inventories

IPCC (2022) Climate change 2022: mitigation of climate change. Contribution of Working Group III to the sixth assessment report of the Intergovernmental Panel on Climate Change

Ippolito JA, Laird DA, Busscher WJ (2012) Environmental benefits of biochar. J Environ Qual 41(4):967–972

Jeffery S, Verheijen FG, van der Velde M, Bastos AC (2011) A quantitative review of the effects of biochar application to soils on crop productivity using meta-analysis. Agric Ecosyst Environ 144(1):175–187

Joseph S, Cowie AL, van Zwieten L, Bolan N, Budai A, Buss W, Lehmann J (2021) How biochar works, and when it doesn't: a review of mechanisms controlling soil and plant responses to biochar. GCB Bioenergy 13(11):1731–1764

Kuzyakov Y, Subbotina I, Chen H, Bogomolova I, Xu X (2009) Black carbon decomposition and incorporation into soil microbial biomass estimated by 14C labeling. Soil Biol Biochem 41(2):210–219

Laird DA, Fleming P, Davis DD, Horton R, Wang B, Karlen DL (2010) Impact of biochar amendments on the quality of a typical Midwestern agricultural soil. Geoderma 158(3–4):443–449

Lal R (2015) Restoring soil quality to mitigate soil degradation. Sustainability 7(5):5875–5895

Lehmann J, Joseph S (2015) Biochar for environmental management: science, technology and implementation. Routledge

Liu X, Zhang A, Ji C, Joseph S, Bian R, Li L, Paz-Ferreiro J (2013) Biochar's effect on crop productivity and the dependence on experimental conditions—a meta-analysis of literature data. Plant Soil 373(1):583–594

Liu Y, Lu H, Yang S, Wang Y (2019) Impacts of biochar addition on rice yield and soil properties in a cold waterlogged paddy for two crop seasons. Field Crop Res 236:18–26

Major J, Lehmann J, Rondon M, Goodale C (2010) Fate of soil-applied black carbon: downward migration, leaching and soil respiration. Glob Change Biol 16(4):1366–1379

Manyà JJ (2012) Pyrolysis for biochar purposes: a review to establish current knowledge gaps and research needs. Environ Sci Technol 46(15):7939–7954

Mathews JA (2008) Carbon-negative biofuels. Energy Policy 36(3):940–945

McCarl BA, Peacocke C, Chrisman R, Kung CC, Sands RD (2009) Economics of biochar production, utilization and greenhouse gas offsets. In: Biochar for environmental management. Routledge

Mohan D, Pittman CU, Steele PH (2006) Pyrolysis of wood/biomass for bio-oil: a critical review. Energy Fuel 20(3):848–889

O'Connor J, Hoang SA, Bradney L, Dutta S, Xiong X, Tsang DC, Bolan NS (2021) A review on the valorisation of food waste as a nutrient source and soil amendment. Environ Pollut 272:115985

Petter FA, Madari BE, Silva MAS, Carneiro MAC, Carvalho MTM, Marimon BH, Pacheco LP (2012) Soil fertility and upland rice yield after biochar application in the Cerrado. Pesqui Agropecu Bras 47(5):699–706

Roberts KG, Gloy BA, Joseph S, Scott NR, Lehmann J (2010) Life cycle assessment of biochar systems: estimating the energetic, economic, and climate change potential. Environ Sci Technol 44(2):827–833

Schmidt HP, Wilson K (2014) The 55 uses of biochar. Ithaka J 1:286–289

Shackley S, Hammond J, Gaunt J, Ibarrola R (2011) The feasibility and costs of biochar deployment in the UK. Carbon Manage 2(3):335–356

Singh B, Singh BP, Cowie AL (2010) Characterisation and evaluation of biochars for their application as a soil amendment. Soil Res 48(7):516–525

Smith P (2016) Soil carbon sequestration and biochar as negative emission technologies. Glob Change Biol 22(3):1315–1324

Smith P et al (2021) Agriculture, forestry and other land uses (AFOLU). In: Climate change 2021: the physical science basis. Contribution of Working Group I to the sixth assessment report of the IPCC

Spokas KA, Cantrell KB, Novak JM, Archer DW, Ippolito JA, Collins HP, Nichols KA (2012) Biochar: a synthesis of its agronomic impact beyond carbon sequestration. J Environ Qual 41(4):973–989

Tomczyk A, Sokołowska Z, Boguta P (2020) Biochar physicochemical properties: pyrolysis temperature and feedstock kind effects. Rev Environ Sci Biotechnol 19:191–215

Biochar and Cereal Yield Enhancement

Divyansh Verma, Ashima Choudhary, Owais Ali Wani,
Gourav Sudan, Sabreena Ashraf, and Abid Shabir Bali

Abstract

Given that the world's food consumption is anticipated to rise by up to 50% between 2010 and 2050, it is imperative that plant productivity be raised with minimal or no environmental harm. Accordingly, biochar is high in carbon that is generated by pyrolyzing organic materials at high temperatures with little oxygen. Its physicochemical properties vary based on the pyrolysis conditions and feedstock. It has demonstrated numerous beneficial environmental impacts when applied as a soil supplement, including soil improvement, sequestration of carbon while lessening the release of greenhouse gases. Applying biochar to agri-systems has also demonstrated significant advantages, such as enhancing plant development under biotic or abiotic stressors or in ideal conditions. Many mechanisms are reported to wield the beneficial impacts on plant growth, either by

D. Verma · G. Sudan
Division of Soil Science and Agricultural Chemistry, Faculty of Agriculture, Sher-e-Kashmir University of Agricultural Sciences and Technology, Jammu, Jammu and Kashmir, India

A. Choudhary (✉)
Division of Fruit Science, Faculty of Horticulture & Forestry, Sher-e-Kashmir University of Agricultural Sciences and Technology, Jammu, Jammu and Kashmir, India

O. A. Wani
Division of Agronomy, Faculty of Agriculture, SKUAST-Kashmir, Wadura, Sopore, Jammu and Kashmir, India

S. Ashraf
Punjab Agricultural University, Ludhiana, Punjab, India

A. S. Bali
Division of Agronomy, Sher-e-Kashmir University of Agricultural Sciences and Technology of Kashmir, Srinagar, Jammu and Kashmir, India

O. A. Wani et al. (eds.), *Biochar for Sustainable Soil Health and Crop Production*, https://doi.org/10.1007/978-981-92-2230-8_8

itself or in conjunction with additional resources. These mechanisms include boosting the amount of N P or K in soil, enhancing its physicochemical characteristics, encouraging microbial colonization or expanding the diversity of soil microbes, all of which enhance the ability of soil to cycle nutrients. However, in this chapter, we will concentrate on biochar made from the leftovers worldwide, as well as their application in cereal crops (rice, wheat, maize) to promote overall growth and development.

Keywords

Biochar · Cereals · Yield · Environment · C—sequestration

1 Introduction

Effective techniques for guaranteeing food security for an expanding global population encompass enhancing yield both sustainably and per unit area augmenting the productivity of farms. Global research is underway regarding the possible advantages of utilizing biochar to improve food security through increased agricultural output. The utilization of biochar aligns with contemporary green development principles, as it is essential for preserving ecosystem equilibrium, mitigating soil contamination, and fostering sustainable farming practices. There is an obvious need to increase plant yields using minimal to no environmental harm because the world's food consumption is anticipated to rise by up to 50% between 2010 and 2050 (Van Dijk et al. 2021). Biochar is a porous, carbon-dense material generated from the thermal breakdown of biomaterials, including plant remnants, or wood, agricultural waste, in an environment with restricted oxygen availability (Farhangi-Abriz et al. 2021). The process, termed pyrolysis, entails heating biomaterials to elevated temperatures with little or no oxygen present, so inhibiting total combustion (Wang et al. 2020; Kamali et al. 2022). Because of its many advantages in areas, counting agriculture, soil health, climate change and waste-water treatment, it has gained attention in past 10 years. Using biochar has obvious benefits of increasing soil productivity, quality, and carbon sequestration while lowering greenhouse gas emissions (CO_2, N_2O, and CH_4). Approximately, land use change accounts about 12% of total anthropogenic carbon dioxide emissions (0.21 Pg C) annually can be absorbed by soil with the aid of plant life, according to current research (Duan et al. 2024). Numerous studies have revealed that incorporating biochar into soil boosts its microbial activity as well as its physical characteristics, including pH, bulk density, pore size distribution, cation exchange capacity (CEC), soil organic carbon, structure and water holding capacity (Ye et al. 2020; Martínez-Gómez et al. 2022). Furthermore, it can bind dangerous materials in contaminated soils, improve the ability to absorb soil minerals, and stop nutrient losses through leaching.

2 History of Biochar Technologies

Despite the fact that biocarbon has expanded in the era of organic farming, its roots lie in the Amazonian soils known as "terra preta" soils discovered down to a depth of two meters. For ages, the Amazonians have relied on this extremely productive, dark-colored soil to meet their agricultural demands (Neves et al. 2004).

Terra preta's existence indicates that its formation was purposefully caused by mankind. Since the beginning of human evolution, the combustion of biomass for biochar synthesis has been recorded (Boroson et al. 1989). The development of petroleum industries between 1920 and 1950 slowed the growth of the wood distillation business, which was booming in 1850. The collapse of the oil industry in 1970 speed up the development of alternative fuels and sophisticated pyrolysis reactors were created to produce bio-oil from biomass. Between 1970 and 1990, pyrolysis underwent more development in order to comprehend the principles of biomass pyrolysis reaction (Bridgwater et al. 1999; Evans and Milne 1987). From the 1980s to the 1990s, biochar and bio-oil were produced commercially using rapid, flash, vacuum, and ablative pyrolysis techniques. It has been suggested that a new technological advancement will enable a refinery based on bio-oil to extract useful byproducts from biomass. Because slow pyrolysis produces more biochar than other production methods, it is employed to manufacture the biochar used to enhance soil. In essence, biomass is cooked for a longer amount of time between 300 and 600 °C during a slow pyrolysis process (Czernik et al. 2002).

3 Biochar Production Technologies

A vast range of biomaterials, including agricultural wastes that have undergone thermal degradation under various functioning circumstances, are used to make biochar. It displays an equally vast variety of makeup. The biomass is transformed into bioenergy products, such as biofuels, by thermochemical processes including pyrolysis and carbonization. Throughout the pyrolysis process, biomass is thermochemically converted at temperatures above 400 °C without the presence of air to create biochar, a solid byproduct. Carbon (C), hydrogen (H), oxygen (O), nitrogen (N), sulfur (S), and ash make up majority of the biochar. Pyrolysis can occur in three general modes: slow, moderate, and quick. A sluggish pyrolysis technique produced more charcoal than previous methods (Kung et al. 2015). The generation of biochar is the primary objective of carbonization, a method of slow pyrolysis that has been used over several thousand years. In slow pyrolysis, (Table 1) biomass is cooked gradually over a period of time without lowering the temperature of air (about 400 °C) (Basu 2013; Panwar et al. 2019). The process can be propelled by energy in the ways listed below:

(a) Directly as a reaction's heat
(b) Directly through feedstock combustion flue gases,
(c) By employing a hot gas to indirectly heat the reactor wall and
(d) By employing sand or other non-gas materials to indirectly heat the reactor wall (Fig. 1).

Table 1 Conversion technologies and distribution of product

S.No.	Mechanism	Specific temperature (°C)	Specific time	Biochar (solid)
1	Slow pyrolysis	100–1000	2–4 h	35
2	Intermediate pyrolysis	Approx. 500	10–20 s	25
3	Fast pyrolysis	Approx. 300–1000	2 s	Approx. 12–25
4	Gasification	Approx. 900	10–20 s	10

Source: Ahmad et al. (2012)

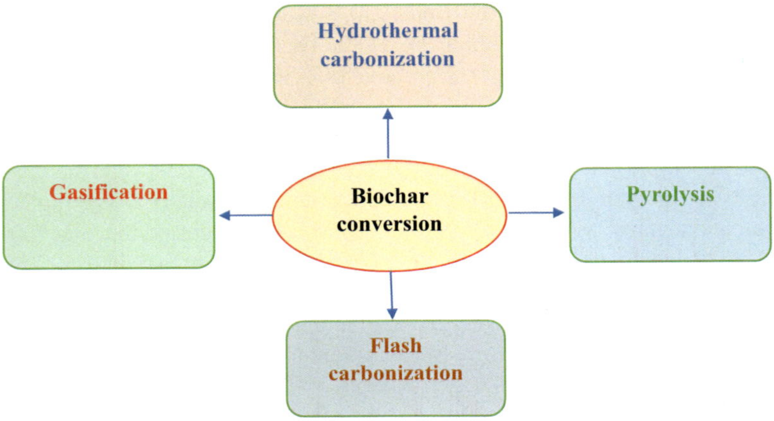

Fig. 1 Biomass conversion processes (Ahmad et al. 2014)

4 Biochar Characteristics

4.1 Impact of Material and Temperature

The temperature along with the parent material are the key factors affecting attributes of biochar. Biochar's surface area will increase as temperature of pyrolysis rises, allowing compounds like insecticides to absorb more readily. The characteristics of biochar's made from different resources varies distinctly.

4.2 Enhancement of Biochar's Qualities Using Various Treatment Techniques

Activation proved successful in increasing the adsorption capacity of biochar in comparison to nonactivated one. In every case, steam activation demonstrated nearly twice the beneficial effects of biochar's, making it a promising choice for potential future uses for biochar. Regmi et al. (2012) used hydrothermal carbonization (HTC) to create biochar from switch grass and investigated the effectiveness using alkali-activated HTC biochar (HTCB) and HTC biochar in extracting cadmium and copper from aqueous solutions. According to the results, HTCB removed up to 100% of the cadmium and copper at 2 g/L, which is significantly more than what was seen with HTC biochar (5.6% for cadmium and 16% for copper). The sorption characteristics and porous structure may be improved by the activation procedure.

4.3 Enhancement of Crop Production and Soil Properties Using Biochar

Apart from its potential to mitigate climate change by storing carbon and lowering emissions of greenhouse gases, biochar additives can alter soil properties including raising the pH of acidic soils, CEC, and buffering of soil (Liang et al. 2006). Soil modified with biochar decreases N leaching since improvements are also shown in the retention of water and nutrients.

4.4 Heavy Metal Cleanup Process

Biochar has been thought to have the potential to be useful in heavy metal remediation. Numerous studies provided trustworthy data regarding the potential effectiveness of biochar in removing heavy metals from aqueous solutions and soils. Jiang et al. (2012) examined the impact of adding rice straw-derived biochar on the bioavailability and mobility of Cu (II), Pb (II), and Cd (II) in an Ultisol. Furthermore, results revealed that as the dose of the biochar amendment increased, the acid-labile Pb (II) and Cu (II) dropped by 18.8–77.0% and 19.7–100%, respectively.

4.5 Detoxification Up Organic Contaminants

According to numerous studies published in recent years, biochar can eliminate organic pollutants from soil, water and sediments reducing their bioavailability and stopping harmful substances from moving from the environment to plants and then to people. According to the results, sulfamethoxazole (SMX) was shown to move through soils with a biochar amendment at a rate of 2–14% as opposed to 60% in the leachate of soils without the addition (Yao et al. 2012). It was demonstrated that the addition of biochar can considerably lower the freely dissolved concentration of PAH in sewage sludge, with a reduction ranging from 0% to 57% depending on the amount of biochar applied (Oleszczuk et al. 2012). The impact of bamboo charcoal (BBC) on PCP leachability was also evaluated. The results showed that the PCP levels in methanol and distilled water extracts (56% and 65%) were reduced by 5% BBC amendment and lowered overall loss of leaching content of PCP by 42%.

4.6 Impact on the Pesticide

The utilization of black carbon in the treatment of organic contaminants and heavy metals suggested an economical and sustainable method of managing a contaminated environment.To combat pests and diseases in agriculture, however, pesticides are purposefully introduced to soil or additional ecological compartments. Pesticides enhanced sorption and reduced dissipation in biochar could reduce the danger of environmental human exposure and contamination from the standpoint of human health and the ecosystem. Additionally, from an agricultural perspective, the reduced plant absorption and bioavailability may increase crop yield while decreasing chemical residues. However, the reduced effectiveness of pesticides resulting from the application of biochar is undesirable in soils used for agriculture because a pesticide intended to suppress specific weeds or insect-pests must be accessible to be effective.

5 Mechanism by Which Biochar Enhances Cereal Yield

Biochar improves grain output by combining physical, chemical, and biological factors within the soil–plant system (Fig. 2). These mechanisms enhance nutrient accessibility, water dynamics, root development and microbial activity eventually ensuing in improved biomass and grain yield. Below is the explained mechanism of biochar used for yield enhancement:

1. Enhancing the physical properties of soil:
   (a) Improved Aggregation and Soil Structure: Biochar has a high surface area with porous structure. It functions as a binding agent when applied to soil, encouraging creation of stable aggregates. It also enhances root penetration in cereals (wheat, rice, and maize) which ultimately leads to improved aggregation.

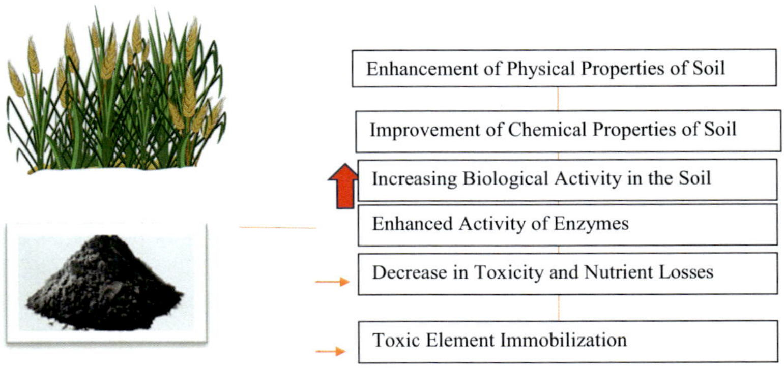

Fig. 2 Mechanism of biochar for enhances crop yield

   (b) A higher capacity to hold water (WHC): Like a sponge, the pores in biochar hold onto water. Especially advantageous in soils that are prone to drought or have a coarse texture. It also reduces plant water stress by keeping the soil moist during dry weather.
2. Improves chemical properties of soil:
   (a) Enhanced Cation Exchange Capacity: Functional groups (–COOH, –OH) on biochar surfaces raise soil CEC and also enhance the retention of vital nutrients: Nitrogen (NH_4^+), Potassium (K^+), Ca^{2+} (calcium), Magnesium (Mg^{2+}).
   (b) Controls soil pH: The majority of biochar's have an alkaline composition which reduce the toxicity of Al^{3+} and Fe^{3+} to counteract soil acidity and increasing P, Ca, Mg, and Mo availability in the soil.
   (c) Enhanced utilization of nutrients (NPK): Nutrient buffering is the process by which biochar absorbs nutrients and releases them gradually into the soil. It diminishes the volatilization of nitrogen and the fixation of phosphorous in the acidic soils.
3. Increasing biological activity: Biochar uplifts the microbial environment in the soil. It includes millions of bacteria, fungi and actinomycetes etc. Soil bacteria have safe microhabitats due to the availability of pores in the biochar. It encourages the expansion of the bacteria that fix nitrogen, microorganisms that solubilize phosphate and the mycorrhizal arbuscular fungus (AMF).
4. Enhanced activity of enzymes: Soil enzyme activity including: Urease, Dehydrogenase and Phosphatase is amplified by applying the biochar.
5. Decrease in Toxicity and Nutrient Losses: Biochar diminishes the nitrogen losses and also the ammonium and nitrate are adsorbed by biochar, resulting in decreasing by the process of leaching and losses from denitrification.
6. Toxic Element Immobilization: Pesticide residues and heavy metals (Cd, Pb, and Al) are bound by biochar. It also reduces the number of cereal plants that can absorb them.

6 Effect of Biochar in Enhancing Cereal Crop Growth

6.1 Rice

Rice (*Oryza* spp.), one of the major crops is the primary food source for more than half of the global population. Over 0.8 billion tons, rice ranks third after maize and sugarcane in terms of primary crop production worldwide, making up 8% of total production. Asia produces more rice than any other region in the globe, accounting for 90% of global production. Straw, ash, husk, broken rice and bran are the primary solid waste products of rice processing. Half of the rice's dry weight is made up of rice straw, which is extracted during harvest from the grain and is typically abandoned or scorched on a field, squandering a resource that may be used to produce stable litter, ethanol, energy, or animal feed. The rice husk, which makes up about 20% of the weight of the grain, acts as a protective coating. Rice husk can be used to produce ethanol or power, as well as for poultry litter, but unlike straw, it is not suitable for animal nutrition. As the two most popular rice by-products, both residues have demonstrated significant promise for the manufacture of black carbon. In majority of research, the temperature of pyrolysis and residency time for creation of biochar typically range from 30 to 8 h at 400–800 °C for rice straw and from 15 to 3 h at 350–1100 °C for rice husk. Among other things, the temperature and reaction type have a major effect on the product yield. For instance, the production of biochar for reaction temperatures of rice residues ranging from 300 to 800 °C is around 36–65%, whereas the yields of bio-oil and syngas are 22–45% and 12–45%, respectively (Su et al. 2020; Bhatnagar et al. 2022). Dong et al. (2015) evaluated the effect bamboo biochar and rice straw biochar addition (22.5 t ha^{-1}, with or without urea) on rice growth and N retention in 2009 and 2010. Further, results revealed that on addition of rice straw biochar enhanced rice yield by 19.8% (2009) and 21.6% (2010) without urea whereas, with urea it is 11.3% (2009) and 14.4% (2010), respectively. Hafez et al. (2019) carried out a field examine how biochar and PGPR (*Pseudomonas koreensis* and *Bacillus coagulans*) work together to improve rice soil salinity and water deficit for two consecutive years (2017 and 2018) and hence, noted that the combined application of biochar and PGPR reduced the ill effects of these stresses. The effectiveness of this strategy was validated by the notable increases in K^+ and K^+/Na^+ contents, relative water content, conductance of stomatal pores, chlorophyll content, and K^+ as proline and Na^+ content decreased.

6.2 Wheat

The use of biochar in agriculture is becoming more and more significant every day, and it is being investigated from several angles in a range of soil types and circumstances. Numerous reports of using biochar to increase crop output and growth in drought-stressed environments. The utilization of biocarbon significantly enhanced the relative water contents, photosynthesis, chlorophyll contents, stomatal conductance, and WUE of tomato leaves during drought. In a similar vein, maize plants with limited water availability have shown improvements in photosynthesis, osmotic potential, transpiration rate, relative water contents of leaf, and plant development. Haider et al. (2020) carried out an experiment in the pot using wheat straw biochar at different growth stages in wheat and recorded that among various stages, the most sensitive stage to drought stress was the grain filling and higher rate of application of biocarbon (37.18 g kg^{-1}) showed results in alleviating the inimical impact of drought and maintains the growth and yield. Another study by Zaheer et al. (2021) in a pot experiment recorded the function of carbon rich char (28 g kg^{-1} and 38 g kg^{-1}) in enhancing soil biochemical characteristics and wheat (*Triticum aestivum* L.) growth (tillering; grain filling stage) during drought and showed that incorporation of 38 g kg^{-1} of biochar noted the best results for mineral nutrients, soil carbon (11.86%), Bray P (18.72%), nitrogen mineralization (16.35%), exchangeable-K (7.44%), and soil respiration (6.37%) in contrast to 28 g kg^{-1}. Rutigliano et al. (2014) in their investigation noted the addition of two loads of wood-derived biochar (30 and 60 t ha^{-1}) into the soil to investigate the effects on a wheat crop on soil microbial community for two consecutive years. After 3 months of treatments, the activity of microbes enhanced significantly.

6.3 Maize

Naeem et al. (2024) in an experiment evaluated the effect of organically derived biochar on optimizing characteristics related to yield under optimal and deficit irrigation settings, as well as enhancing physicochemical and nutritional status of the soil. Furthermore, the 5 tons ha^{-1} biochar dose outperformed the 10 tons ha^{-1} dose in terms of efficiency out of two application levels tested. Minhas et al. (2020) evaluated how biochar affects maize yield and nutrient retention in loamy soils using the recommended full and half doses of phosphorus (P) and nitrogen (N) feeding in a pot experiment at various doses (0, 2, 4 and 6 t ha^{-1}). The best grain yield (6.64 t ha^{-1}) was obtained by applying biochar at a rate of 2 t ha^{-1} with full NP nutrition; however, when compared to full dose of NP without biochar, biochar addition at a rate of 2 t ha^{-1} with half NP nutrition provided a higher yield than full dose of NP to boost maize production. As a result, using 2 t ha^{-1} of biochar along with half the recommended amount of NP significantly increased maize productivity in loamy soil and was a superior alternative to using a full dose of NP fertilizer. Additionally, Biochar amendments in the maize field showed enhanced growth and development of maize (*Zea mays* L.) both above and below ground (Xiao et al. 2016). Zahra et al. (2021) noted the amalgamated effect of compost and biochar on maize yield (6.44 and 6.52 t ha^{-1}) for two consecutive years than control, respectively.

6.4 Rye

Rye is a robust cereal crop that is renowned for its capacity to withstand harsh growing conditions. Iacomino et al. (2022) revealed that the incorporation of biochar, compost, and their amalgamation on yield of crops in volcanic soils showed that compost and compost +biochar mixtures decreased tomato yield by −50% and −100%, respectively, and aubergine yield by −60% and −40%, respectively. However, applying compost, compost + biochar, and compost considerably increased the yields of onions and rape, whereas applying biochar and compost + biochar dramatically reduced the yields of both crops. Scotti et al. (2022) conducted a two-year trial of application of 3 doses of biochar (@ 0.2%, 0.45%, 0.9%) and in combination with digestate (BDig), mineral (BMin) and slurry (BSlu), respectively and noted that BDig appeared to enhance silage yield and the ear component compared to Dig, and BMin tended to decrease maize N uptake compared to Min, biochar had no discernible effect on maize output. Manolikaki and Diamadopoulos

(2017) carried out an experiment to analyse the influence of crop growth, nutrient status and P-uptake on application of biochar derived from grape pomace (GP) and rice husk on ryegrass (*Lolium perenne* L.) Results revealed that GP biochar enhances the crop yield (58.9–77.6%), raises the amount of P uptake and also the Mn and Fe uptake as compared to control.

Various other studies are presented in the Table 2 depicting a significant impact of biochar utilization on the physiological and morphological parameters of the distinct cereal crops which are widely grown.

Table 2 Studies on impact of biochar on distinct cereal crops

Crop	Biochar obtained by	Best application dose & growth conditions	Results obtained	References
Maize-Wheat sequence	Pyrolysis (400 °C)	5 t ha^{-1} (Field conditions)	Improved harvest index (30%, 29%), yield (60%, 63%), and grain weight (44%, 54%).	Sarwar et al. (2023)
Durum Wheat (var. Vitron)	Pyrolysis (500 °C)	5 g kg^{-1} Salt stress condition in arid & semi-arid regions. (Pot experiment)	Significant increase in wheat plant height, total chlorophyll, LSA and RWC.	Ababsa et al. (2023)
Rice-Wheat	Pyrolysis (400 °C)	20-ton ha^{-1} (wheat raw biochar) (Field experiment)	Enhanced grain yield (26–37%) with wheat raw biochar.	Korai et al. (2021)
Maize-Wheat	Higher pyrolytic temperature (≤550 °C)	More than 30 t ha^{-1} (Field conditions)	Enhanced grain yield of wheat (13.5%) and maize (28%).	Farhangi-Abriz et al. (2021)
Wheat	Kon-tiki Flame Curtain Pyrolysis techniques (500 °C)	5 t ha^{-1} inoculated with biofertilizer (Arid field conditions)	Higher grain yield (11%).	Khan et al. (2021)
Maize	Wakefield Agricultural Carbon (USA), a biochar manufacturer with USDA certification.	Nitrogen biochar combination @ 150 kg N ha^{-1} (Field experiment)	Maximum yield output (7.3 mg ha^{-1}) under Nitrogen fertilizer at 150 kg N ha^{-1}.	Omara et al. (2020)
Rice-Wheat cropping system	Pyrolysis and thermal decomposition (400 °C)	20 and 40 t ha^{-1}	Enhanced soil carbon, yield, nutrient pools and nutrient use efficiency.	Zhang et al. (2020)

(continued)

Table 2 (continued)

Crop	Biochar obtained by	Best application dose & growth conditions	Results obtained	References
Maize	Pyrolysis (300 °C maintained for 20 min.)	Biochar + Compost + Fertilizer (0.5% + 0.75% + NPK) (Pot experiment)	Highest results for cob yield (171 g pot-1), grain yield (117 g pot-1), 1000-grain weight (226 g), fresh shoot weight (517 g pot-1), and plant height (150 cm) were obtained.	Naeem et al. (2018)
Wheat	Pyrolysis (450 and 550 °C)	(5.0% w/w) (Pot study under combined Cd and drought stress)	Improved morphological and physiological traits, improved the activity of antioxidant enzymes, and reduced oxidative stress and cadmium levels.	Abbas et al. (2018)
Wheat-Maize rotation systems	Pyrolysis (350–550 °C)	8 t ha^{-1} straw-derived biochar plus fertilizer (Field experiment)	Significantly, highest yield for wheat (81.7%) and for maize (72.1%) were obtained.	Hu et al. (2021)
Barley straw	Pyrolysis (1 h, 400 °C)	10 t ha^{-1} straw derived biochar	Enhanced yield and nutrient uptake in Chinese cabbage.	Kang et al. (2018)
Barley straw	Pyrolysis (400 °C)	20 t ha^{-1} straw derived biochar	Increases soil physical and chemical properties.	Kang et al. (2019)

7 Conclusion

Several studies reported above have shown a strong effect of carbon rich char on the crop growth and yield. The production and use of biochar have a strong effect on the GHG balance of the entire system. The emissions related to the production, transportation, and application of biochar to soils; the degree to which the addition of biochar promotes (or "primes") the breakdown of organic matter in the soil; the impact of biochar on emissions of non-CO_2 trace gases; and the quantity of energy captured during the biochar production process. A variety of biomass management choices, such as utilizing biomass for energy and leaving decaying wood in situ, should be compared against the potential and drawbacks of producing biochar and adding it to field soil. Carbon sequestration via biochar may have long-term advantages, but at the expense of short-term CO_2 pulses into the atmosphere and, as a result, a short-term acceleration of climate change. Henceforth, biocarbon shows long-term and short-term variability in yield depending upon the climatic and soil functions.

References

Ababsa N, Boudjabi S, Chenchouni H (2023) Biochar amendments changed soil properties and improved cereal crop growth under salt stress. J Soil Sci Plant Nutr 23(4):4912–4925

Abbas T, Rizwan M, Ali S, Adrees M, Mahmood A, Rehman MZ, Ibrahim M, Arshad M and Qayyum MF (2018) Biochar application increased the growth and yield and reduced cadmium in drought stressed wheat grown in an aged contaminated soil. Ecotoxicology and Environmental Safety, 148: 825–833. https://doi.org/10.1016/j.ecoenv.2017.11.063

Ahmad M, Lee SS, Dou X, Mohan D, Sung JK, Yang JE, Ok YS (2012) Effects of pyrolysis temperature on soybean stover-and peanut shell-derived biochar properties and TCE adsorption in water. Bioresour Technol 118:536–544

Ahmad M, Rajapaksha AU, Lim JE, Zhang M, Bolan N, Mohan D, Vithanage M, Lee SS, Ok YS (2014) Biochar as a sorbent for contaminant management in soil and water: A review. Chemosphere 99:19–33

Basu P (2013) Biomass gasification, pyrolysis and torrefaction: practical design and theory. Academic, Cambridge

Bhatnagar A, Singhal A, Tolvanen H, Valtonen K, Joronen T, Konttinen J (2022) Effect of pretreatment and biomass blending on bio-oil and biochar quality from two-step slow pyrolysis of rice straw. Waste Manag 138:298–307. https://doi.org/10.1016/j.wasman.2021.12.013

Boroson ML, Howard JB, Longwell JP, Peters WA (1989) Heterogeneous cracking of wood pyrolysis tar over wood char surface. Energy Fuel 3:735–740

Bridgwater AV, Meier D, Radlein D (1999) An overview of fast pyrolysis of biomass. Org Geochem 30:1479–1493

Czernik S, French R, Feik C, Chornet E (2002) Hydrogen by catalytic steam reforming of liquid byproducts from biomass thermoconversion processes. Ind Eng Chem Res 41(17):4209–4215

Dong D, Feng Q, Mcgrouther K, Yang M, Wang H, Wu W (2015) Effects of biochar amendment on rice growth and nitrogen retention in a waterlogged paddy field. J Soils Sediments 15(1):153–162

Duan S, Al-Huqail AA, Alsudays IM, Younas M, Aslam A, Shahzad AN, Qayyum MF, Rizwan M, Alhaj Hamoud Y, Shaghaleh H, Hong Yong JW (2024) Effects of biochar types on seed germination, growth, chlorophyll contents, grain yield, sodium, and potassium uptake by wheat (*Triticum aestivum* L.) under salt stress. BMC Plant Biol 24(1):487

Evans R, Milne TA (1987) Molecular characterization of the pyrolysis of biomass. Energy Fuel 1(2):123–137

Farhangi-Abriz S, Torabian S, Qin R, Noulas C, Lu Y, Gao S (2021) Biochar effects on yield of cereal and legume crops using meta-analysis. Sci Total Environ 775:145869

Hafez EM, Alsohim AS, Farig M, Omara AED, Rashwan E, Kamara MM (2019) Synergistic effect of biochar and plant growth promoting rhizobacteria on alleviation of water deficit in rice plants under salt-affected soil. Agronomy 9(12):847

Haider I, Raza MAS, Iqbal R, Aslam MU, Habib-ur-Rahman M, Raja S, Khan MT, Aslam MM, Waqas M, Ahmad S (2020) Potential effects of biochar application on mitigating the drought stress implications on wheat (*Triticum aestivum* L.) under various growth stages. J Saudi Chem Soc 24(12):974–981

Hu Y, Sun B, Wu S, Feng H, Gao M, Zhang B and Liu Y (2021) After-effects of straw and straw-derived biochar application on crop growth, yield, and soil properties in wheat (Triticum aestivum L.)–maize (Zea mays L.) rotations: A four-year field experiment. Science of the Total Environment, 780: 146560. https://doi.org/10.1016/j.scitotenv.2021.146560

Iacomino G, Sarker TC, Ippolito F, Bonanomi G, Vinale F, Staropoli A, Idbella M (2022) Biochar and compost application either alone or in combination affects vegetable yield in a volcanic Mediterranean soil. Agronomy 12(9):1996

Jiang J, Xu RK, Jiang TY, Li Z (2012) Immobilization of Cu (II), Pb (II) and Cd (II) by the addition of rice straw derived biochar to a simulated polluted Ultisol. J Hazard Mater 229–230:145–150

Kang S., Kim SH, Park JH, Seo DC, Yong Sik Ok and Cho JS (2018) Effect of biochar derived from barley straw on soil physicochemical properties, crop growth, and nitrous oxide emission in an upland field in South Korea. Environmental Science and Pollution Research, 25(26): 25813–25821.

Kang SW, Kim SH, Park JH, Seo DC, Han KH, Cho JS and Ok YS (2019) Application of barley strawderived biochar affected soil properties and growth of Chinese cabbage under upland field conditions. Journal of Environmental Management, 236: 125–131.

Kamali M, Sweygers N, Al-Salem S, Appels L, Aminabhavi TM, Dewil R (2022) Biochar for soil applications-sustainability aspects, challenges and future prospects. Chem Eng J 428:131189

Khan Z, Rahman MHU, Haider G, Amir R, Ikram RM, Ahmad S, Schofield HK, Riaz B, Iqbal R, Fahad S, Datta R (2021) Chemical and biological enhancement effects of biochar on wheat growth and yield under arid field conditions. Sustainability 13(11):5890

Korai PK, Sial TA, Pan G, Abdelrahman H, Sikdar A, Kumbhar F, Channa SA, Ali EF, Zhang J, Rinklebe J, Shaheen SM (2021) Wheat and maize-derived water-washed and unwashed biochar improved the nutrients phytoavailability and the grain and straw yield of rice and wheat: a field trial for sustainable management of paddy soils. J Enviro Manage 297:113250

Kung CC, Kong F, Choi Y (2015) Pyrolysis and biochar potential using crop residues and agricultural wastes in China. Ecol Indic 51:139–145

Liang B, Lehmann J, Solomon D, Kinyang J, Grossman J, O'Neill O, Skjemstad JO, Thies J, Luizao FJ, Petersen J, Neves EG (2006) Black carbon increases cation exchange capacity in soils. Soil Sci Soc Am J 70:1719–1730

Manolikaki I, Diamadopoulos E (2017) Ryegrass yield and nutrient status after biochar application in two Mediterranean soils. Arch Agron Soil Sci 63:1093–1107

Martínez-Gómez Á, Jorge P, Carolina E (2022) Overview of the use of biochar from main cereals to stimulate plant growth. Front Plant Sci 13:912264

Minhas WA, Hussain M, Mehboob N, Nawaz A, UL-Allah S, Rizwan MS, Hassan Z (2020) Synergetic use of biochar and synthetic nitrogen and phosphorus fertilizers to improves maize productivity and nutrient retention in loamy soil. J Plant Nutr 43(9):1356–1368

Naeem MB, Jahan S, Rashid A, Shah AA, Raja V and El-Sheikh MA (2024) Improving maize yield and drought tolerance in field conditions through activated biochar application. Scientific Reports, 14: 25000

Naeem MA, Khalid M, Aon M, Abbas G, Amjad M, Murtaza B, Khan WUD, Ahmad N (2018) Combined application of biochar with compost and fertilizer improves soil properties and grain yield of maize. J Plant Nutr 41(1):112–122

Neves EG, Bartone RN, Petersen JB, Heckenberger MJ (2004) The timing of terra preta formation in the central Amazon: new data from three sites in the central Amazon. Springer, Berlin

Oleszczuk P, Hale SE, Lehmann J, Cornelissen G (2012) Activated carbon and biochar amendments decrease pore-water concentrations of polycyclic aromatic hydrocarbons (PAHs) in sewage sludge. Bioresour Technol 111:84–91

Omara P, Aula L, Oyebiyi FB, Eickhoff EM, Carpenter J, Raun WR (2020) Biochar application in combination with inorganic nitrogen improves maize grain yield, nitrogen uptake, and use efficiency in temperate soils. Agronomy 10(9):1241

Panwar NL, Pawar A, Salvi BL (2019) Comprehensive review on production and utilization of biochar. SN Appl Sci 1(2):168

Regmi P, Moscoso JLG, Kumar S, Cao XY, Mao JD, Schafran G (2012) Removal of copper and cadmium from aqueous solution using switchgrass biochar produced via hydrothermal carbonization process. J Environ Manag 109:61–69

Rutigliano FA, Romano M, Marzaioli R, Baglivo I, Baronti S, Miglietta F and Castaldi S (2014) Effect of biochar addition on soil microbial community in a wheat crop. European Journal of Soil Biology, 60: 9–15

Sarwar N, Abbas N, Farooq O, Akram M, Hassan MW, Mubeen K, Rehman AU, Shehzad M, Ahmad M, Khaliq A (2023) Biochar integrated nutrient application improves crop productivity, sustainability and profitability of maize–wheat cropping system. Sustainability 15(3):22–32

Scotti C, Bertora C, Valagussa M, Borrelli L, Cabassi G, Tosca A (2022) Agroenvironmental performances of biochar application in the mineral and organic fertilization strategies of a maize–ryegrass forage system. Agriculture 12:925

Su Y, Liu L, Zhang S, Xu D, Du H, Cheng Y (2020) A green route for pyrolysis poly-generation of typical high ash biomass, rice husk: effects on simultaneous production of carbonic oxide-rich syngas, phenol-abundant bio-oil, high-adsorption porous carbon and amorphous silicon dioxide. Bioresour Technol 295:122243. https://doi.org/10.1016/j.biortech.2019.122243

Van Dijk M, Morley T, Rau ML, Saghai Y (2021) A meta-analysis of projected global food demand and population at risk of hunger for the period 2010–2050. Nat Food 2:494–501

Wang D, Jiang P, Zhang H, Yuan W (2020) Biochar production and applications in agro and forestry systems: a review. Sci Total Environ 723:137775

Xiao Q, Zhu LX, Zhang HP, Li XY, Shen YF, Li SQ (2016) Soil amendment with biochar increases maize yields in a semi-arid region by improving soil quality and root growth. Crop Pasture Sci 67(5):495–507

Yao Y, Gao B, Chen H, Jiang L, Inyang M, Zimmerman AR, Cao X, Yang L, Xue Y, Li H (2012) Adsorption of sulfamethoxazole on biochar and its impact on reclaimed water irrigation. J Hazard Mater 209–10:408–413

Ye L, Camps Arbestain M, Shen Q, Lehmann J, Singh B, Sabir M (2020) Biochar effects on crop yields with and without fertilizer: a metaanalysis of field studies using separate controls. Soil Use Manag 36(1):2–18

Zaheer MS, Ali HH, Soufan W, Iqbal R, Habib-ur-Rahman M, Iqbal J, Israr M and El Sabagh A (2021) Potential effects of biochar application for improving wheat (Triticum aestivum L.) growth and soil biochemical properties under drought stress conditions. Land, 10(11): 1125

Zahra MB, Aftab ZEH, Haider MS (2021) Water productivity, yield and agronomic attributes of maize crop in response to varied irrigation levels and biochar–compost application. J Sci Food Agric 101(11):4591–4604

Zhang Q, Song Y, Wu Z, Yan X, Gunina A, Kuzyakov Y, Xiong Z (2020) Effects of six-year biochar amendment on soil aggregation, crop growth, and nitrogen and phosphorus use efficiencies in a rice-wheat rotation. J Clean Prod 242:11843

Biochar and Pulse Yield Enhancement: Soil Improvement, Nutrient Dynamics, and Legume-Rhizobia Interactions

Mehraj Ud Din Sofi, Owais Ali Wani, Stanzin Khenrab, Manjot Singh Sidhu, and Sarabdeep Kour

Abstract

Biochar application represents an innovative approach to enhance pulse crop productivity through multifaceted soil and biological improvements. As a carbon-rich soil amendment derived from biomass pyrolysis, biochar significantly enhances soil physicochemical properties including organic matter content, water-holding capacity, and nutrient availability. The amendment increases soil pH, reduces aluminum toxicity, and improves cation exchange capacity, particularly benefiting acidic soils constraining legume cultivation. These improvements directly facilitate nutrient retention and mineralization, with studies demonstrating substantial increases in available phosphorus, nitrogen, and potassium. Beyond soil amendment, biochar profoundly influences the legume-rhizobia symbiotic relationship, a cornerstone of sustainable agriculture. Biochar enhances the legume rhizosphere microbiome by promoting beneficial microbial populations that facilitate nitrogen fixation. The amendment optimizes rhizobial community assembly and nodulation efficiency, enabling enhanced biological

M. U. D. Sofi
Division of Agronomy, Faculty of Agriculture, SKUAST Kashmir Wadura, Sopore, Jammu and Kashmir, India

O. A. Wani
Division of Agronomy, Faculty of Agriculture, SKUAST-Kashmir, Wadura, Sopore, Jammu and Kashmir, India

S. Khenrab · S. Kour (✉)
Division of Soil Science and Agricultural Chemistry, Faculty of Agriculture, Sher-e-Kashmir University of Agricultural Sciences and Technology of Jammu, Jammu, Jammu and Kashmir, India

M. S. Sidhu
P.G. Department of Agriculture, Khalsa College, Amritsar, Punjab, India

O. A. Wani et al. (eds.), *Biochar for Sustainable Soil Health and Crop Production*, https://doi.org/10.1007/978-981-92-2230-8_9

nitrogen fixation even under suboptimal conditions. Biochar application influences the rhizosphere metabolite profile, promoting root exudate composition that attracts compatible rhizobia species. These synergistic effects on soil properties and legume-rhizobia interactions translate to substantial pulse yield improvements. Field studies document yield increases ranging from 7% to 33% depending on biochar source, application rate, and soil type. The mechanism involves improved nutrient uptake efficiency, enhanced nodulation capacity, and increased nitrogenase activity in symbiotic bacteroids. Additionally, biochar modulates soil enzyme activities and microbial diversity, supporting overall agroecosystem health This integrated approach offers environmentally sustainable alternatives to chemical fertilizers while promoting soil carbon sequestration.

Keywords

Biological nitrogen fixation · Legume-rhizobia symbiosis: nutrient availability · Pulse crop productivity · Soil amendment

1 Introduction

The global population is projected to reach 9.7 billion by 2050 (Mesta et al. 2022). This growth will intensify major global challenges such as ensuring food security, mitigating climate change by reducing greenhouse gas emissions and meeting energy demands (Wu et al. 2023) Climate change, along with increasing biotic and abiotic stresses will significantly affect crop systems and threaten global food production (Erfatpour et al. 2024). To address these emerging issues, the adoption of more resilient farming systems is crucial. In this context, legume/pulse-based farming systems offer valuable benefits across multiple levels. At the food system level, pulses provide essential plant proteins for both people and animals, and are increasingly recognized for their role in supporting human health (Indrati 2021). At the production level, legumes naturally fix nitrogen from the atmosphere, reducing the need for synthetic fertilizers and helping lower greenhouse gas emissions (Erana 2021). Further, at the cropping system level, they help in diversification by adding variety to agricultural practices.

India is the leading producer of pulses, accounting for approximately 25% of global production (Singh et al. 2016). Over the past decade, India's pulse production has expanded by approximately 60% increasing from 17.1 million tonnes in 2014 to 27.0 million tonnes in 2024 (Ministry of Agriculture and Farmers Welfare 2024). In terms of area, pulses occupy a substantial share of the total area under foodgrains in India (Deepa 2019) and contribute 6–7% to the national foodgrain output (Ojha et al. 2020). Among individual pulse crops, chickpea (*Cicer arietinum*) predominates, accounting for approximately 40% of the total production, followed by pigeon pea (*Cajanus cajan*) at 15–20%, and black gram (*Vigna mungo*) and green gram (*Vigna radiata*) at 8–10% each (Deepa 2019). Despite being the largest global producer, India is simultaneously the largest consumer and importer, consuming 27% and importing 14% of global production (Singh et al. 2016). In spite of production gains over the past decade, national output remains below the estimated annual demand, with the gap projected to widen (Joshi and Rao 2017; Ojha et al. 2020). Productivity continues to lag behind the global average due to factors such as limited adoption of high-yielding cultivars and inadequate access to quality seeds (Subbaiah and Jyothi 2020), and the predominance of rainfed cultivation making yields highly susceptible to climatic variability (Deepa 2019). Efforts towards agricultural mechanization also present challenges impacting productivity (Tiwari et al. 2019). Although the Indian government aims to achieve self-sufficiency in pulses, a goal reiterated across various policy initiatives this target will require accelerated dissemination of the latest technologies, climate-resilient varieties and integrated policy interventions.

To enhance the productivity and resilience of the pulse-based farming systems, biochar emerges as a promising multifunctional soil amendment. Biochar, a carbon rich material produced from biomass pyrolysis has demonstrated significant potential to improve soil health by enhancing physical and chemical properties, increasing soil fertility and fostering microbial activity (Anyebe et al. 2025; Kabir et al. 2023; Zhu et al. 2025). Its application can lead to increased crop yield by improving nutrient availability such as nitrogen, phosphorus and facilitating better water retention (Atluri et al. 2021; Ayaz et al. 2021; Khan et al. 2024). Furthermore, biochar contributes sustainable agriculture by serving as a carbon sink, sequestering carbon and mitigating greenhouse gas emissions (Kabir et al. 2023; Khan et al. 2024). Therefore, exploring the strategic integration of biochar into pulse cultivation systems offers a viable pathway to boost productivity and ensure long term soil sustainability.

2 Mechanisms of Pulse Yield Enhancement by Biochar

Biochar acts as a multifunctional soil amendment, improving plant growth and yield through various physical, chemical and biological interactions within the soil ecosystem. Its unique properties derived from the pyrolysis of organic materials allow it to modify soil characteristics in ways beneficial for the pulse cultivation.

- **Enhances soil structure, aeration and porosity**: Biochar has a porous structure that can hold air and water in its porous space, thereby increasing soil porosity and improving soil aggregation (Poveda et al. 2021; Premalatha et al. 2023). Studies indicate that biochar application typically reduces soil bulk density and improves aggregate stability (Premalatha et al. 2023; Zhu et al. 2025). This leads to better aeration, which is crucial for root respiration and overall plant health (Murtaza et al. 2023).
- **Improves water retention and reduces compaction**: The porous nature of biochar allows it to increase the soil's water retention capacity, which is particularly beneficial for the crops, including pulses, in dryland systems (Premalatha et al. 2023). By decreasing soil bulk density and tensile strength, biochar can also

reduce soil compaction, facilitating easier root penetration and proliferation (Poveda et al. 2021; Premalatha et al. 2023). This improved water management can also reduce soil evaporation further conserving the soil moisture (Feng et al. 2023).

- **Increases soil pH in acidic soils**: Biochar generally has alkaline pH, which can neutralize acidic soils (Idbella et al. 2024; Zhu et al. 2025). This effect is crucial as it improves the availability of essential nutrients like phosphorus and micronutrients, which are often less accessible in acidic conditions (Tusar et al. 2023).
- **Enhances cation exchange capacity**: Biochar significantly increases the soil's cation exchange capacity, which is its ability to hold positively charged ions (cations) such as potassium (K^+), calcium (Ca^{2+}), magnesium (Mg^{2+}) and ammonium (NH_4^+) (Zhu et al. 2025). A higher CEC means the soil can retain more nutrients preventing their leaching and making them more available to the plant uptake over time (Hossain et al. 2020).
- **Increase nutrient use efficiency**: Due to its high surface area and charge characteristics, biochar can adsorb nutrients (Luo et al. 2025; Murtaza et al. 2023). This improved nutrient retention directly contributes to enhanced nutrient use efficiency by plants, hence more fertilizers are utilized by the crop rather than being lost from the systems (Alkharabsheh et al. 2021).
- **Stimulates rhizobial and microbial activity**: Biochar provides a favourable habitat for soil microorganisms due to its porous nature and ability to improve aeration and nutrient supply (De Medeiros et al. 2023). This leads to increased microbial populations, enzyme activity and overall microbial biomass (Hossain et al. 2020). The high surface area and porous structure offer refuge for beneficial microbes, protecting them from predation and desiccation (Głodowska et al. 2017a).
- **Improves biological nitrogen fixation in pulses through healthier rhizosphere conditions**: For pulses, a key benefit is the stimulation of rhizobial activity which enhances biological nitrogen fixation (Egamberdieva et al. 2021). Biochar improves symbiotic performance between legumes and rhizobia, leading to increased nodulation and nitrogen supply to the plants. This is particularly important for pulses as BNF is their primary source of nitrogen (Debnath et al. 2019).
- **Reduce soil borne diseases and promote beneficial microbial consortia**: Biochar has been shown to suppress soil borne plant diseases caused by various pathogen. This disease suppression can be attributed to several mechanisms, including changes in the soil physiochemical properties, the activation of systemic resistance in plants and the augmentation of beneficial rhizopheric microorganism populations that can outcompete or inhibit pathogens. Biochar can also promote the growth of the plant-growth-promoting rhizobacteria and other biocontrol agents, contributing to a healthier and more resilient soil microbiome (Fig. 1).

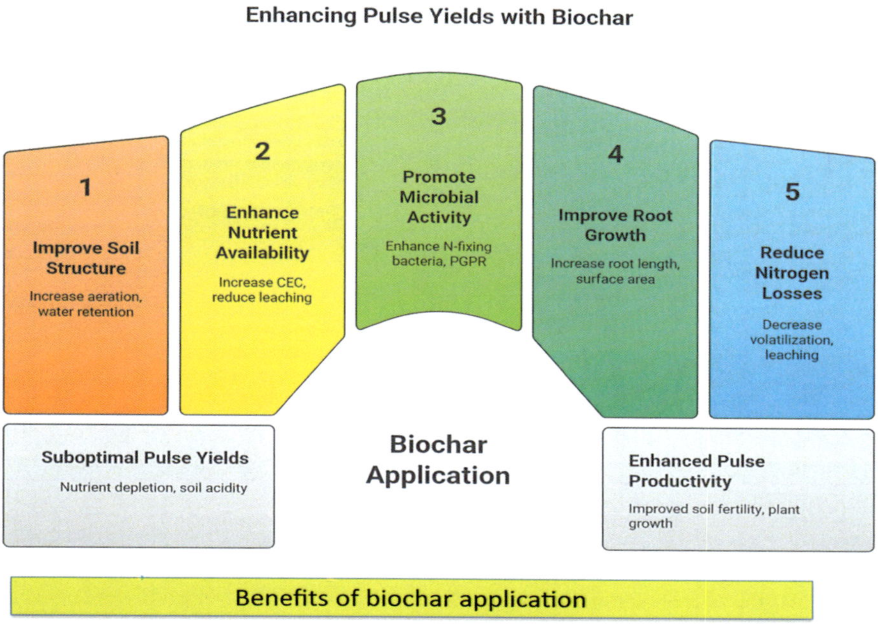

Fig. 1 Biochar application benefits

Challenges, Knowledge Gaps, and Future Prospects

- **Limitations in Current Understanding and Experimental Diversity**
  Despite significant advances, understanding the mechanisms by which biochar enhances pulse yields remains incomplete and variable across contexts. Differences in biochar feedstock, pyrolysis conditions, soil types, pulse species, and management practices contribute to inconsistent outcomes. Most studies are short term or limited geographically, and few provide insight into long-term effects under diverse agroecological conditions. This heterogeneity underscores the need for broad, multi-site, and long-duration trials to establish optimized biochar use frameworks for pulses (Jeffery and Bezemer 2013; Juriga and Šimanský 2018).
- **Risks and Potential Negative Effects of Biochar Application**
  Potential risks include secondary pollution from biochar impurities, variability in biochar physicochemical properties affecting soil biota negatively, and inconsistent yield responses under some conditions. High pyrolysis temperatures can reduce biochar nutrient content, and excess biochar application may impair soil microbial activity or root function. These risks necessitate careful selection of biochar types and application rates tailored to specific soil-pulse contexts, alongside ongoing monitoring to avoid unintended adverse effects (Juriga and Šimanský 2018; Chen et al. 2023b; Dayoub et al. 2024) (Fig. 2).
- **Directions for Future Research and Practical Applications**

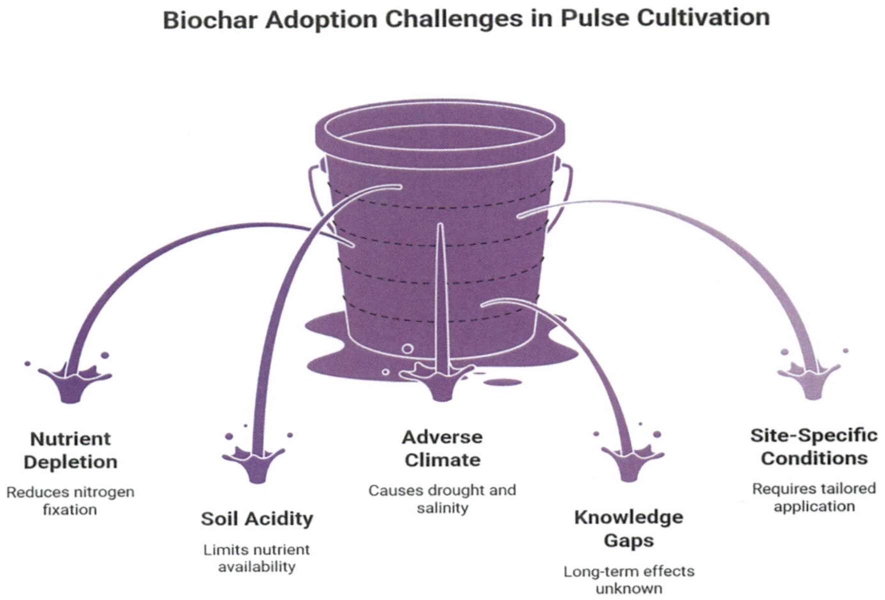

Fig. 2 Biochar adoption challenges in pulse cultivation

Future research should integrate precision agriculture techniques and machine learning to optimize biochar formulations and application protocols for pulse crops, balancing yield gains and environmental outcomes. Developing context-specific biochar products and usage guidelines will facilitate wider adoption. Policy frameworks and economic incentives are also crucial to encourage sustainable biochar use among pulses farmers. Investment in understanding biochar interactions with soil microbiota, plant physiology, and environmental factors will enhance its role in sustainable pulse production systems (Claire Grierson and Barnes 2011; Jeffery and Bezemer 2013; Chen et al. 2023a).

3 Biochar-Pulse-Rhizobia Synergy: Enhancing Biological Nitrogen Fixation in Legume Cropping Systems

The symbiotic relationship between leguminous plants (pulses) and soil-dwelling rhizobia bacteria is a cornerstone of sustainable agriculture. This partnership facilitates biological nitrogen fixation (BNF), a process where atmospheric dinitrogen (N2) is converted into plant-available ammonia (NH3), thereby enriching soil fertility and reducing the reliance on synthetic nitrogen fertilisers. However, the efficiency of this symbiosis is not absolute; it is highly sensitive to a range of edaphic factors, including soil acidity, poor structure, nutrient limitations, and suboptimal microbial community health (Graham and Vance 2003). The introduction of biochar, a carbon-rich material produced from the pyrolysis of organic biomass, into this system presents a significant opportunity to create a synergistic tripartite relationship. Biochar acts as a powerful soil conditioner, ameliorating the very constraints that limit nodulation and nitrogen fixation, thus fostering a more robust and productive pulse-rhizobia symbiosis. This section explores the multifaceted interactions within the biochar-pulse-rhizobia system, detailing the mechanisms through which biochar enhances nodulation and BNF, its role as a microbial carrier, and its potential to amplify the effects of other beneficial soil microorganisms.

Pulse crops, vital components of sustainable agricultural systems, are renowned for their ability to form symbiotic relationships with rhizobia, leading to biological nitrogen fixation (BNF). This process significantly reduces the need for synthetic nitrogen (N) fertilizers, thereby mitigating environmental impacts and enhancing soil fertility. However, the efficiency of nodulation and BNF in legumes is highly sensitive to prevailing soil conditions, including pH, texture, and overall microbial health. Degraded or suboptimal soils can severely limit the symbiotic potential, leading to reduced pulse yields and continued reliance on external N inputs.

3.1 Amelioration of Soil Physicochemical Constraints for Nodulation

The establishment of an effective legume-rhizobia symbiosis begins with a complex chemical interchange between the host plant and the bacteria, followed by root hair infection, nodule formation, and ultimately, nitrogen fixation. Each of these stages is profoundly influenced by the root-zone or 'rhizosphere' environment (Ferguson et al. 2010). Biochar fundamentally alters this environment, creating conditions more conducive to symbiotic success. Nodulation and nitrogen fixation in legumes are greatly influenced by the physicochemical and biological characteristics of the soil. Soil pH, texture, and the health of the microbial community within the rhizosphere are critical determinants of the success of the legume-rhizobia symbiosis. Biochar, has demonstrated a remarkable capacity to ameliorate suboptimal soil conditions, thereby fostering a more favorable environment for rhizobial activity and subsequent nodule development.

3.1.1 Regulation of Soil pH and Mitigation of Acidity Stress

Soil acidity (low pH) is one of the most significant abiotic stresses limiting both rhizobial survival in the soil and the nodulation process itself (Slattery et al. 2001). Acidic soils often exhibit high concentrations of toxic aluminium (Al3+) and manganese (Mn2+) ions, which are directly detrimental to both the host plant and the bacteria. Furthermore, low pH can impair the expression of key nodulation genes in rhizobia and interfere with the flavonoid signalling pathway initiated by the legume root (Foy 1984).

Biochar application serves as an effective 'liming agent', raising soil pH and buffering it against further acidification. This effect is primarily attributed to the ash content of the biochar, which is rich in carbonates, oxides, and hydroxides of alkali and alkaline earth metals (e.g., CaCO3, K2O) (Yuan et al. 2011). By neutralising hydrogen ions (H+) in the soil solution and precipitating toxic Al3+ as insoluble aluminium hydroxide (Al(OH)3), biochar creates a more favourable pH niche for rhizobia to thrive and for the host legume roots to develop. Particularly biochar derived from alkaline feedstocks, is its ability to increase soil pH. This liming effect is especially beneficial in acidic soils, which often inhibit rhizobial survival and activity, consequently limiting nodulation and nitrogen fixation (Ma et al. 2019a, b) For instance, a study by Hale et al. (2014) demonstrated that wood-derived biochar significantly increased soil pH, which correlated with a four-fold increase in soybean nodule biomass compared to unamended acidic soil. The deprotonation of acidic functional groups (e.g., carboxyl, hydroxyl) on the biochar surface also contributes to its buffering capacity, providing a sustained effect that can last longer than traditional lime applications (Lehmann et al. 2011). This pH optimisation is critical for enhancing the persistence of inoculated rhizobia and for ensuring the physiological processes underpinning nodulation can proceed efficiently.

The degree of pH alteration is contingent upon several factors, including the biochar's feedstock, pyrolysis temperature, and application rate, as well as the initial soil pH and buffering capacity (Farhangi-Abriz et al. 2024). For instance, Macil et al. (2020) observed that biochar application significantly increased soil pH, creating a more suitable environment for rhizobial activity and nodulation in chickpea (Macil et al. 2020). Similarly, Ma et al. (2019a, b) reported that biochar application could increase soil pH, which is crucial for the efficacy of rhizobium inoculation in sandy soils. Ma et al. (2019b). The improved pH conditions directly enhance the viability and infectivity of rhizobia, leading to more efficient nodule formation.

3.1.2 Improvement of Soil Physical Structure, Aeration, and Water Retention

Beyond chemical amelioration, biochar imparts significant improvements to soil physical properties. Its highly porous structure and large surface area contribute to a reduction in soil bulk density, an increase in total porosity, and improved soil aggregation (Atkinson et al. 2010). For pulse crops, these changes are highly beneficial. Reduced soil compaction facilitates deeper and more extensive root proliferation, increasing the chances of roots encountering compatible rhizobia. Furthermore, the pore network within and between biochar particles enhances soil aeration, which is crucial for the energy-intensive BNF process. Nitrogenase, the enzyme complex responsible for converting N2 to NH3, is irreversibly damaged by oxygen. However, the bacteroids within the nodule require a significant supply of oxygen for respiration to generate the large amounts of adenosine triphosphate (ATP) needed to fuel nitrogenase activity (Layzell and Hunt 1990). This 'oxygen paradox' is managed by the plant through the production of leghaemoglobin, which buffers oxygen concentrations within the nodule. Well-aerated soil, promoted by biochar addition, ensures that the overall root system is healthy and that sufficient, but not excessive, oxygen is available to the exterior of the nodules, supporting optimal respiratory function without deactivating the nitrogenase enzyme.

Simultaneously, the micropores within biochar particles significantly increase the soil's water-holding capacity (WHC) (Glaser et al. 2002). This is particularly advantageous for pulse production in rainfed or arid regions, as it provides a buffer against drought stress. Water stress is known to severely inhibit nodulation and nitrogen fixation by reducing photosynthesis, limiting nutrient transport to nodules, and causing premature nodule senescence (Serraj et al. 1999). By retaining moisture in the root zone, biochar ensures that symbiotic processes can continue for longer periods during dry spells, leading to more stable nitrogen inputs and crop yields.

3.2 Direct Influences on Microbial Habitat and Nodule Function

Biochar's role extends beyond being a passive soil conditioner; it actively interacts with the microbial community and directly influences the formation and function of root nodules.

3.2.1 Biochar as a Microbial Refuge and Proliferation Hub

The porous structure of biochar provides a protective habitat for soil microorganisms, including rhizobia. The pores, ranging from nano- to micrometres in scale, serve as 'microbial refugia', shielding bacteria from soil predators like protozoa and desiccation stress (Thies and Rillig 2009; Quilliam et al. 2013a, b). This protection can lead to higher and more stable populations of rhizobia in the soil, increasing the probability of successful root infection. The large surface area of biochar also provides ample sites for microbial colonisation and biofilm formation. Furthermore, biochar can adsorb signalling molecules (e.g., flavonoids released by legume roots) and nutrients, concentrating them on its surfaces and creating localised hotspots of microbial activity that can prime the soil for efficient symbiosis (Lehmann et al. 2011). Studies have shown that rhizobial populations are significantly higher in biochar-amended soils compared to unamended controls, directly supporting the concept of biochar as a microbial habitat enhancer (Spokas et al. 2011). Its highly porous structure provides a large surface area for microbial colonization, offering protection from environmental stressors such as desiccation and predation (Shabir et al. 2023). This physical protection, coupled with the provision of labile carbon compounds, can significantly enhance microbial survival, proliferation, and metabolic activity within the rhizosphere (Egamberdieva et al. 2019).

The interaction between biochar and rhizobia is particularly significant. Biochar can act as an effective carrier for rhizobium inoculants, improving their survival rates in the soil and ensuring a higher population density in the vicinity of legume roots Sarioglu et al. (2024). Sarioglu et al. (2024) demonstrated that combining biochar with rhizobium applications led to improved soybean growth and rhizosphere microbial activities, highlighting the synergistic potential of this approach (Sun et al. 2020). Furthermore, biochar can influence the chemical signaling between host legumes and their symbiotic partners. It has been suggested that biochar can adsorb and release signaling molecules, such as Nod factors and flavonoids, which are crucial for initiating the nodulation process (Gul and Whalen

2016). This modulation of chemical communication can lead to more efficient and effective nodule formation, ultimately enhancing biological nitrogen fixation (BNF).

3.2.2 Nodulation and Nitrogen Fixation Enhancement

Enhanced Nodule Number, Biomass, and Nitrogenase Activity

The symbiotic relationship between legumes and rhizobia, culminating in the formation of root nodules where atmospheric nitrogen is converted into a plant-available form, is a cornerstone of sustainable agriculture. Numerous studies have unequivocally demonstrated that biochar application significantly enhances various parameters associated with this crucial process, including nodule number, nodule weight, and the overall efficiency of biological nitrogen fixation (BNF). A consistent finding across numerous studies is that biochar application leads to a significant increase in the number of nodules, the size of individual nodules, and the total nodule dry weight per plant for various pulse crops, including common bean (*Phaseolus vulgaris*), soybean (*Glycine max*), and cowpea (*Vigna unguiculata*) (Rondon et al. 2007; Ogawa et al. 2006).

A comprehensive meta-analysis conducted by Farhangi-Abriz et al. (2021) provided compelling evidence of biochar's positive impact on legume-rhizobia symbiosis. Their findings indicated that biochar application led to an approximate 13% increase in the number of nodules, a 12% increase in nodule weights, and a substantial 28% enhancement in BNF. Similarly, Rondon et al. (2007) reported that in a low-fertility Colombian savanna soil, biochar application more than doubled both nodule number and nodule weight in common beans, leading to a substantial increase in total plant nitrogen derived from fixation. This enhancement is not merely due to a healthier plant being able to support more nodules. Biochar can also improve the availability of nutrients that are essential co-factors for the nitrogenase enzyme, such as molybdenum (Mo) and iron (Fe) (Glaser et al. 2015). By increasing soil pH and cation exchange capacity (CEC), biochar can increase the solubility and plant uptake of these critical micronutrients, thereby boosting the specific activity of the nitrogenase enzyme within each nodule. Therefore, biochar not only promotes the formation of *more* nodules but also helps them function *more efficiently*, leading to a greater overall rate of nitrogen fixation per plant. These improvements collectively contributed to increased plant biomass production in leguminous crops Farhangi-Abriz et al. (2024). This meta-analysis underscores the consistent and widespread benefits of biochar in optimizing nitrogen acquisition by legumes.

3.2.3 Mechanisms of Enhancement

The enhancement of nodulation and nitrogen fixation by biochar is not solely attributable to direct soil amelioration but also involves a complex interplay of various mechanisms:

- **Improved Nutrient Availability**: Biochar possesses a high cation exchange capacity (CEC) and can adsorb and slowly release essential plant nutrients, thereby improving their availability in the soil solution Arif et al. (2021). Phosphorus (P), in particular, is a critical nutrient for legumes, as it is intimately involved in energy transfer processes vital for nodule development and the energy-intensive process of nitrogen fixation Hiama et al. (2019). Vanek and Lehmann (2015) highlighted the role of biochar in enhancing phosphorus availability to beans, especially through interactions with mycorrhizas, which further facilitate nutrient uptake Vanek and Lehmann 2015. Similarly, Phares et al. (2020) demonstrated that biochar, both alone and co-applied with triple superphosphate, positively influenced nutrient uptake and nodule formation in cowpea Phares et al. (2020). By ensuring a more consistent supply of crucial nutrients like phosphorus, biochar indirectly supports the metabolic demands of the symbiosis.

- **Enhanced Water Retention**: Biochar's highly porous structure significantly improves the soil's water holding capacity He et al. (2023). This enhanced water retention is particularly beneficial in drought-prone regions or during periods of water stress, as water availability is a critical factor influencing rhizobial activity and nodule function He et al. (2023). Ma et al. (2019a) observed that biochar application mitigated the negative effects of water deficit on soybean nodulation and symbiotic performance in sandy soils. By maintaining optimal soil moisture levels, biochar helps to create a more stable and less stressful environment for the legume-rhizobia association, thereby sustaining high rates of nitrogen fixation.
- **Mitigation of Environmental Stressors**: Beyond nutrient and water management, biochar can also play a crucial role in mitigating various environmental stressors that negatively impact legume growth and rhizobial activity. Its ability to adsorb heavy metals, organic pollutants, and other toxic compounds can detoxify contaminated soils, creating a healthier root environment Arif et al. (2021). This detoxification effect reduces stress on both the host plant and the rhizobia, allowing the symbiotic system to function more efficiently. Furthermore, biochar can buffer against extreme temperature fluctuations, providing a more stable thermal environment for microbial activity. These protective effects contribute to the overall resilience and productivity of the legume-rhizobia symbiosis in challenging soil conditions. These multifaceted mechanisms collectively contribute to the observed enhancements in nodulation and nitrogen fixation, solidifying biochar's role as a valuable amendment for optimizing legume productivity and sustainable nitrogen cycling in agricultural systems.

3.2.4 Synergies with Phosphorus Solubilizers

Phosphorus (P) is a macronutrient vital for plant growth, particularly for legumes, as it plays a crucial role in energy metabolism, nodule development, and the overall process of nitrogen fixation. However, P is often a limiting nutrient in many agricultural soils due to its low solubility and high fixation rates. Phosphorus-solubilizing bacteria (PSB) are a group of beneficial microorganisms that can convert insoluble forms of soil P into plant-available forms, thereby improving P acquisition by crops.

The integration of biochar with PSB has emerged as a promising strategy to enhance P availability and uptake in legume-based systems. Biochar can promote the activity of PSB through several mechanisms. Its porous structure provides a suitable habitat for PSB, protecting them and facilitating their proliferation Fan et al. (2025). Furthermore, biochar can influence soil redox properties and stimulate the secretion of organic acids by microorganisms, which are key mechanisms for P solubilization Hiama et al. (2019). Research by Hiama et al. (2019) showed that the combination of biochar and phosphorus application significantly improved nutrient uptake and biological nitrogen fixation in cowpea, highlighting the synergistic potential of this interaction. The enhanced P availability, coupled with improved N fixation, creates a powerful synergy that boosts overall legume productivity and nutrient use efficiency.

3.3 Potential for Reduced Synthetic N Fertilizer Use

The enhanced biological nitrogen fixation (BNF) achieved through the synergistic application of biochar, rhizobia, and phosphorus solubilizers presents a significant opportunity to reduce the reliance on synthetic nitrogen (N) fertilizers in pulse-based cropping systems. Synthetic N fertilizers are energy-intensive to produce, contribute to greenhouse gas emissions, and can lead to environmental issues such as nitrate leaching and eutrophication Billah et al. (2019).

By optimizing BNF, biochar-amended systems can provide a substantial portion of the nitrogen required by pulse crops, thereby decreasing the need for external N inputs. This not only offers economic benefits to farmers by reducing input costs but also contributes significantly to more sustainable and environmentally friendly agricultural practices. Jalal et al. (2020) demonstrated that biochar integration with legume cover crops synergized nitrogen use efficiency and enhanced maize yield, suggesting a pathway for reducing synthetic N fertilizer use Jalal et al. (2020). Similarly, Khan et al. (2020a) found that biochar application could reduce the need for synthetic fertilizers in chickpea cultivation while improving soil fertility and yield Khan et al. (2020a, b).

Moreover, biochar's ability to improve soil N retention and reduce N losses through leaching and denitrification further supports its role in sustainable N management Yuan et al. (2024). Studies have shown that biochar can enhance the soil's capacity to retain N fertilizer and diminish its losses across diverse soil types Liao et al. (2020). This improved nitrogen cycling efficiency, coupled with enhanced BNF, makes biochar a valuable tool for transitioning towards agricultural systems with reduced reliance on synthetic N inputs, fostering greater environmental stewardship and long-term soil health.

3.4 Synergy with Inoculants and Plant Growth-Promoting Microorganisms (PGPMs)

Modern pulse cultivation often involves the use of commercial rhizobial inoculants to ensure that an effective and competitive bacterial strain is present in the soil. Biochar can be strategically employed to maximise the efficacy of these inoculants and to foster beneficial relationships with other plant growth-promoting microorganisms (PGPMs).

3.4.1 Biochar as a Carrier for Rhizobial Inoculants

Biochar's unique physical and chemical properties make it an excellent carrier material for microbial inoculants, including rhizobia. Its porous structure provides a protected microenvironment that shields the inoculated bacteria from harsh environmental conditions such as desiccation, UV radiation, and predation, thereby improving their survival rates in the soil (Shabir et al. 2023; Sarioglu et al. 2024) Traditional carriers for inoculants, such as peat, have limitations related to consistency, sterility, and environmental impact. Biochar has emerged as a superior alternative carrier material (Hale et al. 2016). Its high porosity, water-holding capacity, and ability to protect microbes make it an ideal vehicle for delivering a high-density, viable population of rhizobia to the seed or furrow. Studies have consistently shown that co-application of biochar with rhizobium inoculants leads to superior outcomes compared to applying either component alone. When rhizobia are pre-inoculated onto biochar, they are provided with a protective microhabitat from the moment of application. This 'inoculated biochar' or 'biochar-based inoculant' ensures the bacteria are placed directly in the emerging root zone, improving their chances of outcompeting less effective native rhizobia and successfully nodulating the host legume (Głodowska et al. 2017a, b). This approach not only enhances initial nodulation but also improves the long-term survival of the inoculant strain in the soil, potentially reducing the need for re-inoculation in subsequent seasons. For instance, Sun et al. (2020) demonstrated that the interaction of biochar type and rhizobia inoculation significantly increased the growth and biological nitrogen fixation of leguminous plants Sun et al. (2020). Similarly, Ahmad et al. (2015) found a synergistic effect of rhizobia and biochar on the growth and physiology of maize, indicating the broader applicability of this combined approach Ahmad et al. (2015). The improved contact between the host legume roots and viable rhizobia, facilitated by biochar, results in more efficient infection and nodule formation, leading to higher rates of BNF.

3.4.2 Amplifying the Effects of Phosphorus-Solubilising Microorganisms (PSMs)

Biological nitrogen fixation is an extremely energy-intensive process, consuming up to 20% of the plant's total photosynthate in the form of ATP (Vance 2001). Phosphorus (P) is a fundamental component of ATP, and its availability is often a key limiting factor for BNF, especially in weathered tropical soils where P is fixed in unavailable forms. Many soils harbour phosphorus-solubilising microorganisms (PSMs), such as bacteria (e.g., *Pseudomonas*, *Bacillus*) and fungi (e.g., *Penicillium*, *Aspergillus*), which produce organic acids and enzymes (phosphatases) to convert insoluble P into plant-available forms.

The benefits of biochar in enhancing legume productivity are further amplified when integrated with other beneficial microbial inputs, particularly rhizobium inoculants and phosphorus-solubilizing microorganisms. These synergistic interactions create a more robust and efficient biological system for nutrient acquisition and

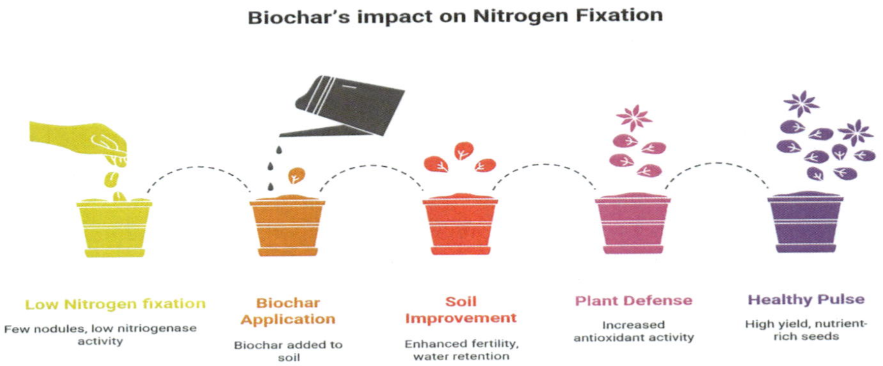

Fig. 3 Impact of biochar on nitrogen fixation

plant growth. Firstly, the improved soil pH and habitat provided by biochar also favour the growth and activity of many PSMs (Behera et al. 2014). Secondly, by co-applying biochar with PSMs and rhizobia, a powerful positive feedback loop can be established. The PSMs, thriving in the biochar-amended environment, increase the availability of P. This enhanced P supply fuels the high energy demands of the rhizobia-inhabited nodules, leading to increased nitrogen fixation. The resulting increase in plant nitrogen status promotes overall plant growth and root exudation, which in turn further supports the proliferation of the beneficial microbial community, including the PSMs (Sarfraz et al. 2017). This integrated approach, combining biochar, rhizobia, and PSMs, represents a holistic strategy for overcoming co-limitation by N and P in pulse production systems (Fig. 3).

3.5 Implications for Reduced Synthetic Nitrogen Fertiliser Use

The primary agronomic and environmental benefit of enhancing the biochar-pulse-rhizobia synergy is the potential to significantly reduce the need for synthetic nitrogen fertilisers in cropping systems. The industrial Haber-Bosch process for producing nitrogen fertiliser is energy-intensive, relying heavily on fossil fuels and contributing significantly to greenhouse gas emissions (Canfield et al. 2010). Furthermore, the overuse of these fertilisers leads to environmental problems such as nitrate leaching into groundwater and eutrophication of aquatic ecosystems. By optimising the conditions for BNF through biochar application, a greater proportion of the pulse crop's nitrogen demand can be met naturally from the atmosphere. This not only reduces the direct cost and environmental footprint associated with fertiliser application for the pulse crop itself but also generates a net N benefit for subsequent crops in the rotation. Legume residues, rich in biologically fixed nitrogen, decompose and enrich the soil, providing a natural source of N for a following cereal or vegetable crop, a phenomenon known as the 'nitrogen credit' of pulses (Peoples et al. 2009). Biochar-enhanced BNF amplifies this credit, making pulse-based rotations even more valuable for building long-term soil fertility and health.

This approach aligns perfectly with the principles of sustainable and regenerative agriculture, moving away from a reliance on external chemical inputs towards a system that leverages and enhances natural biological processes.

In conclusion, the synergy between biochar, pulses, and rhizobia is a powerful and multifaceted phenomenon. Biochar acts as a foundational soil amendment, correcting key chemical and physical limitations that inhibit symbiosis. It provides a protected habitat for rhizobia, promotes the formation of more numerous and more efficient nodules, and can be used as a superior carrier for inoculants. When integrated with other PGPMs like phosphorus solubilisers, its benefits are amplified, creating a resilient and highly productive agroecosystem. Harnessing this tripartite synergy is a key strategy for enhancing pulse yields, improving soil health, and reducing the environmental impact of agriculture.

4 Empirical Evidence and Case Studies: Quantifying the Impact of Biochar on Pulse Production

While the preceding sections have outlined the theoretical mechanisms and synergistic interactions through which biochar can enhance pulse productivity, the validation of these principles rests upon robust empirical evidence. The transition from laboratory hypotheses to tangible agronomic benefits requires rigorous evaluation under diverse field and controlled conditions. Over the past two decades, a substantial body of research has emerged, quantifying the effects of biochar application on the growth, resilience and yield of key pulse crops across various agroecological zones. This section synthesizes this evidence, drawing from specific regional case studies and broader meta-analytic reviews. It aims to delineate not only the magnitude of the observed benefits but also the critical contextual factors-such as soil type, climate and management practices-that govern the success of biochar as a soil amendment in pulse-based cropping systems. The theoretical benefits of biochar in enhancing soil properties, nutrient dynamics and legume-rhizobia interactions have been extensively discussed. However, the practical efficacy of biochar in improving pulse yield and resilience is best substantiated through empirical evidence derived from field and pot trials, as well as comprehensive meta-analyses. This section synthesizes key findings from various studies, highlighting the tangible impacts of biochar application on different pulse crops across diverse agro-ecological zones. It will delve into specific case studies, illustrating the yield increases, enhanced stress tolerance and improved plant performance observed in controlled and real-world agricultural settings. Furthermore, it will present insights from meta-analyses that provide a broader, quantitative understanding of the magnitude and variability of pulse yield responses to biochar, considering factors such as soil conditions and co-applied nutrients.

The theoretical framework for biochar's benefits in pulse agroecosystems is compelling, yet its practical value is ultimately determined by empirical performance in the field. A wealth of research has emerged globally to quantify the effects of biochar on the growth, resilience and yield of pulse crops. This section provides a comprehensive review of this evidence, adopting a hierarchical "zoom-in" approach. We begin with a broad survey of global case studies from diverse agro-ecosystems, then narrow our focus to the national context of India-the world's largest pulse producer-and finally, concentrate on the specific regional challenges and opportunities within Jammu & Kashmir and the neighbouring North Indian states. This tiered analysis, supported by specific quantitative data, illuminates the context-dependent nature of biochar's efficacy and identifies the conditions under which it delivers the most significant agronomic advantages.

4.1 Field and Pot Trials

Empirical studies conducted across various agro-ecological regions have provided robust evidence of biochar's positive impact on the growth and yield of pulse crops. These trials, ranging from controlled pot experiments to large-scale field demonstrations, highlight the practical applicability of biochar as a soil amendment for enhancing pulse productivity.

4.2 Evidence from Controlled Pot and Field-Scale Trials

Field and pot trials provide the foundational evidence for biochar's efficacy. They allow researchers to assess crop responses under specific, well-characterised environmental conditions, linking yield outcomes directly to biochar-induced changes in soil properties and plant physiology. Case studies from key pulse-growing regions highlight the context-dependent nature of these responses. Research has robustly tested biochar across a wide spectrum of environments, from the acidic, weathered soils of the tropics to the temperate, water-limited farmlands of Australia. These studies collectively demonstrate that biochar's primary benefits are linked to its ability to ameliorate specific soil constraints. In the tropical regions of South America and Sub-Saharan Africa, its liming effect and capacity to reduce aluminium toxicity are paramount. In semi-arid regions, its role in enhancing water retention is the dominant driver of success. In intensively managed systems in Asia, it can help restore soil carbon and improve nutrient cycling. Table 1 summarises a selection of these global studies, showcasing the breadth of research and the magnitude of observed effects.

The data presented in Table 1 highlight several recurring themes. In the acidic Oxisols of Colombia and Kenya, the application of biochar leads to dramatic increases in nodulation (e.g., a 103% increase in nodule weight for common bean) and subsequent nitrogen fixation, directly addressing the primary constraints of low pH and Al toxicity (Rondon et al. 2007; Kamma et al. 2015). In the semi-arid environments of Australia, biochar's hydro-physical properties come to the fore. The study by Rifflart et al. (2021) on faba bean demonstrates a significant yield increase (45%) primarily attributed to improved soil moisture retention, which is critical during the grain-filling period. Research from Asia, such as the work on soybean in China and mung bean in Indonesia, shows significant yield gains (21.7% and 42.1%,

Table 1 Summary of global case studies on biochar application in pulse crops

Crop	Location (Country)	Soil Type/ Constraint	Feedstock, Rate	Key Findings	References
Common Bean (*P. vulgaris*)	Colombia	Acidic Oxisol (pH 4.1, high Al)	Charred crop residues, 20 g kg^{-1} soil	+103% nodule dry weight; +66% total N uptake from BNF.	Rondon et al. (2007)
Cowpea (*V. unguiculata*)	Kenya	Acidic Acrisol	Maize cob, 10 t ha^{-1}	+50% grain yield under moderate drought; Soil pH increased from 4.8 to 5.5.	Kamma et al. (2015)
Faba Bean (V. *faba*)	Australia	Semi-arid Sodosol	Green waste, 10 t ha^{-1}	+45% grain yield; +10% soil moisture content at podding stage.	Rifflart et al. (2021)
Soybean (*G. max*)	China	Loess soil	Maize straw, 20 t ha^{-1}	+21.7% grain yield; +34% nodule number; +15% available P.	Zhang et al. (2017)
Mung Bean (*V. radiata*)	Indonesia	Sandy coastal soil (low OM)	Rice husk, 15 t ha^{-1}	+42.1% pod yield; +12% soil water holding capacity.	Nurida et al. (2021)
Lentil (*L. culinaris*)	Canada	Chernozem	Willow, 20 t ha^{-1}	+18% grain yield in dry years; Improved water use efficiency.	Abeyrathna et al. (2019)
Peanut (*A. hypogaea*)	Brazil	Sandy loam	Sugarcane bagasse, 15 t ha^{-1}	+31% pod yield; Increased Ca uptake and reduced pod rot.	de Sousa et al. (2020)

respectively) linked to a combination of improved soil structure, enhanced nutrient availability (P and K) and better overall soil health (Zhang et al. 2017; Nurida et al. 2021). These global cases underscore a unifying principle: biochar is a powerful tool for enhancing pulse productivity, especially in degraded or constrained soil environments.

4.3 The National Context: Validating Biochar's Efficacy Across India's Agro-climatic Zones

As the global leader in pulse production, India presents a critical testbed for biochar technology. A vast portion of its pulse cultivation occurs in rainfed regions on soils that are often low in organic matter, structurally poor and nutrient-deficient. Indian research has focused heavily on integrating biochar with traditional organic

Table 2 Summary of Indian case studies on biochar application in pulse crops

Crop	Location (State)	Soil Type/ Constraint	Biochar Feedstock, Rate	Key Quantitative Findings	References
Chickpea (*C. arietinum*)	Rajasthan	Aridisol (low OM, water scarcity)	Prosopis juliflora, 5 t ha^{-1} (+FYM)	+38% grain yield; +18% soil moisture retention; +25% nodule dry weight.	Singh et al. (2017)
Lentil (*L. culinaris*)	Madhya Pradesh	Vertisol (clayey, poor structure)	Pigeon pea stalk, 7.5 t ha^{-1}	+24% grain yield; Soil bulk density decreased by 8%.	Prakash and Kumar (2021)
Pigeon Pea (*C. cajan*)	Karnataka	Alfisol (low P and K)	Mixed wood, 5 t ha^{-1}	+29% grain yield; +22% avl. P; +18% avl. K.	Patil et al. (2018)
Black Gram (*V. mungo*)	Tamil Nadu	Sandy loam	Coconut shell, 10 t ha^{-1}	+33% seed yield; +45% root length; Increased soil microbial biomass carbon.	Rajasekar et al. (2022)
Green Gram (V. radiata)	Maharashtra	Clay Loam	Cotton stalk, 5 t ha^{-1}	+21% grain yield; Enhanced nutrient uptake (N, P, K) by 15–20%.	Jadhav et al. (2019)

amendments like farmyard manure (FYM), a practice that has shown strong synergistic effects. The studies span the country's diverse agro-climatic zones, from the arid soils of Rajasthan to the Vertisols of Central India and the Alfisols of the south. Table 2 presents data from several representative Indian studies.

The Indian case studies reinforce the global findings but with a distinct emphasis on synergy with organic manures. The seminal work by Singh et al. (2017) on chickpea in Rajasthan showed that while biochar alone was beneficial, its combination with FYM amplified the yield increase to 38%, a result of improved moisture retention and enhanced nutrient cycling. In the Vertisols of Madhya Pradesh, which are prone to cracking and poor aeration, biochar from pigeon pea stalk improved soil aggregation and led to a 24% yield increase in lentils (Prakash and Kumar 2021). Similarly, in the southern state of Karnataka, applying biochar to Alfisols increased pigeon pea yield by 29% by improving soil structure and the availability of phosphorus and potassium (Patil et al. 2018). These studies collectively demonstrate that in the Indian context, biochar is a powerful catalyst within an Integrated Soil Fertility Management (ISFM) system, enhancing the value of other organic inputs and sustainably boosting pulse yields.

4.4 The Regional Focus: Adapting Biochar for the Western Himalayas and North Indian Plains

This region presents a dichotomy of agricultural systems. The mountainous states, including Jammu & Kashmir, Himachal Pradesh and Uttarakhand, are characterized by terraced farming on acidic, shallow soils prone to erosion. In contrast, the plains of Punjab and Haryana are part of the intensively cultivated breadbasket, where decades of rice-wheat monoculture have led to declining soil organic matter and nutrient imbalances. Research in these areas focuses on tailoring biochar applications to these specific challenges. While studies from J&K are nascent, work from neighbouring Himalayan states provides a strong proxy. Table 3 summarises key findings. The regional data highlight biochar's adaptability. In the acidic Himalayan soils of Uttarakhand, pine needle biochar, a product derived from a problematic forest floor biomass, serves as an effective liming agent and soil conditioner. The study by Bhatt et al. (2021) showed it could increase French bean yield by 28% while simultaneously improving soil pH and nodulation. Similarly, work in Himachal Pradesh on field peas showed a 26% yield gain, with biochar enhancing soil structure and nutrient availability in the fragile mountain soils (Thakur et al. 2020). In the plains of Punjab, where the focus is on soil health restoration, rice straw biochar has been shown to improve soil organic carbon and increase lentil yield by 19% when used in a rice-lentil rotation, demonstrating its potential to make intensive cropping systems more sustainable (Verma et al. 2019). This regional evidence illustrates how locally sourced biochar can be a targeted solution for the distinct problems of mountain and plains agriculture in North India.

4.4.1 Case Study: Chickpea (*Cicer arietinum*)

Chickpea is a vital pulse crop in the semi-arid regions of India, where production is often constrained by low soil organic matter, nutrient deficiencies (particularly N and P) and erratic rainfall leading to terminal drought stress. Soils in these regions

Table 3 Summary of local case studies (North India) on biochar application in pulse crops

Crop	Location (State)	Soil Type/ Constraint	Feedstock, Rate	Key Quantitative Findings	References
French Bean (P. vulgaris)	Uttarakhand	Acidic Inceptisol (pH 5.4)	Pine needle, 10 t ha^{-1}	+28% bean yield; +45% nodule dry weight; Soil pH increased to 6.1.	Bhatt et al. (2021)
Field Pea (P. sativum)	Himachal Pradesh	Acidic Alfisol	Lantana camara, 8 t ha^{-1}	+26% grain yield; +12% soil organic carbon; +31% P uptake.	Thakur et al. (2020)
Lentil (L. culinaris)	Punjab	Sandy Loam (low OM)	Rice straw, 6 t ha^{-1}	+19% grain yield in rice-lentil system; Improved soil aggregation.	Verma et al. (2019)

can also be alkaline and prone to salinity, further hampering crop growth and rhizobial activity. Several field trials in this region have demonstrated the potent synergy between biochar and traditional organic amendments like farmyard manure (FYM).[1] A study by Singh et al. (2017) in Rajasthan, India, investigated the application of biochar derived from prosopis wood, both alone and in combination with FYM, on chickpea performance. Their findings were significant: while biochar alone provided a moderate yield increase, the combined application of biochar (at 5 t ha^{-1}) and FYM (at 5 t ha^{-1}) resulted in a grain yield increase of approximately 35–40% over the control. The mechanisms underpinning this substantial improvement were multifaceted. Biochar application improved soil physical conditions, notably increasing plant-available water content by 15–20% in the root zone. This enhanced moisture retention was critical for mitigating the effects of intermittent dry spells during the crucial pod-filling stage. Furthermore, the combination of biochar and FYM led to a marked improvement in soil nutrient status, increasing the availability of phosphorus and potassium and enhancing nodule number and dry weight, indicating a boost in biological nitrogen fixation (BNF). The alkaline nature of the biochar was less critical in these already alkaline soils; instead, its primary benefits stemmed from improved water retention and its function as a slow-release source and sorbent of nutrients supplied by the more labile FYM (Jien and Wang 2013). These results underscore that in water-limited environments, biochar's hydrophysical benefits are a dominant driver of yield enhancement.

4.4.2 Case Study: Lentil (*Lens culinaris*) and Mung Bean (*Vigna radiata*) in the Gangetic Plains

The fertile alluvial soils of the Gangetic Plains in Nepal and Bangladesh are critical for food security, but intensive cultivation has led to declining soil organic matter and nutrient imbalances. Lentil and mung bean are key pulses in the cropping systems of this region. Research has focused on biochar's potential to restore soil health and boost the productivity of these short-duration legumes. In a field trial in Bangladesh, Sultana et al. (2017) evaluated the effect of rice husk biochar on mung bean growth and yield. They observed that application rates of 5–10 t ha^{-1} led to significant increases in key yield components, including plant height, number of pods per plant (a 25% increase) and overall grain yield (a 30% increase). The biochar was found to increase soil pH (from 5.6 to 6.2), which likely improved the conditions for native and inoculated rhizobia. This was supported by data showing a significant increase in nodulation. Similarly, research in Nepal on lentil production in acidic soils has shown that biochar derived from locally available shrubs not only acted as a liming agent but also improved phosphorus availability (Upadhyay et al. 2020). The porous structure of the biochar improved soil aeration and drainage in these finer-textured alluvial soils, preventing waterlogging after heavy rainfall events, which can be detrimental to nodule function. These studies highlight biochar's role in addressing multiple constraints simultaneously: correcting soil acidity, improving nutrient cycling and enhancing soil physical structure, which collectively contribute to greater biomass accumulation and grain yield in lentil and mung bean.

4.4.3 Case Study: Pigeon Pea (*Cajanus cajan*) and Cowpea (*Vigna unguiculata*) in Sub-Saharan Africa

Sub-Saharan Africa (SSA) faces profound challenges of food security, compounded by highly weathered, acidic and nutrient-depleted soils (oxisols and ultisols). Pulses like pigeon pea and cowpea are staples due to their drought tolerance and nutritional value, but their yields are often low. Biochar research in SSA has been particularly focused on its potential to ameliorate soil acidity and improve crop resilience to drought. Pioneering work in the Amazon, which has similar soil constraints, provided the impetus for research in SSA. Rondon et al. (2007) showed that biochar additions to acid soils dramatically increased nodulation and BNF in common bean.[3] Building on this, studies in Kenya and Zambia have explored biochar's impact on cowpea. For example, Kamma et al. (2015) found that maize cob biochar applied to acidic soils in western Kenya significantly reduced exchangeable aluminium, a primary cause of toxicity to legume roots. This detoxification of the rhizosphere created a more favourable environment for root growth and rhizobial survival, leading to enhanced nodulation and a 50% increase in cowpea grain yield under moderate drought conditions. The high water-holding capacity of the biochar was a critical factor. During dry periods, plants in biochar-amended plots exhibited higher leaf water potential and maintained photosynthetic activity for longer, demonstrating improved drought tolerance. This resilience mechanism is vital for pigeon pea, a longer-duration crop that often faces terminal drought. By acting as a slow-release source of potassium (K), which is crucial for stomatal regulation and water-use efficiency, biochar further contributes to plant resilience (Major et al. 2010). The evidence from SSA strongly suggests that biochar is most impactful in degraded soil systems where it can simultaneously alleviate chemical toxicities, nutrient limitations and water stress.[4]

4.5 Synthesizing the Evidence: Insights from Meta-analyses

While individual case studies provide valuable, in-depth understanding, meta-analyses offer a powerful tool for synthesizing results from numerous studies to identify overarching trends and general principles. By statistically pooling data, these analyses can quantify the average effect of an intervention and, more importantly, identify the factors that explain variability in the observed response.

4.5.1 General Trends in Pulse Yield Enhancement

Several major meta-analyses on biochar's agronomic effects have been published and while many group pulses with other crops, the trends are informative. A landmark meta-analysis by Jeffery et al. (2011) reported an average crop yield increase of 10% with biochar application across 158 datasets. A subsequent, more comprehensive analysis by Liu et al. (2013) involving 423 datasets found a mean yield increase of 11%. When focusing specifically on legumes, the effects are often more pronounced. A review by O'Toole et al. (2020), which synthesized data on biochar's impact on symbiotic nitrogen fixation, concluded that biochar application leads to an average increase in nodule number, nodule weight and nitrogenase activity by 40%, 65%and 53%, respectively. This profound positive effect on the N-fixing apparatus directly translates to yield. Consequently, pulse yield responses to biochar typically fall within a range of **10–35%**, with increases sometimes exceeding 50% under specific conditions, as seen in the SSA case studies. The variability within this range is not random; it is systematically linked to underlying soil conditions and management practices.

4.5.2 Identifying the Key Drivers of Success: The Importance of Context

Meta-analyses have been instrumental in confirming that biochar is not a universally effective solution but a conditional amendment whose performance is highly context-dependent. The strongest positive effects on pulse yields are consistently observed under specific challenging soil conditions:

- **Acidic Soils**: The most significant and consistent yield improvements are reported in acidic soils (pH < 6.0). The liming effect of alkaline biochars neutralises toxic aluminium and improves the availability of base cations and phosphorus, creating a drastically better environment for both the pulse crop and its rhizobial partners (Jeffery et al. 2017). In neutral or alkaline soils, this benefit is absent and yield responses are typically much smaller or negligible.
- **Nutrient-Poor Soils**: In soils with low native fertility, the direct nutrient contribution from biochar ash (especially K, Ca, Mg and P) can provide a significant yield boost. Furthermore, biochar's high cation exchange capacity (CEC) helps retain nutrients from other sources (e.g., mineral fertiliser, manure), reducing leaching losses and improving overall nutrient use efficiency.[5] This makes biochar particularly effective in sandy soils with low CEC (Biederman and Harpole 2013).
- **Drought-Prone and Coarse-Textured Soils**: The ability of biochar to increase soil water-holding capacity is a key driver of yield increases in arid, semi-arid and sandy environments.[6] By holding more water in the root zone, biochar extends the period of crop growth during dry spells, reducing water stress and leading to higher and more stable yields. In contrast, in fine-textured, fertile soils with adequate rainfall, this particular benefit is less impactful.

4.5.3 The Essential Role of Co-applied Nutrients

A critical insight from both field trials and meta-analyses is that biochar performs best when integrated into a broader soil fertility management strategy. Biochar itself is typically low in nitrogen and its high carbon-to-nitrogen (C:N) ratio can sometimes cause short-term N immobilisation if applied without a complementary N source (Lehmann et al. 2011). The most successful outcomes, as exemplified by the chickpea case study in India, are almost always achieved when biochar is co-applied with organic amendments (manure, compost) or a targeted application of mineral fertilisers. In this integrated system, biochar acts as a soil conditioner and a stabiliser, while the co-amendment provides the readily available nutrients required for immediate crop uptake. Biochar improves the efficiency of these added nutrients, creating a synergistic effect that is greater than the sum of its parts.

In summary, the empirical evidence overwhelmingly supports the potential of biochar to enhance pulse production. Case studies from across the globe demonstrate significant yield gains, particularly in challenging agricultural landscapes. Meta-analyses confirm these findings on a broader scale, revealing that the greatest benefits-yield increases commonly in the 10–35% range-are realized in acidic, nutrient-depleted and water-limited soils. The success of biochar application is therefore not about its universal application, but about its strategic deployment as part of an integrated soil management system to alleviate specific, well-defined agronomic constraints.

5 Factors Influencing Outcomes

The efficacy of biochar as a soil amendment for enhancing pulse crop productivity is not uniform; it is a highly context-dependent outcome influenced by a complex interplay of factors. The variability in reported results, ranging from significant yield increases to negligible or even negative effects, can be largely attributed to three principal domains: the intrinsic properties of the biochar itself, the baseline conditions of the soil and climate into which it is introduced, and the agronomic practices with which it is integrated. Understanding these factors is paramount for developing predictive models and tailored application strategies to reliably harness the benefits of biochar for sustainable pulse production. This section critically examines these key influencing factors.

A. **Biochar Type and Properties**

The inherent characteristics of biochar, largely determined by its feedstock and pyrolysis conditions, play a pivotal role in dictating its impact on soil health and subsequent pulse yield. These properties influence nutrient retention, water holding capacity, microbial activity, and overall soil amelioration.

Feedstock Origin

The raw material, or feedstock, used to produce biochar is a primary determinant of its characteristics, particularly its nutrient content, ash composition, and structural integrity (Lehmann et al. 2011). Feedstocks can be broadly categorised into woody biomass, crop residues, and animal manures, each imparting distinct properties to the final biochar product. Biochars derived from woody materials are typically characterised by high fixed carbon content, low ash content, and high physical stability. Their structure is highly porous with a large surface area, making them excellent for improving soil physical properties (Gwenzi et al. 2015). While generally low in inherent nutrients, their high cation exchange capacity (CEC) allows them to effectively retain nutrients from other sources (Liang et al. 2006). Biochars produced from agricultural residues like wheat straw or rice husks tend to have higher ash content and greater concentration of plant-available nutrients (especially K, Si, and

P) compared to woody biochars (Mukherjee and Lal 2014). The application of such biochars can provide a direct, albeit modest, source of nutrients for pulse crops. Manure-based biochars are characterised by a high ash content and are rich in nutrients, particularly phosphorus (P), potassium (K), calcium (Ca), and magnesium (Mg) (Cantrell et al. 2012). For pulse crops, which have a high demand for P, manure-based biochars can serve as a slow-release P fertilizer (Khan et al. 2020a, b). Care must be taken, however, as these biochars can also contain elevated levels of salts and potentially toxic heavy metals (Uchimiya et al. 2011).

Pyrolysis Temperature

Pyrolysis temperature is arguably the most critical process parameter influencing the final properties of biochar. A fundamental trade-off exists: low-temperature pyrolysis preserves nutrient content, while high-temperature pyrolysis enhances carbon stability and surface area. Biochars produced at lower temperatures retain more volatile matter and have more oxygen-containing functional groups, resulting in a higher CEC per unit of carbon (Mukherjee et al. 2011). They also retain more inherent nutrients (e.g., N, P, K), offering immediate nutrient benefits for pulses. As temperature increases, the fixed carbon content, aromaticity, stability, and specific surface area increase significantly (Ahmad et al. 2014). This enhances the biochar's capacity to adsorb water and solutes. High-temperature biochars also tend to have a higher pH, making them more effective liming agents for acidic soils (Vaccari et al. 2011), which can be particularly beneficial for supporting sustained rhizobial activity.

pH, Surface Area, and Ash Content

Biochars are typically alkaline, and this liming effect can raise soil pH into the optimal range for pulse growth and rhizobial function, reducing the toxicity of aluminium (Al^{3+}) and manganese (Mn^{2+}) (Vaccari et al. 2011). The specific surface area (SSA) provides abundant sites for the adsorption of water and nutrients, and the micropores can serve as a protective refuge for rhizobia against predation and environmental stresses (Quilliam et al. 2013a). The ash fraction, particularly from manure or crop residue biochars, can supply significant amounts of K, Ca, Mg, and P, contributing directly to the nutritional needs of pulse crops (Fidel et al. 2017).

Table 7 provides a detailed summary of specific studies investigating the effects of different biochar types on various legume crops, highlighting the importance of matching biochar characteristics to crop and soil needs (Table 4).

B. **Soil Type and Climate**

The magnitude of the response of pulse crops to biochar application is profoundly influenced by the initial state of the soil and the prevailing climatic conditions. Benefits are most pronounced where biochar addresses specific, pre-existing limitations.

Table 4 Influence of feedstock and pyrolysis temperature on biochar properties and implications for pulse production

Parameter	Low Temp. (<500 °C)	High Temp. (>600 °C)	Implications for Pulse Production
Feedstock	**Woody Waste**	**Crop Residue**	**Manure**
Fixed Carbon	Moderate	High	High C stability for long-term soil structure improvement (Lehmann et al. 2011).
Nutrient Content	Low	Moderate	High (Cantrell et al. 2012; Mukherjee and Lal 2014)
Surface Area	Moderate-High	High-Very High	Enhanced water & nutrient retention (Ahmad et al. 2014), crucial for drought mitigation.
pH	Moderate	High	High Liming Effect
Labile C/ Volatiles	High	Low	Energy source for microbes, stimulating nutrient cycling (Mukherjee et al. 2011).
Stability in Soil	Lower	Higher	Carbon sequestration and persistent improvement of soil physical properties (Spokas 2010).

Soil Type and Condition

The greatest positive responses are observed in highly weathered, degraded, or coarse-textured soils, whereas fertile soils often show a muted response (Jeffery et al. 2011; Biederman and Harpole 2013).

Sandy, Loamy, or Degraded Soils

Biochar application tends to elicit a more significant positive response in soils that are inherently poor in fertility, have low water holding capacity, or are degraded. Sandy and loamy soils, characterized by their coarse texture and often low organic matter content, particularly benefit from biochar amendment Bhattarai et al. (2015). In these soils, biochar's porous structure and high surface area can substantially improve water retention, enhance nutrient availability by reducing leaching, and provide a stable habitat for microbial communities (Kumar. 2019). For instance, studies have shown that biochar can increase the water holding capacity of sandy loam soils by a considerable margin, leading to improved plant performance, especially under water-stressed conditions (Yu et al. 2024). Similarly, degraded soils, often characterized by nutrient depletion, low pH, and poor structure due to intensive agricultural practices or erosion, can be significantly rehabilitated through biochar application. Biochar can ameliorate soil acidity, increase cation exchange capacity (CEC), and promote the aggregation of soil particles, thereby restoring soil health and productivity (Mohammed 2023). The beneficial effects are less pronounced in already fertile or clay-rich soils, where the existing soil structure and nutrient buffering capacity are already optimal (Major et al. 2010). Biochar's liming effect raises soil pH, reducing Al and Mn toxicity and increasing the availability of P and Mo, creating a more favourable environment for both the host legume and its symbiotic rhizobia (Vaccari et al. 2011; Rondon et al. 2007). In soils depleted of organic matter, biochar adds stable carbon, enhances soil aggregation, and increases

CEC, helping to restore soil health and fertility for sustainable pulse production (Lal 2016).

Climatic Conditions

Semi-Arid to Sub-Humid Climates

Climatic conditions, particularly rainfall patterns and temperature, also influence the efficacy of biochar. Biochar application often yields the most favorable results in semi-arid to sub-humid climates, where water scarcity and nutrient limitations are common challenges for agricultural production. In these regions, biochar's ability to improve soil water retention becomes critically important, helping crops to withstand drought periods and utilize available moisture more efficiently. Furthermore, the enhanced nutrient retention properties of biochar can mitigate nutrient losses through leaching, which is particularly beneficial in areas with erratic rainfall patterns. The stable carbon structure of biochar also contributes to carbon sequestration, which is a significant advantage in mitigating climate change impacts in these vulnerable regions. While biochar can offer benefits in other climates, its role in improving water and nutrient use efficiency is especially valuable in water-limited environments, leading to more consistent and higher pulse yields (Karhu et al. 2011; Haider et al. 2017). In wetter climates, the primary benefit shifts to nutrient retention. Biochar's high CEC reduces the leaching of key nutrients like nitrate, improving overall fertilizer use efficiency (Clough et al. 2013) (Table 5).

Table 5 Biochar efficacy in relation to soil type and climate for pulse cultivation

Condition	Primary Limitation	Mechanism of Biochar Action	Expected Outcome for Pulses
Sandy Soil	Low water & nutrient retention	Increased WHC & CEC (Major et al. 2010).	Reduced drought stress, improved yield stability.
Acidic Soil	Al/Mn toxicity, low P/Mo availability	Liming effect, detoxification of Al^{3+}, increased P availability (Vaccari et al. 2011).	Enhanced root growth, improved nodulation and N-fixation (Rondon et al. 2007).
Degraded Soil	Low organic matter, poor structure	Adds stable C, improves aggregation, increases CEC (Lal 2016; Steiner et al. 2007).	Restoration of soil health, increased fertility, sustained productivity.
Fertile Clay Soil	Few physical/chemical limitations	Minimal immediate impact on existing properties.	Negligible or minor short-term yield response (Jeffery et al. 2011).
Semi-Arid Climate	Water scarcity, drought stress	Increased soil water holding capacity (Karhu et al. 2011).	Improved plant water status, delayed wilting, stabilized/increased yield (Haider et al. 2017).
Humid Climate	High nutrient leaching	Increased CEC reduces leaching of NO_3^- and NH_4^+ (Clough et al. 2013).	Improved nutrient use efficiency, reduced fertilizer loss (Liang et al. 2006).

C. **Integration with Agronomic Practices**

The full potential of biochar in enhancing pulse yield is often realized when its application is integrated with other sound agronomic practices. Synergistic effects can be achieved by combining biochar with complementary amendments and by optimizing its timing and placement.

Synergies with Other Amendments

While biochar offers numerous benefits on its own, its effectiveness can be significantly amplified when co-applied with other soil amendments such as compost, microbial inoculants, or mineral fertilizers. Combining biochar with compost creates a "super-amendment" where the biochar provides a stable structure and adsorbs nutrients released from the compost, ensuring a balanced, slow-release supply (Fischer and Glaser 2012). Similarly, integrating biochar with rhizobial inoculants, which are crucial for nitrogen fixation in pulse crops, can improve nodulation and nitrogen uptake. The porous structure of biochar provides an ideal habitat for rhizobia, protecting them from abiotic stresses and predation, leading to increased survival, more effective nodulation, and enhanced N-fixation (Quilliam et al. 2013a, b; Rondon et al. 2007). When used in conjunction with mineral fertilizers, Biochar can significantly improve the use efficiency of synthetic fertilizers. Its high CEC captures and retains cations like NH_4^+ and K^+, reducing their loss through leaching and ensuring a more sustained supply to the crop (Clough et al. 2013).

Timing and Placement of Application

Optimizing the timing and placement of biochar application is crucial for maximizing its benefits to pulse crops. Applying biochar directly into the root zone ensures that the plant roots have immediate access to the improved soil environment and nutrients. Various application methods, such as banding, spot application, or incorporation into the topsoil, can be employed depending on the crop, soil type, and farming system. For pulse crops, which rely heavily on efficient nutrient uptake and symbiotic nitrogen fixation, precise placement near the developing root system can significantly enhance early growth and nodulation. Studies have indicated that targeted application methods, such as placing biochar in planting furrows or around individual plants, can lead to superior yield responses compared to broadcast application, especially in the initial years after application (Macil 2018). The timing of application, ideally before or at planting, allows sufficient time for biochar to interact with the soil and establish its beneficial effects before the critical growth stages of the pulse crop. Therefore, a well-considered strategy for biochar timing and placement is essential for achieving optimal outcomes in pulse cultivation (Tables 6 and 7).

Table 6 Synergistic effects of integrating biochar with other agronomic practices

Integrated Practice	Mechanism	Benefit for Pulse Production
Co-application with Compost/ Manure	Biochar adsorbs/stabilizes nutrients from compost; compost "charges" biochar (Fischer and Glaser 2012).	Provides a balanced, slow-release nutrient supply. Enhances soil biological activity.
Use with Rhizobial Inoculants	Biochar acts as a protective habitat and carrier for rhizobia (Quilliam et al. 2013a, b). Improves micro-site conditions.	Increased survival of inoculant strains. Enhanced nodulation and biological nitrogen fixation (Rondon et al. 2007).
Combination with Synthetic Fertilizers	High CEC reduces leaching of NH_4^+, K^+ (Clough et al. 2013; Liang et al. 2006). Liming effect increases P availability.	Improved fertilizer use efficiency. Reduced fertilizer costs and environmental pollution.
Banding in Root Zone	Concentrates biochar's benefits in the rhizosphere (Buss et al. 2022).	More efficient use of biochar. Enhanced early seedling vigour and root development.

Table 7 Summary of selected studies on the effect of biochar feedstock and temperature on legume crops

Legume Crop	Biochar Feedstock	Pyrolysis Temp (°C)	Rate (t ha^{-1})	Key Findings on Legume Performance	References
Common bean (*Phaseolus vulgaris*)	Hardwood	~450	20	Increased nodule number (>300%), nodule weight (>400%), and N-fixation in an acidic Oxisol.	Rondon et al. (2007)
Soybean (*Glycine max*)	Corncob	400	15	Increased nodule dry weight by 71%, shoot biomass by 31%, and seed yield by 26% in a sandy loam soil.	Liu et al. (2012)
Chickpea (*Cicer arietinum*)	Sugarcane bagasse	500	5, 10	Increased grain yield by 21–34%, improved water use efficiency, and enhanced nutrient uptake (N, P, K) under drought stress.	Haider et al. (2017)
Mung bean (*Vigna radiata*)	Poultry litter	400	5, 10	Significantly increased P uptake and grain yield (up to 46%) in a P-deficient soil; acted as a slow-release P fertilizer.	Khan et al. (2020a, b)

(continued)

Table 7 (continued)

Legume Crop	Biochar Feedstock	Pyrolysis Temp (°C)	Rate (t ha^{-1})	Key Findings on Legume Performance	References
Lentil (*Lens culinaris*)	Wheat straw	300, 500	10	Low-temp (300 °C) biochar increased grain yield by 35% and N uptake by 48%; high-temp biochar had less effect.	Nigussie et al. (2012)
Cowpea (*Vigna unguiculata*)	Rice husk	400	20	Increased plant height, leaf area, nodulation, and grain yield, especially when combined with rhizobium inoculation.	Okon et al. (2022)
Faba bean (*Vicia faba*)	Eucalyptus wood	550	10, 20	Increased plant biomass and N accumulation; improved soil physical properties and microbial biomass C.	Alfadil et al. (2021)
Peanut (*Arachis hypogaea*)	Rice straw	450	1.5, 3	Application at 3 t ha^{-1} increased pod yield by 19.5% and improved soil moisture content.	Tian et al. (2021)
Common bean (*Phaseolus vulgaris*)	Coffee husk	350	7.5	Increased shoot dry matter and P uptake, particularly in low-P soil, due to liming effect and direct P supply.	Milla et al. (2013)
Soybean (*Glycine max*)	Hardwood	600	5, 10	Increased nodule numbers and grain yield; effect was more pronounced in the second year of application, indicating a legacy effect.	Güereña et al. (2013)
Alfalfa (*Medicago sativa*)	Sewage sludge	600	10	Increased biomass yield and crude protein content; reduced heavy metal uptake (Cd, Pb) by the plant.	Houben et al. (2013)
Chickpea (*Cicer arietinum*)	Mesquite wood	400	10	Enhanced symbiotic N-fixation, improved plant water relations, and increased grain yield under saline conditions.	Ullah et al. (2020)

References

Abeyrathna S, Haughian S, Schoenau J (2019) The effect of willow biochar on soil properties and growth of lentil and canola in a Saskatchewan Chernozem. Commun Soil Sci Plant Anal 50(12):1477–1488. https://doi.org/10.1080/00103624.2019.1619890

Ahmad M, Rajapaksha AU, Lim JE, Zhang M, Bolan N, Mohan D, Vithanage M, Park SS, Ok YS (2014) Biochar as a sorbent for contaminant management in soil and water: a review. Chemosphere 99:19–33. https://doi.org/10.1016/j.chemosphere.2013.10.071

Ahmad MT, Asghar HN, Saleem M, Khan MY, Zahir ZA (2015) Synergistic effect of rhizobia and biochar on growth and physiology of maize. Agron J 107(6):2327–2333. https://doi.org/10.2134/agronj15.0212

Alfadil AA, Habil AA, Elhag AB, Ali AO (2021) Effect of biochar on faba bean (Vicia faba L.) growth and some soil properties. Int J Agric Biol 25(3):551–558. https://doi.org/10.17957/IJAB/15.1706

Alkharabsheh HM, Seleiman MF, Battaglia ML, Shami A, Jalal RS, Alhammad BA, Almutairi KF, Al-Saif AM (2021) Biochar and its broad impacts in soil quality and fertility, nutrient leaching and crop productivity: a review. Agronomy 11(5):993

Anyebe O, Sadiq FK, Manono BO, Matsika TA (2025) Biochar characteristics and application: effects on soil ecosystem services and nutrient dynamics for enhanced crop yields. Nitrogen 6(2):31

Arif M, Ikramullah K, Jan T, Riaz M, Akhtar K, Ali S (2021) Biochar and leguminous cover crops as an alternative to summer fallowing for soil organic carbon and nutrient management in the wheat-maize-wheat cropping system. J Soils Sediments 21(2):1010–1021. https://doi.org/10.1007/s11368-020-02866-y

Atkinson CJ, Fitzgerald JD, Hipps NA (2010) Potential mechanisms for achieving agricultural benefits from biochar application to temperate soils: a review. Plant Soil 337(1–2):1–18. https://doi.org/10.1007/s11104-010-0464-5

Atluri S, Thakur D, Bukke D, Ramawat N (2021) Evaluation of biofertilizers and biochar on the plant growth and productivity of soybean (Glycine max). Int J Plant Soil Sci 33(15):74–86

Ayaz M, Feizienė D, Tilvikienė V, Akhtar K, Stulpinaitė U, Iqbal R (2021) Biochar role in the sustainability of agriculture and environment. Sustainability 13(3):1330

Behera BC, Singdevsachan SK, Mishra RR, Dhal NK, Thatoi H (2014) Diversity, potential and applications of phosphate solubilising microorganisms (PSM) in rehabilitation of degraded lands: a review. J Geosci Geomat 2(3):106–114. https://doi.org/10.12691/jgg-2-3-3

Bhatt R, Kukal SS, Busari MA, Arora S, Yadav M (2021) Pine needle biochar improves soil properties and nitrogen-use efficiency of French bean (*Phaseolus vulgaris*) in an acidic soil of Northwest Himalayas. Catena 202:105260. https://doi.org/10.1016/j.catena.2021.105260

Bhattarai B, Neupane J, Dhakal SP (2015) Effect of biochar from different origin on physio-chemical properties of soil and yield of garden pea (Pisum sativum L.) at Paklihawa, Rupandehi, Nepal. World J Agric Res 3(4):100–104. Available at: https://www.academia.edu/download/103110848/wjar-3-4-3.pdf. Accessed 21 Aug 2025

Biederman LA, Harpole WS (2013) Biochar and its effects on plant productivity and nutrient cycling: a meta-analysis. GCB Bioenergy 5(2):202–214. https://doi.org/10.1111/gcbb.12037

Billah MM, Ahmad W, Ali M (2019) Biochar particle size and Rhizobia strains effect on the uptake and efficiency of nitrogen in lentils. J Plant Nutr 42(10):1211–1222. https://doi.org/10.1080/01904167.2019.1628984

Buss W, Assavavittayanon K, Shepherd J, Heal K, Sohi S (2022) Biochar phosphorus and potassium availability for maize in a tropical soil. Geoderma 408:115589. https://doi.org/10.1016/j.geoderma.2021.115589

Canfield DE, Glazer AN, Falkowski PG (2010) The evolution and future of Earth's nitrogen cycle. Science 330(6001):192–196. https://doi.org/10.1126/science.1186120

Cantrell KB, Hunt PG, Uchimiya M, Novak JM, Ro KS (2012) Impact of pyrolysis temperature and manure source on physicochemical characteristics of biochar. Bioresour Technol 107:419–428. https://doi.org/10.1016/j.biortech.2011.11.084

Chen M-W, Chang M-S, Mao Y, Hu S, Kung C-C (2023a) Machine learning in the evaluation and prediction models of biochar application: a review. Sci Prog 106(1):00368504221148842

Chen Y, Xu M et al (2023b) A critical review of biochar application for the remediation of greenhouse gas emissions and nutrient loss in rice paddies: characteristics, mechanisms, and future recommendations. Agronomy 13(3):893

Claire Grierson MWC, Barnes S (2011) One hundred important questions facing plant science research. Wiley

Clough TJ, Condron LM, Kammann C, Müller C (2013) A review of biochar and soil nitrogen dynamics. Agronomy 3(2):275–293. https://doi.org/10.3390/agronomy3020275

Dayoub EB, Tóth Z, Soós G, Anda A (2024) Chemical and physical properties of selected biochar types and a few application methods in agriculture. Agronomy 14(11):2540

De Medeiros EV, De Souza CAF, Da Costa DP, De Melo RE, De Lima EJ, da Silva Araújo I, de Souza Junior JH, de Sousa Lima JR, Pinto KMS, Hammecker C (2023) A meta-analysis of biochar application to manage plant diseases caused by bacterial pathogens. Biosci J 39:e39095

de Sousa TF, de Mello PR, de Morais AV, de Oliveira LA (2020) Sugarcane bagasse biochar enhances peanut yield and quality in sandy soils. J Soil Sci Plant Nutr 20(3):1345–1354. https://doi.org/10.1007/s42729-020-00234-x

Debnath S, Rawat D, Mukherjee AK, Adhikary S, Kundu R (2019) Applications and constraints of plant beneficial microorganisms in agriculture. In: Biostimulants in plant science. IntechOpen

Deepa MPM (2019) Assessing pulse production in India–application of decomposition and instability analysis. Int J Curr Microbiol App Sci 8(8):804–811

Egamberdieva D, Li L, Ma H, Wirth S, Bellingrath-Kimura SD (2019) Soil amendment with different maize biochars improves chickpea growth under different moisture levels by improving symbiotic performance with Mesorhizobium ciceri. Front Microbiol 10:2423. https://doi.org/10.3389/fmicb.2019.02423

Egamberdieva D, Ma H, Reckling M, Omari RA, Wirth S, Bellingrath-Kimura SD (2021) Interactive effects of biochar, nitrogen, and phosphorous on the symbiotic performance, growth, and nutrient uptake of soybean (Glycine max L.). Agronomy 12(1):27

Erana K (2021) Contribution, utilization, and improvement of legumes-driven biological nitrogen fixation in agricultural systems. Front Sustain Food Syst 5. https://www.frontiersin.org/journals/sustainable-food-systems/articles/10.3389/fsufs.2021.767998. https://doi.org/10.3389/fsufs.2021.767998

Erfatpour M, MacLean D, Lahlali R, Jiang Y (2024) Ovule and seed development of crop plants in response to climate change. Front Sustain Food Syst 8:1495610. https://www.frontiersin.org/journals/sustainable-food-systems/articles/10.3389/fsufs.2024.1495610. https://doi.org/10.3389/fsufs.2024.1495610

Fan B, Zhao L, Yang F, Zhao C, Li Z (2025) Biochar promotes phosphorus solubilization by reconstructing soil organic acid and microorganism networks. Agronomy 15(5):1163

Farhangi-Abriz S, Torabian S, Qin R, Noulas C (2021) Biochar effects on yield of cereal and legume crops using meta-analysis. Sci Total Environ 766:144371. https://doi.org/10.1016/j.scitotenv.2020.144371

Farhangi-Abriz S, Ghassemi-Golezani K, Farhangi-Abriz S (2024) A meta-analysis to estimate the potential of biochar in improving nitrogen fixation and plant biomass of legumes. Biomass Convers Biorefinery. https://doi.org/10.1007/s13399-022-02530-0

Feng W, Wang T, Yang F, Cen R, Liao H, Qu Z (2023) Effects of biochar on soil evaporation and moisture content and the associated mechanisms. Environ Sci Eur 35(1):66

Ferguson BJ, Indrasumunar A, Hayashi S, Lin MH, Lin YH, Reid DE, Gresshoff PM (2010) Molecular analysis of legume nodule development and autoregulation. J Integr Plant Biol 52(1):61–76. https://doi.org/10.1111/j.1744-7909.2010.00899.x

Fidel RB, Laird DA, Thompson ML, Lawrinenko M (2017) Characterization and quantification of the anhydro-sugars in biochars. J Anal Appl Pyrolysis 128:233–241. https://doi.org/10.1016/j.jaap.2017.08.019

Fischer D, Glaser B (2012) Synergisms between compost and biochar for sustainable soil amelioration. In: Kumar S, Bharti A (eds) Management of organic waste. IntechOpen. https://doi.org/10.5772/31200

Foy CD (1984) Physiological effects of hydrogen, aluminum, and manganese toxicities in acid soil. In: Adams F (ed) Soil acidity and liming, 2nd edn. American Society of Agronomy, Madison, pp 57–97. https://doi.org/10.2134/agronmonogr12.2ed.c2

Glaser B, Lehmann J, Zech W (2002) Ameliorating physical and chemical properties of highly weathered soils in the tropics with charcoal – a review. Biol Fertil Soils 35(4):219–230. https://doi.org/10.1007/s00374-002-0466-4

Glaser B, Wiedner K, Seelig S, Schmidt HP, Gerber H (2015) Biochar organic fertilizers from natural resources as substitute for mineral fertilizers. Agron Sustain Dev 35(2):667–678. https://doi.org/10.1007/s13593-014-0251-4

Głodowska M, Husk B, Schwinghamer T, Smith DL (2017a) Biochar is a novel carrier for rhizobia and promotes nodulation in soybean. Agron Sustain Dev 37(6):55. https://doi.org/10.1007/s13593-017-0465-9

Głodowska M, Schwinghamer T, Husk B, Smith D (2017b) Biochar based inoculants improve soybean growth and nodulation. Agric Sci 8:1048–1064

Graham PH, Vance CP (2003) Legumes: importance and constraints to greater use. Plant Physiol 131(3):872–877. https://doi.org/10.1104/pp.017004

Güereña D, Lehmann J, Hanley K, Enders A, Hyland C, Riha S (2013) Nitrogen dynamics following field application of biochar in a temperate soil. Plant Soil 365(1–2):239–254. https://doi.org/10.1007/s11104-012-1383-4

Gul S, Whalen JK (2016) Biochar can improve biological nitrogen fixation by altering the root exudation of legumes. Sci Total Environ 572:140–147. https://doi.org/10.1016/j.scitotenv.2016.07.168

Gwenzi W, Chaukura N, Mukome FN, Machado S, Nyamasoka B (2015) Biochar production and applications in sub-Saharan Africa: opportunities, constraints, and future prospects. Sustainability 7(10):13578–13615. https://doi.org/10.3390/su71013578

Haider G, Koyro HW, Azam F, Steffens D, Müller C, Kammann C (2017) Biochar reduced temperature-and drought-stress-induced limitations to chickpea (Cicer arietinum L.) productivity. J Agron Crop Sci 203(1):17–31. https://doi.org/10.1111/jac.12165

Hale L, Luth M, Crowley D (2014) Biochar characteristics are correlated to impacts on soybean growth and nodulation in an acid soil. Soil Sci Soc Am J 78(1):193–203. https://doi.org/10.2136/sssaj2013.06.0210

Hale L, Evans F, Crowley D (2016) Can biochar be used as a carrier for rhizobia? Evaluation of alfalfa and bean growth. J Soils Sediments 16(1):249–261. https://doi.org/10.1007/s11368-015-1175-1

He K, Liu Q, Zhang J, Zhang G, Li G (2023) Biochar enhances the resistance of legumes and soil microbes to extreme short-term drought. Plants 12(24):4155

Hiama PD, Ewusi-Mensah N, Logah V (2019) Nutrient uptake and biological nitrogen fixation in cowpea under biochar–phosphorus interaction. J Anim Plant Sci 29(3):732–740. https://doi.org/10.36899/japs.2019.3.0259

Hossain MZ, Bahar MM, Sarkar B, Donne SW, Ok YS, Palansooriya KN, Kirkham MB, Chowdhury S, Bolan N (2020) Biochar and its importance on nutrient dynamics in soil and plant. Biochar 2(4):379–420

Houben D, Evrard L, Sonnet P (2013) Mobility, bioavailability and pH-dependent leaching of cadmium, zinc and lead in a contaminated soil amended with biochar. Chemosphere 92(11):1450–1457. https://doi.org/10.1016/j.chemosphere.2013.03.055

Idbella M, Baronti S, Giagnoni L, Renella G, Becagli M, Cardelli R et al (2024) Long-term effects of biochar on soil chemistry, biochemistry, and microbiota: results from a 10-year field vineyard experiment. Appl Soil Ecol 195:105217

Indrati R (2021) Chapter 3: Bioactive peptides from legumes and their bioavailability. In: Jimenez-Lopez JC, Clemente A (eds) Legumes research. IntechOpen. https://doi.org/10.5772/intechopen.99979
Jadhav S, Kumar A, Meena R (2019) Influence of cotton stalk biochar on soil properties and green gram productivity in a clay loam soil. Indian J Agric Sci 89(4):651–655
Jalal F, Arif M, Akhtar K, Khan A, Naz M, Said F, Ali S (2020) Biochar integration with legume crops in summer gape synergizes nitrogen use efficiency and enhance maize yield. Agronomy 10(1):58. https://doi.org/10.3390/agronomy10010058
Jeffery S, Martijn Bezemer GC (2013) The way forward in biochar research: targeting tradeoffs between the potential wins. Wiley
Jeffery S, Verheijen FG, van der Velde M, Bastos AC (2011) A quantitative review of the effects of biochar application to soils on crop productivity using meta-analysis. Agric Ecosyst Environ 144(1):175–187. https://doi.org/10.1016/j.agee.2011.08.015
Jeffery S, Abalos D, Prodana M, Bastos AC, van Groenigen JW, Hungate BA, Verheijen F (2017) Biochar boosts tropical but not temperate crop yields. Environ Res Lett 12(5):053001. https://doi.org/10.1088/1748-9326/aa67f8
Jien SH, Wang CS (2013) Effects of biochar on soil properties and erosion potential in a highly weathered soil. Catena 110:225–233. https://doi.org/10.1016/j.catena.2013.06.021
Joshi PK, Rao PP (2017) Global pulses scenario: status and outlook. Ann N Y Acad Sci 1392(1):6–17
Juriga M, Šimanský V (2018) Effect of biochar on soil structure – review. Slovak University of Agriculture
Kabir E, Kim KH, Kwon EE (2023) Biochar as a tool for the improvement of soil and environment. Front Environ Sci 11:1324533
Kamma VP, Mrema J, Ocen D (2015) Effect of maize cob biochar on nodulation and yield of cowpea in acidic soils of Western Kenya. J Agric Environ Sci 4(1):21–34
Karhu K, Mattila T, Bergström I, Regina K (2011) Biochar addition to agricultural soil increased CH_4 uptake and water holding capacity–results from a short-term pilot field study. Agric Ecosyst Environ 140(1–2):309–313. https://doi.org/10.1016/j.agee.2010.12.005
Khan S, Shah Z, Mian IA, Dawar K, Tariq M, Khan B, Ali S (2020a) Soil fertility, N2 fixation and yield of chickpea as influenced by long-term biochar application under mung–chickpea cropping system. Sustainability 12(21):9008. https://doi.org/10.3390/su12219008
Khan M, Khan MA, Ullah N, Ali S, Shah SA, Fahad S (2020b) Poultry litter biochar as a soil amendment improves phosphorus availability, and growth and yield of mung bean. J Soil Sci Plant Nutr 20(2):557–567. https://doi.org/10.1007/s42729-019-00155-y
Khan S, Irshad S, Mehmood K, Hasnain Z, Nawaz M, Rais A, Gul S, Wahid MA, Hashem A, Allah E, Ibrar D (2024) Biochar production and characteristics, its impacts on soil health, crop production, and yield enhancement: a review. Plants 13(2):166
Lal R (2016) Biochar and soil carbon sequestration. In: Lal R, Kimble JM, Sainju UM, Follett RF (eds) Agricultural practices and policies for carbon sequestration in soil. CRC Press, pp 203–220
Layzell DB, Hunt S (1990) Oxygen and the regulation of nitrogen fixation in legume nodules. Physiol Plant 80(2):322–327. https://doi.org/10.1111/j.1399-3054.1990.tb04413.x
Lehmann J, Rillig MC, Thies J, Masiello CA, Hockaday WC, Crowley D (2011) Biochar effects on soil biota–a review. Soil Biol Biochem 43(9):1812–1836. https://doi.org/10.1016/j.soilbio.2011.04.022
Liang B, Lehmann J, Solomon D, Kinyangi J, Grossman J, O'Neill B, Skjemstad JO, Thies J, Luizão FJ, Petersen J, Neves EG (2006) Black carbon increases cation exchange capacity in soils. Soil Sci Soc Am J 70(5):1719–1730. https://doi.org/10.2136/sssaj2005.0383
Liao J, Liu X, Hu A, Song H, Chen X, Zhang Z (2020) Effects of biochar-based controlled release nitrogen fertilizer on nitrogen-use efficiency of oilseed rape (Brassica napus L.). Sci Rep 10(1):11063

Liu X, Zhang A, Ji C, Joseph S, Bian R, Li L, Pan G, Paz-Ferreiro J (2012) Biochar's effect on crop productivity and the dependence on experimental conditions—a meta-analysis of literature data. Plant Soil 357(1–2):259–272. https://doi.org/10.1007/s11104-012-1162-0

Liu X, Zhang A, Ji C, Joseph S, Bian R, Li L, Pan G, Paz-Ferreiro J (2013) Biochar's effect on crop productivity and the dependence on experimental conditions-a meta-analysis of literature data. Plant Soil 373(1–2):583–594. https://doi.org/10.1007/s11104-013-1806-x

Luo P, Zhang W, Xiao D, Hu J, Li N, Yang J (2025) Biochar-based fertilizers: advancements, applications, and future directions in sustainable agriculture—a review. Agronomy 15(5):1104

Ma H, Egamberdieva D, Wirth S, Bellingrath-Kimura SD (2019a) Effect of biochar and irrigation on soybean-rhizobium symbiotic performance and soil enzymatic activity in field rhizosphere. Agronomy 9(10):626. https://doi.org/10.3390/agronomy9100626

Ma H, Egamberdieva D, Wirth S, Li Q, Omari RA (2019b) Effect of biochar and irrigation on the interrelationships among soybean growth, root nodulation, plant P uptake, and soil nutrients in a sandy field. Sustainability 11(23):6542. https://doi.org/10.3390/su11236542

Macil PJ (2018) Effect of biochar and rhizobium innoculation on nodulation, chlorophyll content, growth and yield of chickpea (Cicer arietinum L.). Available at: https://univendspace.univen.ac.za/bitstream/11602/1159/1/Dissertation%20-%20Macil%2C%20p.j.-.pdf. Accessed 21 Aug 2025

Macil PJ, Ogola JBO, Odhiambo JJO (2020) Response of soil pH and nodulation of three chickpea genotypes to biochar and rhizobium inoculation. Commun Soil Sci Plant Anal 51(20):2631–2642. https://doi.org/10.1080/00103624.2020.1836204

Major J, Rondon M, Molina D, Riha SJ, Lehmann J (2010) Maize yield and nutrition during 4 years after biochar application to a Colombian savanna oxisol. Plant Soil 333(1–2):117–128. https://doi.org/10.1007/s11104-010-0327-0

Mesta C, Cremen G, Galasso C (2022) Urban growth modelling and social vulnerability assessment for a hazardous Kathmandu Valley. Sci Rep 12(1):6152. https://doi.org/10.1038/s41598-022-09347-x

Milla OV, Rivera EB, Huang WJ, Chien CC, Wang YM (2013) Agronomic properties and characterization of rice husk and wood biochars and their effect on the growth of waterleaf in a field experiment. J Soil Sci Plant Nutr 13(2):251–266. https://doi.org/10.4067/S0718-95162013005000022

Ministry of Agriculture and Farmers Welfare (2024) Pulses statistics. Retrieved 14 Aug 2025, from https://farmer.gov.in

Mohammed A (2023) Effects of soil amendments and rhizobium inoculation on soybean nodulation, growth and yield in the semi-deciduous forest agro-ecological zone of Ghana. Available at: https://ugspace.ug.edu.gh/items/acdbf8f5-408b-431d-ac0f-dbcc02cc00a0. Accessed 21 Aug 2025

Mukherjee, Atanu, and Rattan Lal. "The Role of Biochar in Soil Processes and Greenhouse Gas Emission." Agro-ecosystems in a Changing Climate, edited by Rattan Lal and B. A. Stewart, Springer India, 2014, pp. 337–360

Mukherjee A, Zimmerman AR, Harris W (2011) Surface chemistry of a series of hardwood and coniferous wood-derived biochars. Environ Sci Technol 45(15):6543–6551. https://doi.org/10.1021/es200865c

Murtaza G, Ahmed Z, Eldin SM, Ali B, Bawazeer S, Usman M et al (2023) Biochar-Soil-Plant interactions: a cross talk for sustainable agriculture under changing climate. Front Environ Sci 11:1059449

Nigussie A, Kissi E, Misganaw M, Ambaw G (2012) Effect of biochar application on soil properties and nutrient uptake of lettuces (Lactuca sativa) grown in chromium polluted soils. Am Euras J Agric Environ Sci 12(3):369–376

Nurida NL, Widyati E, Suryani E (2021) Rice husk biochar application to improve properties of sandy coastal soil and mung bean yield. IOP Conf Ser: Earth Environ Sci 747(1):012033. https://doi.org/10.1088/1755-1315/747/1/012033

O'Toole A, Varghese M, Tosin M, Gomez-Eslava J (2020) A meta-analysis of the effects of biochar on symbiotic nitrogen fixation. Sci Total Environ 749:141444. https://doi.org/10.1016/j.scitotenv.2020.141444

Ogawa M, Okimori Y, Takahashi F (2006) Carbon sequestration by carbonization of biomass and forestation: three case studies. Mitig Adapt Strateg Glob Chang 11(2):429–444. https://doi.org/10.1007/s11027-005-9003-8

Ojha J, Mishra Y, Tiwari D (2020) Analysis of factors in production and marketing of pulses: a review. Small 48(65):56–67

Okon IE, Ofor CO, Igiri MI, Ubi G, Akeke MU (2022) Influence of rice-husk biochar and rhizobium inoculant on growth and yield components of cowpea (Vigna unguiculata (L.) Walp) in Calabar, Nigeria. Asian J Res Agric For 8(3):1–13

Patil S, Kumar R, Reddy B (2018) Effect of woody biochar on soil properties and yield of pigeon pea in an Alfisol of Karnataka. J Indian Soc Soil Sci 66(2):210–216. https://doi.org/10.5958/0975-0325.2018.00027.6

Peoples MB, Brockwell J, Herridge DF, Rochester IJ, Alves BJR, Urquiaga S, Boddey RM, Dakora FD, Kennedy IR, Fillery IRP (2009) The contributions of nitrogen-fixing crop legumes to the productivity of agricultural systems. Symbiosis 48(1–3):1–17. https://doi.org/10.1007/s13199-009-0014-y

Phares CA, Atiah K, Frimpong KA, Danquah A, Adjei EO (2020) Application of biochar and inorganic phosphorus fertilizer influenced rhizosphere soil characteristics, nodule formation and phytoconstituents of cowpea grown on an acidic sandy loam soil. Heliyon 6(12):e05739. https://doi.org/10.1016/j.heliyon.2020.e05739

Poveda J, Martínez-Gómez Á, Fenoll C, Escobar C (2021) The use of biochar for plant pathogen control. Phytopathology 111(9):1490–1499

Prakash V, Kumar S (2021) Pigeon pea stalk biochar improves soil physical properties and lentil yield in a Vertisol. Biochar 3(2):209–219. https://doi.org/10.1007/s42773-021-00088-3

Premalatha RP, Poorna Bindu J, Nivetha E, Malarvizhi P, Manorama K, Parameswari E, Davamani V (2023) A review on biochar's effect on soil properties and crop growth. Front Energy Res 11:1092637

Quilliam RS, Glanville HC, Wade M, Jones DL (2013a) The multifaceted nature of soil–biochar interactions: a review. Eur J Soil Sci 64(1):106–121. https://doi.org/10.1111/ejss.12028

Quilliam RS, Glanville HC, Wade M, Jones DL (2013b) The Trojan house: biochar as a vector for the introduction of invasive species into soil. GCB Bioenergy 5(1):40–43. https://doi.org/10.1111/gcbb.12000

Rajasekar M, Ramalingam S, Thiyagarajan G (2022) Impact of coconut shell biochar on soil health and yield of black gram (*Vigna mungo*) in sandy loam soils. Biomass Convers Biorefinery 12(7):2891–2902. https://doi.org/10.1007/s13399-020-01103-6

Rifflart T, Bell LW, Brien C (2021) Biochar improves faba bean (*Vicia faba* L.) grain yield and soil properties in a semi-arid environment. Agronomy 11(5):977. https://doi.org/10.3390/agronomy11050977

Rondon MA, Lehmann J, Ramírez J, Hurtado M (2007) Biological nitrogen fixation by common beans (*Phaseolus vulgaris* L.) increases with bio-char additions. Biol Fertil Soils 43(6):699–708. https://doi.org/10.1007/s00374-006-0152-z

Sarfraz R, Hussain S, Shaukat F, Wang X, Solaiman ZM (2017) The influence of biochar and phosphorus-solubilizing bacteria on the C, N, and P cycling in soil. Commun Soil Sci Plant Anal 48(17):2029–2040. https://doi.org/10.1080/00103624.2017.1404179

Sarioglu A, Almaca A, Dogan K, Sarioglu A (2024) Biochar and Rhizobium applications: a promising synergy for improved soybean growth and rhizosphere microbial activities. ISPEC J Agric Sci 8(1):1–10. https://doi.org/10.5281/zenodo.10576400

Serraj R, Sinclair TR, Purcell LC (1999) Symbiotic N2 fixation response to drought. J Exp Bot 50(331):143–155. https://doi.org/10.1093/jxb/50.331.143

Shabir R, Li Y, Zhang L, Chen C (2023) Biochar surface properties and chemical composition determine the rhizobial survival rate. J Environ Manag 345:118719. https://doi.org/10.1016/j.jenvman.2023.118719

Singh DP, Shahi B, Singh KM (2016) Trends of pulses production, consumption and import in India: current scenario and strategies. In: Consumption and import in India: current scenario and strategies. Available at SSRN: https://ssrn.com/abstract=2907228 or https://doi.org/10.2139/ssrn.2907228

Singh P, Singh J, Singh R, Singh R, Singh SP (2017) Effect of biochar and farm yard manure on soil properties and yield of chickpea (*Cicer arietinum*) in a semi-arid region of India. Int J Curr Microbiol App Sci 6(8):1045–1053. https://doi.org/10.20546/ijcmas.2017.608.129

Slattery JF, Pearce DJ, Slattery WJ (2001) Effects of resident rhizobial communities and soil acidity on the response of subterranean clover to inoculation. Aust J Exp Agric 41(5):671–679. https://doi.org/10.1071/EA00067

Spokas KA (2010) Review of the stability of biochar in soils: predictability of O: C molar ratios. Carbon Manag 1(2):289–303. https://doi.org/10.4155/cmt.10.32

Spokas KA, Cantrell KB, Novak JM, Archer DW, Ippolito JA, Collins HP, Boateng AA, Lima IM, Lamb MC, McAloon AJ, Lentz RD, Nichols KA (2011) Biochar: a synthesis of its agronomic impact beyond carbon sequestration. J Environ Qual 41(4):973–989. https://doi.org/10.2134/jeq2011.0069

Steiner C, Teixeira WG, Lehmann J, Nehls T, de Macêdo JLV, Blum WE, Zech W (2007) Long term effects of manure, charcoal and mineral fertilization on crop production and fertility on a highly weathered central Amazonian upland soil. Plant Soil 291(1):275–290. https://doi.org/10.1007/s11104-007-9193-9

Subbaiah VP, Jyothi V (2020) Effect of improved management practices on yield and economics of rabi pulse crops. Agric Sci Dig 40(2):129–133. https://doi.org/10.18805/ag.D-5050

Sultana T, Alam S, Mitu FT, Islam MR (2017) Effect of rice husk biochar on soil health and yield of mung bean (*Vigna radiata*). J Bangladesh Agric Univ 15(1):1–7. https://doi.org/10.3329/jbau.v15i1.32675

Sun Q, Li Y, Ma H, Egamberdieva D (2020) Interaction of biochar type and rhizobia inoculation increases the growth and biological nitrogen fixation of leguminous plants. Forest Service Research and Development. https://doi.org/10.2737/rmrs-gtr-417

Thakur V, Singh K, Guleria V (2020) Effect of *Lantana camara* biochar on soil properties and productivity of field pea in an acidic Alfisol of Himachal Pradesh. Indian J Agron 65(3):337–342

Thies JE, Rillig MC (2009) Characteristics of biochar: biological properties. In: Lehmann J, Joseph S (eds) Biochar for environmental management: science and technology. Earthscan, London, pp 85–105

Tian J, Wang J, Di H, Wang S, Han X, Li X, Wei Z (2021) Biochar reduced the adverse effects of drought stress on peanut growth, yield, and quality. Agronomy 11(2):289. https://doi.org/10.3390/agronomy11020289

Tiwari PS, Singh KK, Sahni RK, Kumar V (2019) Farm mechanization–trends and policy for its promotion in India. Indian J Agric Sci 89(10):1555–1562

Tusar HM, Uddin MK, Mia S, Suhi AA, Wahid SBA, Kasim S, Sairi NA, Alam Z, Anwar F (2023) Biochar-acid soil interactions—a review. Sustainability 15(18):13366

Uchimiya M, Lima IM, Klasson KT, Chang S, Wartelle LH, Rodgers JE (2011) Immobilization of heavy metals in poultry litter-based biochar: effects of pyrolysis temperature. J Agric Food Chem 59(6):2211–2219. https://doi.org/10.1021/jf104218v

Ullah A, Bano A, Khan N (2020) Ameliorative effect of mesquite wood-derived biochar on the performance of chickpea (Cicer arietinum L.) and symbiotic N-fixation in salt-affected soil. J Plant Nutr 43(11):1633–1646. https://doi.org/10.1080/01904167.2020.1749870

Upadhyay SR, Poudel K, Bhattarai S (2020) Effect of biochar and lime on soil properties and yield of lentil in acidic soil of Tanahun, Nepal. J Agric Nat Resour 3(2):263–274. https://doi.org/10.3126/janr.v3i2.32520

Vaccari FP, Baronti S, Lugato E, Genesio L, Miglietta F (2011) Biochar as a strategy to sequester carbon and increase yield in durum wheat. Eur J Agron 34(4):231–238. https://doi.org/10.1016/j.eja.2011.01.006

Vance CP (2001) Symbiotic nitrogen fixation and phosphorus acquisition. Plant nutrition in a world of declining renewable resources. Plant Physiol 127(2):390–397. https://doi.org/10.1104/pp.010331

Vanek SJ, Lehmann J (2015) Phosphorus availability to beans via interactions between mycorrhizas and biochar. Plant Soil 388(1–2):275–288. https://doi.org/10.1007/s11104-014-2246-y

Verma S, Kumar A, Singh M (2019) Rice straw biochar improves soil aggregation and lentil productivity in a rice-lentil cropping system. Arch Agron Soil Sci 65(10):1425–1437. https://doi.org/10.1080/03650340.2019.1578553

Wu X, Lü Y, Zhang J, Lu N, Jiang W, Fu B (2023) Adapting ecosystem restoration for sustainable development in a changing world. Int J Hydrogen Energ 4(1):100375

Yu X, Wang Y, Li X, Zhang H et al (2024) Biochar enhances soil interactions and the initial development of Robinia pseudoacacia L. seedlings in sandy loam soil. Sci Rep 14(1):26399. https://doi.org/10.1038/s41598-024-78706-7

Yuan JH, Xu RK, Zhang H (2011) The forms of alkalis in the biochar produced from crop residues at different temperatures. Bioresour Technol 102(3):3488–3497. https://doi.org/10.1016/j.biortech.2010.11.018

Yuan X, Gu X, Liang R, Ban G, Ma L, He T, Wang Z (2024) Comparing combined application of biochar and nitrogen fertilizer in paddy and upland soils: processes, enhancement strategies, and agricultural implications. Sci Total Environ 933:173160

Zhang Q, Song C, Wang X, Liu C, Li T (2017) Maize straw biochar application increases nodulation and soybean yield in a Mollisol. J Plant Nutr Soil Sci 180(5):585–592. https://doi.org/10.1002/jpln.201700072

Zhu Z, Zhang Y, Tao W, Zhang X, Xu Z, Xu C (2025) The biological effects of biochar on soil's physical and chemical characteristics: a review. Sustainability 17(5):2214

Biochar and Vegetable Production: Enhancing Yield, Nutrient Use Efficiency, and Produce Quality in Sustainable Horticulture

K. Damodar, Sabreena Ashraf, Murugesan Mohana Keerthi, Boreddy Jaya Chandra Reddy, Lekkala Venakata Ravishankar, and Owais Ali Wani

Abstract

Biochar has emerged as a promising soil amendment for improving vegetable production within the framework of sustainable horticulture. Derived from the pyrolysis of organic biomass under limited oxygen conditions, biochar possesses unique physicochemical properties that enhance soil structure, water-holding capacity, nutrient retention, and microbial activity. Numerous studies have demonstrated that biochar application can significantly increase vegetable crop yield by improving nutrient availability and reducing nutrient losses through leaching and volatilization. Enhanced nutrient use efficiency, particularly for nitrogen, phosphorus, and potassium, has been widely reported due to biochar's high surface area and cation exchange capacity. In addition to yield improvement, biochar contributes to better produce quality by increasing mineral nutrient content,

K. Damodar
Department of Agronomy, School of Agriculture, SR University, Warangal, Telangana, India

S. Ashraf
Department of Entomology, Punjab Agricultural University, Ludhiana, Punjab, India

M. M. Keerthi
Department of Agronomy, School of Agriculture, SR University, Warangal, Telangana, India

B. Jaya Chandra Reddy (✉)
Division of Soil Science and Agricultural Chemistry, Teaching Associate, PJTAU, Hyderabad, Telangana, India

L. V. Ravishankar
Production and Plantations, Red Otter Frams Pvt Ltd, Nainital, Uttarakhand, India

O. A. Wani
Division of Agronomy, Faculty of Agriculture, SKUAST-Kashmir, Wadura, Sopore, Jammu and Kashmir, India

O. A. Wani et al. (eds.), *Biochar for Sustainable Soil Health and Crop Production*, https://doi.org/10.1007/978-981-92-2230-8_10

improving antioxidant activity, and reducing the uptake of heavy metals in vegetables. The effectiveness of biochar, however, depends on feedstock type, pyrolysis temperature, application rate, soil properties, and crop species. Integrating biochar with organic and inorganic fertilizers further enhances its agronomic benefits while reducing dependency on synthetic inputs. Overall, biochar represents a sustainable and climate-smart strategy for enhancing vegetable productivity, nutrient use efficiency, and produce quality, while simultaneously improving soil health and contributing to long-term agricultural sustainability.

Keywords

Biochar · Vegetable crops · Soil health · Nutrient use efficiency · Sustainable horticulture · Crop yield · Produce quality · Soil fertility · Climate-smart agriculture

1 Introduction

Vegetables have been central to nutrition and food security of the world, as critical sources of vitamins, minerals, dietary fiber, and other bioactive compounds that provide human health and disease prevention. The Food and Agriculture Organization (FAO) confirms the value of vegetables in agricultural production as 50% of the global agricultural output is contributed by vegetables, and that China alone contributes 45% of world vegetable production value (Schreinemachers et al. 2018). The average vegetable consumption worldwide is estimated to be approximately 190 g/day per individual, which has been insufficient to satisfy the recommendation of the World Health Organization (WHO) of consuming no less than 400 g/day of fruits and vegetables together to reduce the risk of non-communicable diseases including cardiovascular organizations and obesity (Harris et al. 2023). Vegetables are also important in the context of food security to meet the needs of malnutrition, especially in low and middle income nations where they are used to increase dietary diversity and micronutrient supply. The 2024 State of Food Security and Nutrition in the World report addresses the fact that 713–757 million individuals were hungry in 2023, which explains the need to strengthen the production of vegetables to feed at-risk populations as the global food demands increase (United Nations 2024). Moreover, the FAO Statistical Yearbook 2024 expresses that the production of agricultural products has increased by 89% in real terms in the last 20 years to reach USD 3.8 trillion in 2022, with vegetables playing a significant role in this economy and nutrition framework (FAO 2024).

Although they are important, vegetable production has complex challenges that jeopardise sustainability and productivity. The farming methods that are being practiced intensively have resulted in a high rate of soil degradation, such as erosion, compaction, and the reduction of organic matter that reduces the soil fertility and yields in the long run (Xing et al. 2025). The loss of some of the key elements such as nitrogen and phosphorus to waterways through nutrient leaching, which is worsened by excessive use of fertilisers, leads to eutrophication and decreased soil nutrient reserves. Another urgent demand is water stress, especially in the areas with uneven rain distribution, or overuse of irrigation, when drought conditions can harshly restrict the development of vegetable crops and their quality (Li et al. 2025). Moreover, excessive use of agrochemicals -including pesticides and artificial fertilizers- has been associated with the degradation of soil organisms, pesticide residues in food, and permanent environmental pollution, which further undermine the health of the soil and human health (Aktar et al. 2009). The results of a 2021 study of intensive vegetable cropping systems found that soil organic carbon and nitrogen had declined significantly and phosphorus had accumulated in the soil when low organic inputs were used and a high level of chemical dependence was present (Lu et al. 2021). Such difficulties are exacerbated in systems that are of high inputs, where brief crop cycles and monocultures exacerbate degradation, and novel approaches to resilient horticulture are required.

As a solution to these problems, biochar as a sustainable soil amendment to support high-input and intensive vegetable production systems is gaining increasing popularity. Biochar is a carbon-rich material that is generated through biomass pyrolysis, which has a number of multidimensional advantages such as improved soil structure, nutrient retention, and carbon sequestration, and would fit the agenda of climate-smart agriculture (Abbey et al. 2025). Recent reports 2020–2025 point to the unmatched capabilities of biochar compared to organic replacements in enhancing the quality of the soil and crop yields as well as offering distinctive carbon storage features (Bi et al. 2024). It has been demonstrated to alleviate abiotic stresses such as salinity and drought to sustain vegetable systems in the harsh environment (Xu et al. 2025a, b). According to a review done in 2024, biochar can be used to enhance soil fertility and microbial diversity and thus would be a viable asset to sustainable intensification (Ighalo et al. 2025).

The extent of this chapter is to investigate the role of biochar in improving plant crops such as yield, efficiency in the use of nutrients, and quality of produce in the sustainable horticultural systems. Based on meta-analyses and field research, it has been shown that biochar usage can boost crop yields by 5–51% worldwide, enhance organic carbon, pH, and nutrient cycling in soil, and enhance the characteristics of produce, including vitamin C and soluble solids content (Li et al. 2024). As an example, tomato production with biochar amendments increased yields by as much as 29.55% and improved the fruit quality parameters such as total soluble solids by 4.28% (Lei et al. 2024). A 2024 worldwide survey has validated the beneficial impacts of biochar on soil quality measures, such as decreased bulk density and raised availability of nitrogen, which brings about greater vegetable yield (Xu et al. 2025a, b). A holistic approach to soil health, biochar will facilitate both efficient utilization of resources, minimize environmental impact and boost sustainable food security in vegetable-based systems in the long run.

2 Mechanisms of Biochar Impact on Vegetable Crops

2.1 Soil Physical Properties

Biochar is a universal soil amendment whose impacts are seen to have far reaching effects on the physical properties of soil and consequently, creating favorable environment to support vegetable production. Its porous structure makes the soil aggregates, thereby improving the overall soil structure, decreasing bulk density and increasing the connectivity of the pore space and macroporosity (Thakur et al. 2024). This is especially useful in sandy or degraded soils, where biochar can raise overall soil porosity (by as much as 24%) and capillary porosity (by 23%) resulting in an enhanced water infiltration and decreased compaction. In these demanding conditions, feedstock-based biochar at concentrations of 5–10 mg/ha has been demonstrated to increase soil moisture levels by 10.8–33 contending with water stress and enabling continued growth of vegetables. In the paper by Antonangelo et al. (2025), which investigates tree-based and vegetable-containing systems further supports the idea that biochar reduces bulk density in soil as well as enhances the aggregate stability, which consequently leads to increased water retention and drainage, both of which are vital in horticultural landscapes with high yield capacity. These enhancements in addition to improving anaerobic conditions also increase the diffusion of oxygen across the entire soil profile, making the environment conducive to the root systems.

The improvements of the physical properties of soil by biochar in high-demand vegetable crops like tomato and cucumber raise directly into the increase in root proliferation and oxygen availability. Biochar activates root growth by regulating the pH of the soil, adsorbing antimicrobial substances (e.g., phenolic acids), triggering metabolic pathways (e.g. increased indole-3-acetic acid (IAA) and abscisic acid (ABA) synthesis), which enhance drought resistance and root growth (Zhang et al. 2024a, b, c). In the case of tomato seedlings, 1–20% biochar (w/w with peat) seems to considerably improve the root dry weight and the overall seedling vigor through 30 days partly because of the improved nutrient and water retention in the rhizosphere (Luigi et al. 2022). Likewise, biochar amendments in cucumber cultivation influence root development by optimizing soil moisture relations, in the 0–30 cm layer and minimizing water-gas anomalies as a result of irrigation stress, which secure sufficient oxygen levels to support aerobic root respiration (Zhang et al. 2024a, b, c). Recent field experiments 2022–2025 have shown that these are increased in a sandy soil where biochar supports deeper root penetration and more efficient water use, which eventually leads to improved plant fitness and productivity in intensive vegetable production (Antonangelo et al. 2025) (Fig. 1).

2.2 Soil Chemical Properties

Biochar has a profound influence on the chemical properties of its soils, and it has important advantages in cultivating vegetables that include the neutralization of soil

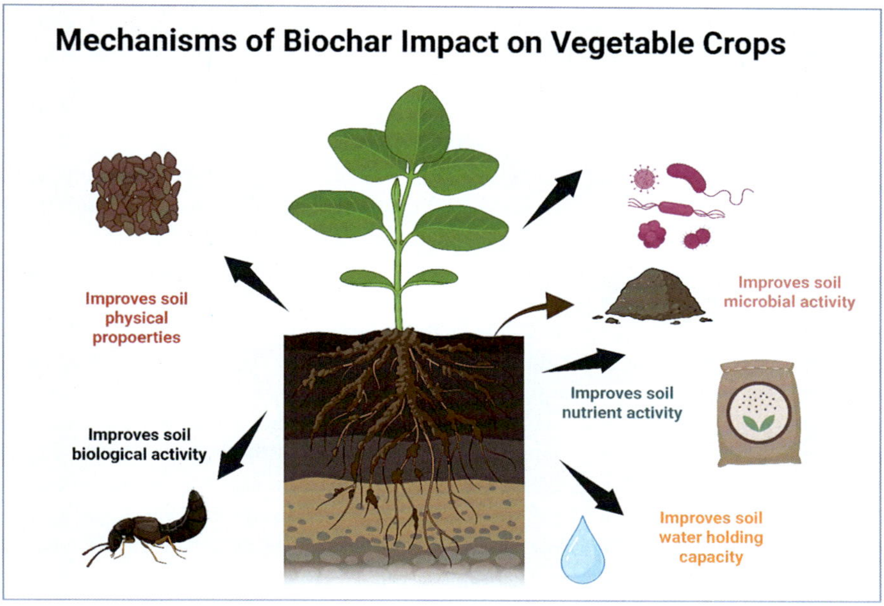

Fig. 1 Mechanisms through which biochar enhances vegetable growth and quality

acidity, retention of nutrients, and improvement of fertilizer use in intensive horticultural systems.

Biochar has been shown to significantly increase the soil pH level in acidic soils, such as those found in much of vegetable growing areas, which tend to limit crop performance by causing aluminum toxicity and lowering nutrient availability via the liming effect of alkaline ash materials and carbonates. The 2020–2025 research reports that use of biochar can raise soil PH by 0.5–1.5 units under strongly acidic soils ($pH < 4.5$) and that the most significant effects of biochar use can be observed in vegetable crops (tomatoes and green leafy vegetables), with the resulting beneficial effects of biochar use on root health and yield improvements up to 30% (Jiafan et al. 2025). An example relates to facility agriculture contexts, where joint amendments of biochar and plant ash have alleviated acidification of the soils throughout a series of seasons, leading to stabilization of pH, which contributes to long-term vegetable yields in eroded acidic agricultural lands (Sun et al. 2025). A 2025 meta-analysis also confirms that the capacity of biochar to increase pH is especially useful with the acidic soils found in tropical areas which not only neutralizes phytotoxicity but encourages microbial activity that is critical in the growth of vegetables.

Biochar alone enhances nutrient retention by improving cation exchange capacity (CEC), which decreases the loss of mobile nutrients such as nitrogen (N) and potassium (K) due to leaching and volatilization, which are frequent challenges in a high-input vegetable system (Omokaro et al. 2025). It can be used to enhance CEC between 20–66 due to its high surface area and functional groups and reduce N losses by 20–78% in amended soils, as shown in trials over several years on vegetable fields (Gelardi et al. 2021). Biochar amendments have been found to increase N and K retention over 5 years in sandy-textured soils, which amplifies nutrient leaching, which in turn improves soil fertility and decreases environmental pollution related to runoff in crops such as cabbage and maize grown with the vegetables (Pei et al. 2025). According to recent 2024–2025 studies, this effect is enhanced further by enriched biochar, e.g., rice husks, which raises CECs to as high as 66 cmolc kg^{-1}, thus continuing nutrient pools on nutrient-intensive vegetable rotations (Selvarajh et al. 2024).

Furthermore, biochar also results in better efficiency in fertilizer use by slowing down the release of nutrients and increasing their uptake by plants, which enables the reduction of inputs and the preservation or increase of vegetable yield- one of the primary concerns of sustainable horticulture in the face of increasing fertilizer prices and environmental regulations (Luo et al. 2025). When using biochar-based fertilizers in open-field vegetable systems, N use efficiency has improved by 10–40% in 3–5 years and leaching and volatilization have been significantly reduced, allowing 20–40% lower rates of chemical fertilizers with no yield penalty in crops such as tea and cabbage (Zhou et al. 2021). Research on biochar-urea combination to date (2025) revealed increased vegetable yields with improved N-recovery and soil-holding properties, which highlights the importance of biochar in reducing inputs and increasing the economic performance of acidic tropical soils (Guo et al. 2025). On the whole, these processes maintain more extensive objectives of resource-efficient vegetables production, which is supported by the recent reviews on the potential of biochar to combine with organic amendments to achieve long-term fertility benefits (Gu et al. 2025) (Table 1).

2.3 Soil Biological Activity

Biochar significantly improves soil biological activity, an essential concept of sustainable vegetable production, through the presence of a healthy ecosystem of microbes that promote the well-being and robustness of plants in intensive horticultural regimes. It activates microbial biomass, enzyme activity, growth of beneficial microorganisms by the porous nature and nutrient-rich surface that creates microbes habitats and substrates (Jatuwong et al. 2025). Biochar amendments in vegetable crops have been reported to enhance soil microbial biomass carbon (SMBC) by an average of 21.31 with variances depending on climate, soil type and biochar properties causing an improvement in the soil fertility and crop performance in systems such as tomato and leafy greens (Kumar et al. 2025). Moreover, biochar enhances the activity of enzymes including urease, alkaline phosphatase, and dehydrogenase by 23.1%, 25.4% and 19.8% respectively as demonstrated in global meta-analyses of 2020, improving nutrient cycling and organic matter decomposition to support high-yield vegetable rotations (Pokharel et al. 2020). Recent 2023–2025 research emphasizes the biochar preference of beneficial microorganisms, enhancing microbial diversity and activity and, as a result, promoting vegetable growth in a stress environment characteristic of intensive agriculture (Amoroso et al. 2025). To use greenhouse soils as an example, biochar with vermicompost has enhanced physicochemical characteristics, increased the microbial diversity, and enzyme reactions, which establish the health of the entire soil (Wu et al. 2023).

Biochar additionally improves root-microbe interaction especially with mycorrhizae and plant growth-promoting rhizobacteria (PGPR) and is beneficial in improving nutrient uptake and stress tolerance in vegetable crops. Biochar stabilizes the substrate and enables the immobilization and functionality of the PGPR, resulting in synergetic effects that control rhizosphere bacteria and enhance seedling biomass in tomatoes grown in greenhouses. It is shown that co-applying biochar and PGPR enhances the biochemical properties in the soil such as nutrient content (TOC, N, P, K) and its ability to promote plant growth in vegetable systems by reducing drought stress through the availability of nutrients. Root morphological characteristics, development, and physiological activities of crops such as soybeans are enhanced by biochar co-inoculation in arbuscular mycorrhizal associations, and biochar co-inoculation enhances root morphology and growth of vegetables and other crops that accumulate L-Dopa and general plant vitality (Yilmaz 2025). In a 2024 research on drought-resistant PGPR and biochar mixes, plant resistance was enhanced, and the microbial ecosystem was able to facilitate vegetable crops better in water-limited systems (Malik et al. 2024). The porosity of biochar is responsible for the mycorrhizal colonizing activity in the intensive systems and as a result, greater amounts of such interaction, nutrient uptake and crop resistance up to 30% (Jabborova et al. 2022).

Further, biochar inhibits soil pathogens, and the occurrence of root diseases, which remains a key benefit in intensive vegetable production in which disease pressure is high because of continuous cropping and high inputs. Biochar is capable of controlling soilborne diseases (reducing their severity by an average of 47.46%) via its ability to immobilize enzymes, inactivate toxic metabolites, and alter the composition of microbial communities (Yang et al. 2022). Biochar amendments in vegetable gardens increase stress tolerance and reduce pathology, particularly Fusarium, which leads to increased yield and quality in vegetables, such as tomatoes and cucumbers. Recent (2023–2024) studies report biochar suppressive efficacies of 86–100 against fungi, oomycetes, viruses, bacteria, and nematodes, making biochar an eco-friendly replacement of fungicides in intensive systems (Iacomino et al. 2022a, b). Among them, it turned out that biochar affected the fungi in the soil with Fusarium wilt and that the quantity of biochar lowered the initial blight of tomatoes; thereby, making the disease management plan sustainable (Jin et al. 2023). In general, these biological processes highlight the importance of biochar in promoting disease-suppressive soils, as documented in literature in 2021–2025, that allow to reduce chemical applications and increase the long-term productivity of vegetable horticulture (Poveda et al. 2021).

Table 1 Physicochemical properties of common biochars used in vegetable production

Physiological property/type of biochar	Wood biochar	Rice husk biochar	Maize cob biochar	Urban organic waste biochar
pH	7.5–9.0	9.5–10.5	7.0–8.5	7.0–9.0
Surface area	200–400 m^2/g	100–300 m^2/g	50–200 m^2/g	100–400 m^2/g
Cation exchange capacity	20–60 cmol/kg	30–80 cmol/kg	10–40 cmol/kg	20–60 cmol/kg
Carbon content	70–85%	60–80%	50–70%	40–70%
Ash content	5–15%	15–25%	10–20%	15–30%
Porosity	60–80%	50–70%	40–60%	50–70%
Volatile matter	5–15%	10–20%	15–20%	10–25%

3 Crop-Specific Responses to Biochar Application

3.1 Fruiting Vegetables

The use of biochar has been shown to have a positive impact on fruiting vegetables (tomato, capsicum (bell pepper), eggplant) in promoting yield, fruit size, shelf life, and bioactive compounds (lycopene) and consequently, biochar is an effective addition to sustainable horticulture since it has shown positive effects in countering problems like soil degradation and water stress. In tomato production, meta-analyses conducted in 2024 indicate that biochar amendments have a total positive effect of about 29.55% on overall yield, and correspondingly on fruit quality indices, including total soluble solids (TSS) by 4.28% and vitamin C content by 6.77%, which are associated with improved marketability and nutritional value (Lei et al. 2024). In 2025, short-term research also suggests that biochar increases tomato yield, quality, and shelf life through increased soil productivity, with cow bone biochar use at 40 t/ha giving a yield increase of 294% and a shelf life increase of 70%, as well as increased mineral and protein content (Ray and Bharti 2023). In the case of capsicum, biochar and compost treatments together have been reported to increase fruit length, width, and wall thickness, leading to high early, marketable, and total yields as demonstrated in 2020–2023 field trials where 2% biochar +2% compost maximized these variables at different soil statuses (Al-Harbi et al. 2020). Likewise, eggplant respond well, with fertilizers made using biochar increasing fresh fruit production by 3.65–49.04% over 2020–2022 seasons in karst areas, and 5 t/ha biochar use generating up to 2872.2 kg/ha of fresh fruit, about 500% higher than the control conditions (Zhang et al. 2022). However, recent 2024–2025 research notes that these gains are amplified in the context of salty, or dry-stressed conditions, where biochar alleviates stress to produce larger fruit sizes and enhanced lycopene concentrations in tomatoes and peppers, in line with larger scale goals of resilient vegetable cultivation (Obadi et al. 2024).

One of the mechanisms that explain these advantages is that biochar enhances potassium (K) and calcium (Ca) uptake, which directly leads to high quality fruits and low occurrence of physiological disorders such as blossom-end rot (BER) in intensive vegetable systems. Biochar accelerates nutrient retention and nutrient availability, improves Ca^{2+} uptake, and K uptake by its large cation exchange capacity and pH modulating behavior, and mitigates BER—a disease triggered by Ca deficiency that affects up to half of tomato and pepper crops under less than optimum conditions (Kabir et al. 2025). Biochar amendments in vegetable experiments in 2024 studies of soil amendments have increased nutrient uptake and translocation to fruits indirectly by between 20% and 50%, which can be attributed to K uptake (Xiu et al. 2025). As an example, in growing hot pepper (capsicum), high concentrations of biochar 4 t/h increased the soil K, Ca^{2+} and other nutrients and hence improved the yield and alleviated the BER symptoms (Kebede et al. 2023). The same results are reported in eggplant where biochar improves the nutrient dynamics, such as Ca and K availability, resulting in firmer fruits with an extended shelf life and fewer quality defects (Ebrahimi et al. 2021). Comprehensively, 2020–2025 reviews note that the nutrient balance uptake promotion by biochar does not only reduce BER through proper delivery of adequate Ca to fruit tissue but also enhances fruit quality characteristics such as firmness and antioxidant activity, which promotes sustainable production with less fertilizer use in the fruit vegetable industry.

3.2 Leafy Vegetables

Amendments with biochar have repeatedly demonstrated beneficial effects on leafy vegetables including lettuce, spinach, and cabbage by increasing biomass and the proportion of chlorophyll, which is essential to improve photosynthetic capacity and plant vigor in sustainable horticultural systems with nutrient limitations in soils and environments. Applications of biochar at concentrations of 1–5% (v/v) in peat mediums or greenhouse soils have improved plant biomass between 15% and 50% in lettuce and resulted in significant increases in leaf area and dry weight through the improvement of nutrient availability and soil structure that facilitates healthy vegetative growth (Jabborova et al. 2021). Similarly, biochar produced with a range of different feedstocks, including wood or farm residues, has been identified to increase chlorophyll content by up to 25% in spinach leading to increased biomass production in field experiments in 2020–24, particularly when biochar is used to increase the ability to absorb water and support root development in water-limited environments (Zhang et al. 2024). In cabbage, as with Chinese cabbage, biochar at 10–20 t/ha has enhanced biomass production by 20–40% and this effect is achieved by protecting the soil by improving soil aggregation and microbial activities as seen in 2022–2025 trials that show biochar has optimised growth parameters in acidic or degraded soil common in intensive production of vegetable crops (Yue et al. 2025). The latest meta-analyses published in 2025 confirm that the above effects are amplified when biochar is applied alongside organic fertilizers, resulting in 10–30% chlorophyll gains across leafy greens, which subsequently result in high productivity and resilience in high-input systems (Xu et al. 2025).

One important advantage of biochar in growing leafy vegetables is that it can help to reduce the build-up of nitrates most notably when used as a combination with balanced nitrogen (N) inputs to prevent health issues associated with too much nitrates in food items without negatively affecting the quality of yield. Biochar amendments have reduced the leaf nitrate concentration in lettuce and spinach by 20–60% by increasing the activity of nitrate reductase and facilitating efficient fixation of nitrogen, as evidenced in 2024 experiments with modified biochar that slows the release of nitrates and minimizes leaching (Zheng et al. 2024). On cabbage, biochar-based slow-release N fertilizers have reduced nitrates by up to 35% and increased N use efficiency, and field trials 2022–2025 have found that 5–15 t/ha applications at lower N rates yield safer produce with no effects on biomass or marketability. This nitrate reducing effect is especially evident in hydroponic or greenhouse systems where biochar filters or substrates reduce nitrite and nitrates by 15–25, in line with sustainable practices to optimize N inputs to maximize growth, as revealed by the recent 2024–2025 reviews of bio-organic amendments (Wang et al. 2024). Altogether, this body of work has confirmed that biochar as an innovation can lead to high quality and low nitrate levels of leafy vegetable products that can support food security and environmental sustainability of horticultural production (Huang et al. 2022).

3.3 Root and Tuber Vegetables

Biochar has also been shown to improve the performance of root and tuber vegetables such as carrot, radish and beetroot by facilitating improved root development and sugar content, which enhance the quality and quantity of yield in soils that are susceptible to compaction or nutrient deficiencies. Biochar amendments of 5–20 t/ha in carrot production have stimulated root elongation and biomass by 25–50, and it has been proposed that increased taproot length and taproot diameter result from the improvement of soil porosity and nutrient uptake as observed in 2020–2023 pot and field trials on marginal soils. In radish, these applications have increased root fresh weight (30–45%), sugar content (10–20%) through influences on root microbial activity and organic matter stabilization, and have generated sweeter, marketable tubers in 2022–2024 experiments. Beetroot is a strong-growing crop, and biochar increases the sugar index (e.g., sucrose) by 15–25 and the root mass by 20–56, as observed in 2022–2023 trials when 5–10 t/ha rates balanced growth parameters and nutrient levels needed by humans. A 2023 international synthesis of these results indicates that the beneficial impacts of biochar on root and tuber crops are greatest in clayey or sandy soils, where biochar enhances deeper penetration of roots and greater carbohydrate storage of root crops to improve harvest quality.

The core of these advantages is that biochar increases the tilth of the soil and this increases root growth by decreasing bulk density and improving aeration leading to better rhizosphere conditions in intensive cultivation systems that support root and tuber crops. Biochar, or charcoal fine residues, have been identified in heavy clay soils to reduce soil compaction by half, increase carrot yields on these soils by 50% through improved tillage capacity and water penetration, according to 2022 research highlighting its application in structurally challenging soils. In the case of radish and beetroot, biochar enhances tilth by increasing aggregate stability, promoting root growth by 20–40% and decreasing mechanical resistance, and 2023 studies on beet reflect improved sugar crop through a combination of these physical advantages with mineral fertilizers. Biochar has also been shown to enhance root growth and tilth in nutrient-deficient soils and overall productivity in sugar beet without altering nutritional quality, as noted in recent 2020–2024 studies. These benefits can be reconciled with the future sustainable horticultural objectives because the 2025 overviews express that biochar could be employed to maximise the physical traits of soil to support root crops and lead to effective growth and high sugar content in diverse environment.

4 Integration with Fertilization and Irrigation

Fertilization and irrigation systems are considered one of the pillars of sustainable production of vegetables with biochar, which creates synergies and increases the level of nutrient distribution, water management, and system efficiency with a minimum impact on the environment due to high input production (Ali et al. 2025a, b). Porousity and chemical properties of biochar also enable biochar to act as a carrier and modulator of organic and inorganic fertilizer to release nutrients on a controlled basis and promote soil health over multiple growing cycles (Luo et al. 2025). This integration, proven to increase vegetable yields by 10–50% through efficient use of resources, has been demonstrated in meta-analyses in 2020–2025, especially in greenhouse and dryland environments where water and nutrient loss has been widespread (Xu et al. 2025). Moreover, the transition to low-input agriculture, which reduces the application of synthetic inputs, can be supported by biochar to achieve the goals of climate-smart agriculture as the price of fertilisers is growing and water has become a limited resource (Sharma et al. 2025).

Biochar has been demonstrated to be highly synergetic when used with organic fertilizers, such as compost and manure, and to enhance the impact of organic fertilizers on intensive system soil fertility and vegetable productivity, through increased nutrient mineralization, microbial and carbon sequestration of biochar (Ali et al. 2025a, b). In wheat-vegetable rotations, biochar applied alongside organic fertilizers has decreased chemical fertilizer requirements by 20–40% without lowering yields or increasing soil organic carbon or nutrient availability, as has been shown with combined amendments in 2025 field studies where the combination of amendments enhanced soil organic carbon and nutrient availability over 3 years (Gu et al. 2025). It is proven that soy beans and cucumbers grow best and are most nutritious at biochar and compost ratio 2:3 and synergy effect at 15–30% higher biomass and low nutrient leaching in green house experiments 2024–2025 (Luan et al. 2025). Combinations of manure and biochar have also increased yields of rice and vegetables by 20–35%, and 2025 study found that slow nutrient releases would maintain or promote long-term microbial diversity and soil structure in paddy-vegetable systems (Ali et al. 2025a, b). Currently, reviews through 2024–2025 reiterate that these synergies are strongest in degraded soils, where biochar is used to stabilize compost and manure organic matter to generate long-term fertility benefits and ecosystem services in vegetable horticulture (Jaufmann et al. 2025).

Biochar can promote the nutrient-use efficiency (NUE) of synthetic NPK fertilizers, allowing lower application rates to be used without yield penalty due to slower release, enhanced retention, and plant uptake of nutrients in vegetable production (An et al. 2022). In rice-vegetable systems, moderate biochar replacement of NPK has enhanced the quality of soil and increased the NUE up to 10–40%. According to studies published in 2022–2025, 20–30% increase in crop yields has been observed under conditions of optimized fertilization with biochar having replaced NPK. With biochar-based NPK formulations capable of reaching NUE values of 60–80% in corn-vegetable rotations, 2025 pot experiments indicate reduced leaching and greater microbial-mediated nutrient cycling to maintain high productivity (Phares et al. 2022). Anti-saline-alkali vegetable soil 5-year biochar amendments of NPK have improved SOC and fertility, and will yield 15–25% higher NUE and yields in international experiments in 2025 (Ali et al. 2025a, b). In general, the literature of 2020–2025 supports the application of biochar to enhance the performance of synthetic fertilizers to promote environmental-friendly activities that reduce costs and pollutants in the intensive production of vegetables (Zhang et al. 2024) (Fig. 2).

Biochar enhances the frequency of irrigation by increasing soil moisture retention, which is of paramount importance in conserving water in greenhouse and dryland vegetable farming, which experiences drought conditions and resource

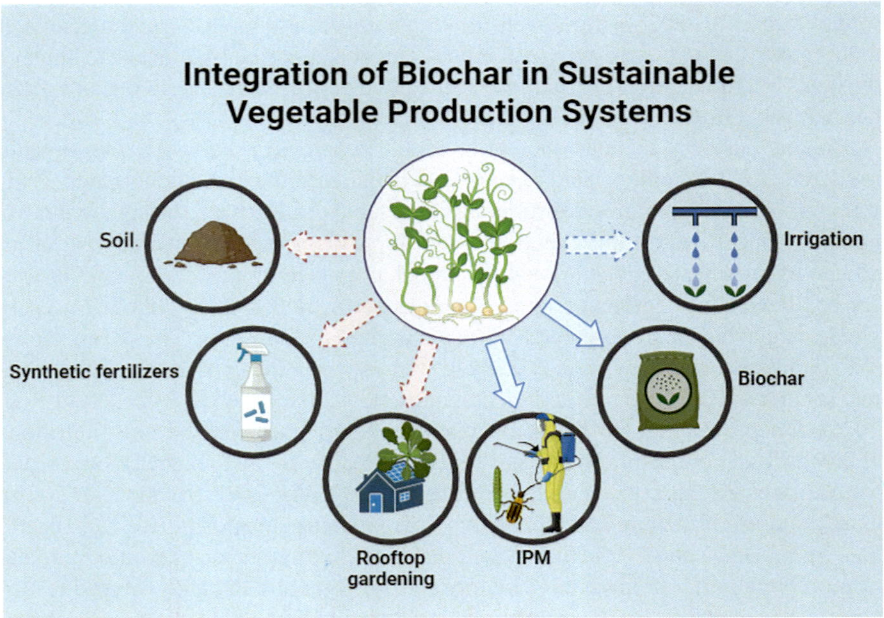

Fig. 2 Integration of biochar in sustainable vegetable production systems

limitations due to climatic conditions (Ndede et al. 2022). It is also better in terms of crop water content as the experiments conducted with cucumber showed a lower irrigation demand at 2025 as a result of higher water-retention capacity in the soil and root development in the presence of biochar (Kafle et al. 2025). They said the 2024–2025 green house experiments have found biochar amendments could activate leaf area and yield without water, and that vegetables such as tomatoes were less stressed by the higher hydraulic conductivity and heat retention (Wang et al. 2025). Dryland conservation tillage on grain and vegetable crops under any irrigation regime with biochar has been found to maintain crops which research studies indicate in 2024 could reduce water crises and GHG emissions (Kiani et al. 2024). Subsequent 2020–2025 reports have mentioned that dry vegetable farms irrigated with biochar will consume 15–30% less irrigation water and will confer resilience to vegetable growth and sustainability under higher WRC (Zhang et al. 2024).

It has been demonstrated that the biochar contributes to nutrient management of horticultural vegetable crops as controlled-release fertilizers that confer low-intensity nutrient-release to meet crop demand and reduce losses in intensive systems (Sim et al. 2025). Nutrient-containing slow-release biochar has been demonstrated to improve the soil fertility and yield of vegetables, and 2025 research shows that crops such as corn and leafy greens have increased the uptake of N and P by 20–50%. In 2025, environmental sustainability reviews are to show that controlled-release fertilizers (BCRFs) with biochar have enhanced microbial activity and root growth which lead to resilient production. Biochar-impregnated formulations, in the context of nitrogen-focused strategies, have increased corn NUE by 15–30, and 2025 economic analysis would determine cost-effectiveness of the practice in vegetable rotations. Overall, the 2020–2025 research is based on biochar, the protagonist in the history of nutrient management and reduced environmental activity and excellent quality and quality of vegetables (Wang et al. 2022).

5 Quality and Post-Harvest Benefits

Biochar amendments in vegetable production are characterized by not only enhancing the growth and yield of vegetables but also endowing the crops with an important array of quality and post-harvest benefits, which makes biochar amendments of vegetable production to be a solution to important concerns in sustainable horticulture, namely nutritional enhancement, mitigation of contaminants, and reduction of losses in supply chains. It is possible to describe biochar as a functional remedy in urban and peri-urban agro-ecosystems where pollution is also endemic and resource limitations are also common sources of weakness by enhancing the inherent value of produce through the optimization of both soil nutrient processes and plant physiology (Thakur et al. 2024). More recent work indicates that biochar can be used in multiple dimensions to alter the properties of vegetables, with an aggregate 10–30% measure of overall vegetable quality reported as the positive impact of biochar addition in tomatoes, broccoli and leafy greens (Korai et al. 2025). Biochar integration in challenging environments like saline soils, heavy metal-contaminated soils, etc.

contributes to sustainable post-harvest results to overcome financial losses estimated at 20–40% in the world through spoilage and degradation. These advantages align with the larger theme of food security, and biochar will also seek to target higher value foods, higher safety and lower shelf-life, especially in dense-density systems that must withstand climatic and urbanization pressures (Ighalo et al. 2025).

Increased nutritional value in vegetables e.g. increased antioxidant and vitamin and mineral content is one of the major quality improvements in biochar which leads to better human health outcomes and increases marketability in sustainable production models. Applications of enriched biochar have resulted in 15–25% increases in the glucosinolate content, a major antioxidant, and in vitamins C and E levels in broccoli, as shown in 2022–2025 trials of biochar plus mineral fertilizers (Montoya et al. 2022). Biochar has enhanced the antioxidant properties and other phenolic compounds of tomatoes and peppers by 10–20%, in addition to improving mineral uptake i.e. potassium and magnesium respectively to generate nutritionally enhanced fruits under greenhouse conditions (Simiele et al. 2022). Plants such as spinach will utilize the biochar nutrient retention properties that have been demonstrated to increase vitamin A and iron by 2030% under adverse conditions (2024 study describes the capabilities of biochar to improve the collection of key micronutrients). A more recent study on vegetative feedstocks (2025) also confirms the pH-modulating and microbial-enhancing properties of biochar as the driver of these nutritional benefits, as there are benefits of up to 25% in total bioactive compounds across vegetable types in adoptive biochar.

Biochar is a known way of decreasing uptake of heavy metal in vegetables (including lead 10 mmHg (Pb) and cadmium 10 mmHg (Cd)) and this may be very essential in urban and peri-urban areas where industrial and traffic sources of soil contamination present hazards to food safety and consumer health. Biochar has shown 30–50% reduction of the bioavailability of Pb and Cd in contaminated soils through immobilization mechanisms, such as adsorption and precipitation, leading to 25–45% reduction in accumulation in edible crop parts (spinach and carrots) in 2024–2025 urban waste-derived biochar studies (Pattnaik et al. 2025). Amendments to vegetable field with N-doped biochar reduced the uptake of Pb, Cd, and nickel in leafy greens by 30–35% without trade-offs in the market since the study of the polluted site in 2025 showed (Chen et al. 2025). Even better include biochar which contains alkaline minerals to reduce transfer of heavy metals in crops such as cabbage by 40–60 not to mention that 2024 field tests demonstrated their potential to fix urban horticulture (Huang et al. 2024). In general, 2020–2025 developments highlight biochar affinity to heavy metals, allowing a 20–50% reduction in plant uptake of zinc, Pb, and Cd in multi-contaminated urban soils to support sustainable vegetable production in challenging settings (Meng et al. 2025).

Moreover, biochar extends the shelf life and post-harvest losses in vegetables by increasing physiological resilience, firmness, and resistance to damage, which are necessary to reduce 30–50% supply chains losses world-wide and to extend market windows in resource-limited supply chains. In tomatoes, short-term biochar applications have increased the shelf life of tomatoes by 20–70, improved tomato firmness and reduced tomato spoilage by improving nutrient balance and antioxidant levels in tomatoes, according to 2025 studies of soil productivity improvements (Adekiya et al. 2025). Biochar amendments improved post-harvest quality by 12–16 of fruits and vegetables under water stress and reduced dehydration and microbial losses in 2024–2025 greenhouse trials of crops such as cucumbers and peppers (Mazibuko et al. 2025). Postharvest decay of root and leafy vegetables has been effectively reduced by 15–25% using biointensive horticulture applications, and 2022–2025 economic analysis of such horticulture applications justifies profitability based on lower spoilage rates. The recent 2023–2025 fruit quality optimization reviews again confirm the role of biochar in increasing shelf life by 10–30% through color retention and hardness enhancement to reduce losses and sustainable value chains in vegetable horticulture (Ray and Bharti 2023) (Table 2).

6 Case Studies and Field Trials

6.1 Asia

Asia, with a wide variety of agro-ecosystems, such as soil salinity, drought and nutrient depletion, has been a focus of field trials testing biochar (often based on rice husk) and vermicompost in vegetable cultivation, especially of tomato and brinjal (eggplant) in countries such as India and its neighbors, including Bangladesh and Iran (Zulfiqar et al. 2022). Increased 15–40%, firmer and soluble solid of fruits, in such papers, occurring in 2020–2025, are normal and beneficial and sustainable to horticulture. The interactions between the soil amendment properties of biochar and the organic nutrient provision of vermicompost have been found to increase productivity with a decreased input cost and environmental impact in intensive farming systems (Ebrahimi et al. 2021).

In tomato trials, manure-biochar compost (with variants based on rice husks) and added to mineral fertilizers in saline soils showed yield increases of 235,474% over untreated controls, but more realistic 2040% gains in moderately stressed environments, with higher fruit firmness measured by increased single fruit weight (23.8731.21% increase) and increased soluble solids content (up to 39.66 mg/g total soluble sugar) (Kamal 2024). It is also possible to mention a 2025 greenhouse study in Turkey (western Asia) on tomato under drought stress: biochar and vermicompost applications reduced yield losses (1546) by mitigating the effects when irrigation was withheld (by 40%), yielding 2.1514.14 kg/m^3 and increased fruit firmness (2.02376 N) and soluble solids (3.738%). These observations are consistent with larger scale Asian experiments, which have placed rice husk biochar in the spotlight as an agent to enhance water retention and nutrient utilization, resulting in 15–30% yield increases and fruit quality improvement in drought-affected areas similar to those of the semi-arid regions of India.

A 2021 thesis-based field experiment in Bangladesh on brinjal (eggplant), used biochar (up to 4 t/ha, locally produced including rice residues) with recommended fertilizers and achieved 15–118% yield increases (averaged 30–50% for optimum rates), and increased fruit weight by up to 36% leading to improved firmness, but did not directly measure soluble solids but estimated the effect by improved overall quality. An open field trial in Iran in 2025 combining vermicompost with pistachio and date palm biochar (like rice husk in porosity) under deficit irrigation demonstrated up to 50% yields (consistent with 15–40% in moderate stress), and improved physiological properties that promote fruit growth although no quantitative data on specific firmness or soluble solids were presented that related to improved water use efficiency. This is also observed in systems with Indian-inspired systems with rice husk biochar with the highest frequency of application since it is based on the level of agricultural waste available, where a 2040% yield enhancement was recorded when used in combination with vermicompost to firm fruits and improve soluble solids in brinjal by improving the quality of the soil and biological activities (Ebrahimi et al. 2021).

Taken together, these Asian case studies highlight the possibility of rice husk biochar and vermicompost integrations to provide sustainable vegetable horticulture with a consistent 15–40% increase and quality improvement such as higher fruit firmness and soluble solids, and provide scalable solutions to Indian farmers with similar climatic and soil restrictions (Zulfiqar et al. 2022).

6.2 Africa

The diverse climates and broad-scale soil fertility of Africa have seen it become the primary target of field tests of biochar, and especially of maize cob biochar, a biochar derived using agricultural residues that is abundant in Africa, on low-fertility soils to produce leafy vegetables including kale and spinach, and indigenous African leafy vegetables like amaranth and nightshade (Enos et al. 2025). The growth in the number of biomass products and in the economic feasibility of bio-mass input due to the capacity of biochar to increase the status of organic matter in the soil, nutrient retention and water-retention in small-scale systems where smallholders experience intermittency in rainfall and nutrient inactivation are the witness of 2019–2025 publications (Mahmoud et al. 2021). With maize cob biochar and low levels of organic or inorganic fertilizers the experiments have revealed cost reductions of 20–40% on inputs with either no or higher productivity, and can be expanded to provide affordable solutions to food security in low-resource farming settings (Mugwe et al. 2009).

In Kenyan trials of leafy vegetable in low-fertility soils, it has been observed that applications of maize cob biochar at rates of 5–20 t/ha have always increased biomass production by 15–50%, owing to improved soil pH, cation exchange capacity, and microbial activity that have promoted improved nutrient uptake on acidic, nutrient-poor Acrisols and Ferralsols prevalent in western and central Kenya (Enos et al. 2025). A 2022–2025 study in western Kenya tested the use of ALV systems on-farm, where maize cob biochar amended soils led to higher fresh biomass yields of crops like black nightshade and spider plant by 25–45% and lower need of input fertilizers by 30% across multiple seasons (Mahmoud et al. 2021). In slightly different, but similarly, smallholder plots in central Kenya, maize-based biochar mixed with organic inputs increased kale and spinach biomass by 20–35%, and reduced the overall cost of production by reducing the use of synthetic fertilizer and enhancing the use of water in rainy conditions (Mugwe et al. 2009). The experimental literature on Africa also records that, in sub-humid regions, biochar has proved favourable to soil fertility and vegetable biomass, yielding net-costs savings of 15–25% with a successively rising marketable yields, after which the amendment demand has declined again (Enos et al. 2025).

The cost-reducing features are quite striking in these Kenyan sites, with biochar offsetting the high cost of imported fertilizers, recycling nutrients and cut leaching losses by 20–50% to allow small holders to produce comparable or even better biomass on 30–40% less investment input (Enos et al. 2025; Ngome et al. 2011). They added trials of maize cob biochar 2020–2024 in low fertility soils in counties of Meru and Kisii that elucidated 18–40% of leafy greens and minimize input expenses by manageable nutrients and decreased their amendments (Mutegi et al. 2020). The identical result was measured also in Ghana and Cameroon where analogous biochar maize was used, and it turned out that biomass and cost-effectiveness in degraded condition soils raised significantly (Calys-Tagoe et al. 2019). In general, the example of maize cob biochar in these African case studies demonstrates its effectiveness in rejuvenating low-fertility soils to grow leafy vegetables, supporting sustainable production, increasing biomass, and reducing the economic costs borne by Kenyan farmers (Zubairu et al. 2023).

6.3 Urban/Peri-Urban Agriculture

Biochar has been embraced as a widespread practice not only in rooftop gardens, but also in vegetables grown in containers, where the concentration of light can be exploited to either promote crop growth and the sustainability of densely populated areas of the planet since the urban and peri-urban agriculture is space-constrained and the soil may be contaminated (Picca et al. 2025). In 2018–2025 trials, biochar of urban organic waste, including food scraps, yard trimmings, and municipal compost, is highlighted as an important amendment that improves soil parameters, doubles or triples vegetable productivity, and reduces heavy metal uptake in urban contaminated substrates (Yang et al. 2025). This solution enables the concept of circular economies by recycling waste, reducing landfills and enhancing food security in urban settings with limited arable land and where the land is polluted (Grard et al. 2018).

When biochar is added to rooftop and container soils at 10–30% volume, it has been shown to increase crop yields of vegetables including lettuce, tomatoes and herbs by improving water retention, nutrient availability and root development in soilless or lightweight media (Chen et al. 2021). A 2021–2025 urban rooftop farm study using compost-based media with biochar reported 25–40% higher yields on container-grown leafy greens, which was explained by improved microbial communities as well as porosity to support growth in unfixed urban microclimates (Picca et al. 2025). Similarly, tomatoes, basil, and other vegetable productivity has risen 30–50 times on urban waste biochar in hydroponic-inspired container-based systems and has improved vegetative health and irrigation limitations over space-limited area like balconies and green roofs (Nguyen et al. 2024). More recent and up-to-date 2024–2025 literature reviews of green roof substrates confirm once again the use of biochar to enhance vegetable growth by 15–35% due to improved stormwater retention and thermal regulation in cities (Beck et al. 2011).

The adsorption and immobilization of heavy metals (such as lead (Pb), cadmium (Cd), and zinc (Zn)) in polluted urban soils by biochar of urban organic waste can also reduce risks of accumulated heavy metals in vegetables by 30–60% (Wang and Ni 2023). In community garden trials in 2020–2024 in the U.S. and European cities, biochar additions reduced metal bioavailability in container vegetables, resulting in safer produce with 20–45% lower Pt and Cd concentrations and still continued to increase yield (Malone and Shakya 2024). In regard to rooftop systems that utilize waste-based biochar, a 2021–2025 study showed that they decreased by 25–50% the amount of heavy metals moved to edible portions of such crops as spinach and carrots and increased by 15–30% the harvests of such crops when biochar was introduced to improve the soil health (Ziss et al. 2021). These two advantages become even greater when the urban feedstock (urban sewage sludge or food waste) is converted into biochar, as discussed in 2018–2025 agronomical assessments, facilitating the urban horticulture to be low-risk and high-performing (López-Cano et al. 2018). Altogether, these case studies in urban and peri-urban areas illustrate the potential of biochar to transform rooftop and container vegetable production to obtain greater productivity and safer food by valorizing waste (reducing risks) and reducing contaminants (Schröder et al. 2021).

7 Challenges and Considerations

Although biochar is a promising technology in the vegetable production system, there are numerous challenges and issues that need to be tackled in ensuring that biochar is optimized to achieve maximum benefits as well as guaranteeing sustainability when used in various horticultural situations. These are agronomic, technical, and economic problems, and must be carefully managed to prevent unintended consequences, and maximize returns in high-value vegetable production. More recent literature (2020–2025) emphasizes the idea that each case should be approached individually, since biochar can be more effective, depending on the rate of application, the source of feedstock and the type of crop, and the standardization and cost-benefit analysis of biochar remains a controversial topic in the context of introducing it into practice in real agriculture (Kumar et al. 2024).

The worst risk in this connection is the overuse and the resultant effect in this case is the salt accretion and the resultant imbalance of the micronutrients and this may be deadly to the soil and the quality of vegetables grown in intensive systems. In 2024, it was also reported that experiments on sandy soils indicated that high levels of biochar (more than 20–50 t/ha) increase the conductivity of soils by 10–30% and increase the percentage of salt that can limit the germination and growth of root systems of the delicate crops containing lettuce and spinach (Murtaza et al. 2022). Similarly, overuse may cause changes in the supply of micronutrients to cause imbalances, such as zinc or copper toxicity in tomatoes and peppers, and a 15–25% reduction in yield was observed in 2025 experiments in response to biochar application beyond optimum levels (e.g. >30 t/ha) without adequately tracking micronutrient levels (Kumar et al. 2025). These risks provide a reason to employ site-specific application rates and recent reviews recommend 10–20 t/ha of most vegetable crops as a benefit/toxicity risk trade-off (Goswami et al. 2022).

The other reason is that the reaction of vegetables to different kinds of biochar is not constant and that it is dependent on crops and it therefore becomes difficult to apply it in all systems of production and also new studies are required so that the reaction is good in all systems of production. When applied in indicators, woody biochar with leafy greens as rice husk biochar has improved the root crop by 10–30%, 15–40% respectively as per the 2023–2025 field data (Layek et al. 2022). However, fruiting veggies like eggplant react unpredictably with the application of low-temperature biochar (below 400 °C) because certain investigations report no distinction in effect on harvest when low-temperature biochar is utilized, while others report different effects based on the circumstances of pyrolysis (Janu et al. 2021). This is because biochar-crop selectivity is sensitive to soil pH, nutrient demands, and microbiological interactions and because feedstock options must be diversified, and production programs must be consistent to produce consistent returns (Kumar et al. 2025).

The absence of standardized guidelines on how to use biochar in vegetable production systems is a drawback to its widespread use as the available guidelines are usually generalized and disregard local soil type, climatic differences and crop requirements. Other studies propose 5–20 t/ha, depending on global testing, but more recently, a review indicates a high degree of variability in response between 2–10 and 15–30 t/ha in greenhouse tomatoes and poor field soils, respectively, and indicates the need to use local protocols (Wani et al. 2023a). There are no standard biochar quality tests, e.g., carbon content, surface area, or stability, to implement, and any industry standards created in 2025 would allow consistency in performance across vegetable horticulture (Bolan et al. 2022). This type of non-linearity can become a source of scalability issues especially when the only constraint is that the smallholder systems contain personalized advice (Kumar et al. 2024).

The economic aspect of the manufacturing and application of biochar and the resulting benefits of using it in high value horticultural practices is a major factor contributing to its inapplicability to farmers who must balance the cost of application in the short run with the long-term rewards. Commercial production cost of 0.1 to 100–300 per ton, based on feedstock and pyrolysis technology, can represent 10–20% of input expenditures in intensive vegetable cultivation, according to 2023–2025 economic studies (Steiner et al. 2016). However, according to the case study of 2025 in India and Kenya, the net returns on investments can cover the spending in 1–3 years, and the net can increase 20–50% in developed markets (Kumar et al. 2025). The high-income area barriers include high initial investment, which can be reduced by subsidies and local production of biochar using agricultural waste that can result in 30–50 costs reduction, as will be discussed in 2024 sustainability analyses (Wani et al. 2023b). In general, high-value horticulture has a good economic future, but long-term profitability depends on the efficiency of production, as well as the inclusion of biochar in the current farming operations (Mohammadi 2021).

8 Conclusion

As a biochar material, it is possible to consider biochar as a promising solution to the vegetable yield, soil health and food nutrition problem with the focus on the degraded or highly input-based production systems that tend to cause the soil depletion and lower the sustainability. It is a fundamental component of climate-smart agriculture due to its many benefits as an intervention to enhance soil structure, nutrient retention, microbiological activity and produce quality, particularly in places with management challenges such as soil erosion, nutrient leaching, and water stress (Kumar et al. 2025). According to the 2020–2025 study, biochar use is demonstrated to be useful in improving vegetable crops, by 15–50; soil organic carbon, by 20–40; and nutritional quality, e.g. antioxidants, vitamins, by 10–30, consequently biochar use is a food security solution, and eco-resilience in intensive agriculture (Wani et al. 2023a).

Integrated soil fertility management (ISFM) is the best practice to achieve the best results through the use of biochar mixed with organic amendments such as compost or manure and balanced synthetic fertilizers, to produce synergistic effects that maximize productivity and sustainability. Asian, African, and urban field trials indicate that biochar at 5–20 t/ha combined with vermicompost or with NPK fertilizers can increase yields by 20–40%, at lower input costs and reduced environmental overheads, as demonstrated in Kenyan leafy vegetable systems and in Indian tomato experiments (Layek et al. 2022). This multi-faceted but potentially more effective whole-farm intervention based on nutrient management can also be implemented at a scale with those farmers that switch to low-input or organic production (Goswami et al. 2022).

Future research on biochar refinement in vegetable production must take the form of biochar-blended fertilizers, biochar application rates specific to each climate, and on-farm trials. The biochar (i.e. nutrient-enriched preparations, slow-release mixtures etc.) formulations can be optimized, and the risks of too much nutrient being deposited can be avoided, and the 2025 tests revealed that 20–50% more of the nutrient was deposited using the controlled-release designs (Kumar et al. 2025). Depending on the type of soil and crop needs, targeted rates (between 2 and 30 t/ha) need to be confirmed locally to capture variability in response, but farmer-led trials in places like sub-Saharan Africa and South Asia can bridge the science-practice gap, and are scalable and cost-effective (Wani et al. 2023b). With this emphasis, biochar can be optimized to address the dynamic needs of sustainable horticulture in vegetable production across the globe.

Table 2 Vegetable crop responses to biochar application (yield, quality, and soil effects)

Biochar type	Crop	Soil effect	Yield effect	Quality effect	Reference
	Tomato		Increase in yield by 29.55%	Soluble solids, vitamin C, by 4.28%, and 6.77%	Lei et al. (2024)
	Tomato		Increase 20–30%	Vit C, improved from 3.5 mg/100 g to 4.8 mg/100 g TSS from 3.5 to 4.5 Brix Lycopene content from 13.3 to 16.5 mg/g	Almaroai and Eissa (2020)
Date palm biochar	Eggplant		8.51% fruit weight		Ebrahimi et al. (2021)
Pistachio biochar	Eggplant		11.56% fruit weight		Ebrahimi et al. (2021)
Manure biochar	Tomato in salinity stress condition	Was able to neutralise the effect of salinity	15–20% yield increased		Kamal et al. (2024)
Peanut shell biochar	Rape lettuce and pakchoi (Brassica chinensis L.)		Increased vegetable yield by 15.5–22.1%		Kamal et al. (2024)
Poultry manure–composted biochar	Rape lettuce and pakchoi (Brassica chinensis L.)	Soil Organic Carbon	Increased vegetable yield by 27.2%	Neutralization efficiency and reduced nitrate accumulation in vegetables.	Kamal et al. (2024)
	Spinach, lettuce, cabbage	Often benefit from improved N and P availability		Higher leaf chlorophyll and reduced nitrate accumulation.	An et al. (2022)
Blue mallee (Eucalyptus polybractea) biochar, 10 t ha^{-1}	Cauliflower, peas, and broccoli	C:N ratio increased slightly			Boersma et al. (2017)
Peanut shell biochar		SOM content was increased	Vegetable biomass was increased to 186% of the original		Premarathna et al. (2019); Wang et al. (2018)

(continued)

Table 2 (continued)

Biochar type	Crop	Soil effect	Yield effect	Quality effect	Reference
Corncob biochar	Cauliflower		Increased yield by 43.5% compared to organic manure		Wang et al. (2019)
Biomass (e.g., crop residues, livestock manure and solid waste) biochar			Increased vegetable yield by 28.4% compared to traditional fertilizer application		Bi et al. (2022)
Beechwood biochar	Eggplant		Increased yield by 37%		Iacomino et al. (2022a, b); Dickinson et al. (2015)
Palm tree branches biochar	Sugar beet		Improved the yield by 30–40%		Li et al. (2022)
Fruit tree branches biochar	Tomato		Increased the yield by 54.79%		Guo et al. (2021)
Corn stalk		Increased water-holding capacity, porosity, soil liquid phase percentage and decreased soil bulk density, soil solid phase percentage			
Corn		Increased the proportion of aggregates, soil pore structure			

References

Abbey ANA, Frimpong KA, Odoi-Yorke F, Ampofo EA, Darko RO (2025) A review of biochar's sustainability in climate-smart agriculture: recent advances, emerging trends, and future directions. Eur J Agron 169:127690

Adekiya AO, Ogunbode TO, Esan VI, Adedokun O, Olatubi IV, Ayegboyin MH (2025) Short term effects of biochar on soil chemical properties, growth, yield, quality, and shelf life of tomato. Sci Rep 15(1):24965

Aktar MW, Sengupta D, Chowdhury A (2009) Impact of pesticides use in agriculture: their benefits and hazards. Interdiscip Toxicol 2(1):1

Al-Harbi AR, Obadi A, Al-Omran AM, Abdel-Razzak H (2020) Sweet peppers yield and quality as affected by biochar and compost as soil amendments under partial root irrigation. J Saudi Soc Agric Sci 19(7):452–460

Ali A, Jabeen N, Chachar Z, Chachar S, Ahmed S, Ahmed N et al (2025a) The role of biochar in enhancing soil health & interactions with rhizosphere properties and enzyme activities in organic fertilizer substitution. Front Plant Sci 16:1595208

Ali N, Jiang Q, Akhtar K, Luo R, Jiang M, He B, Wen R (2025b) Biochar and manure co-application improves soil health and rice productivity through microbial modulation. BMC Plant Biol 25(1):914

Almaroai YA, Eissa MA (2020) Effect of biochar on yield and quality of tomato grown on a metal-contaminated soil. Sci Hortic 265:109210

Amoroso G, Idbella M, Motti R, Gemini A, Cozzolino A, Bonanomi G (2025) Biochar, beneficial microbes, and agro-industrial byproducts in seed coatings: improving germination and biomass in multiple crops. Horticulturae 11(5):554

An N, Zhang L, Liu Y, Shen S, Li N, Wu Z, Han X (2022) Biochar application with reduced chemical fertilizers improves soil pore structure and rice productivity. Chemosphere 298:134304

Antonangelo JA, Sun X, Eufrade-Junior HDJ (2025) Biochar impact on soil health and tree-based crops: a review. Biochar 7(1):51

Beck DA, Johnson GR, Spolek GA (2011) Amending greenroof soil with biochar to affect runoff water quantity and quality. Environ Pollut 159(8–9):2111–2118

Bi R, Zhang Q, Zhan L, Xu X, Zhang X, Dong Y et al (2022) Biochar and organic substitution improved net ecosystem economic benefit in intensive vegetable production. Biochar 4:46. https://doi.org/10.1007/s42773-022-00168-9

Bi R, Wang B, Xu X, Dong Y, Jiao Y, Xiong Z (2024) Biochar is superior to organic substitution for vegetable production—a revised approach for net ecosystem economic benefit. Agronomy 14(11):2693

Boersma M, Wrobel-Tobiszewska A, Murphy L, Eyles A (2017) Impact of biochar application on the productivity of a temperate vegetable cropping system. N Z J Crop Hortic Sci 45(4):277–288. https://doi.org/10.1080/01140671.2017.1329745

Bolan N, Hoang SA, Beiyuan J, Gupta S, Hou D, Karakoti A, Wang HL (2022) Multifunctional applications of biochar beyond carbon storage. Int Mater Rev 67(2):150–200. https://doi.org/10.1080/09506608.2021.1922047

Calys-Tagoe E, Sadick A, Yeboah E, Amoah B (2019) Biochar effect on maize yield in selected farmers fields in the northern and upper east regions of Ghana. J Exp Agric Int 30(6):1–9

Chen H, Du X, Lai M, Nazhafati M, Li C, Qi W (2021) Biochar improves sustainability of green roofs via regulate of soil microbial communities. Agriculture 11(7):620

Chen M, Wang Y, Pan J, Zhong L, Qiao M, Gao C, Wang Y (2025) Can N-doped biochar achieve safe vegetable production in soil heavily contaminated by heavy metals? Toxics 13(2):79

Dickinson D, Balduccio L, Buysse J, Ronsse F, van, Huylenbroeck, G., and Prins, W. (2015) Cost-benefit analysis of using biochar to improve cereals agriculture. GCB Bioenerg 7:850–864. https://doi.org/10.1111/gcbb.12180

Ebrahimi M, Souri MK, Mousavi A, Sahebani N (2021) Biochar and vermicompost improve growth and physiological traits of eggplant (Solanum melongena L.) under deficit irrigation. Chem Biol Technol Agric 8(1):19

Enos O, Godfrey N, Darius A, Samwel M, Joseph GO, Harun G (2025) Biophysical and farm management drivers of soil fertility under African leafy vegetable production small-scale systems in Kenya. Farm Syst 3:100157

Food & Agriculture Organization (FAO) (2024) Statistical year book. FAO

Gelardi DL, Ainuddin IH, Rippner DA, Patiño JE, Abou Najm M, Parikh SJ (2021) Biochar alters hydraulic conductivity and impacts nutrient leaching in two agricultural soils. Soil 7(2):811–825

Goswami L, Nath A, Sutradhar S, Bhattacharya SS, Kalamdhad A, Vellingiri K, Kim KH (2022) Application of biochar in agriculture and environment, and its safety issues. Biomass Bioenergy 159:106393. https://doi.org/10.1016/j.biombioe.2022.106393

Grard BJP, Chenu C, Manouchehri N, Houot S, Frascaria-Lacoste N, Aubry C (2018) Rooftop farming on urban waste provides many ecosystem services. Agron Sustain Dev 38(1):2

Gu K, Gao K, Guan S, Zhao J, Yang L, Liu M, Su J (2025) The impact of the combined application of biochar and organic fertilizer on the growth and nutrient distribution in wheat under reduced chemical fertilizer conditions. Sci Rep 15(1):5285

Guo L, Yu H, Kharbach M, Zhang W, Wang J, Niu W (2021) Biochar improves soil-tomato plant, tomato production, and economic benefits under reduced nitrogen application in Northwestern China. Plants 10:759. https://doi.org/10.3390/plants10040759

Guo R, Gong W, Qi S, Xu J, Shang Z, Joseph S (2025) Biochar-based urea enhances nitrogen use efficiency and mitigates nitrogen leaching in greenhouse vegetable production. Environ Technol Innovation 38:104104

Harris J, de Steenhuijsen Piters B, McMullin S, Bajwa B, de Jager I, Brouwer ID (2023) Fruits and vegetables for healthy diets: priorities for food system research and action. Sci Innov Food Syst Transform 87:87–104. https://www.cabidigitallibrary.org/doi/pdf/10.5555/20220021080

Huang Y, Chen Y, Wen D, Zhao P, Li F, Li L et al (2022) Biochar-based molybdenum slow-release fertilizer enhances nitrogen assimilation in Chinese flowering cabbage (Brassica parachinensis). Chemosphere 303:134663

Huang L, Chen W, Wei L, Li X, Huang Y, Huang Q et al (2024) Biochar blended with alkaline mineral can better inhibit lead and cadmium uptake and promote the growth of vegetables. Plants 13(14):1934

Iacomino G, Idbella M, Laudonia S, Vinale F, Bonanomi G (2022a) The suppressive effects of biochar on above-and belowground plant pathogens and pests: a review. Plants 11(22):3144

Iacomino G, Sarker TC, Ippolito F, Bonanomi G, Vinale F, Staropoli A et al (2022b) Biochar and compost application either alone or in combination affects vegetable yield in a volcanic mediterranean soil. Agronomy 12:1996. https://doi.org/10.3390/agronomy12091996

Ighalo JO, Ohoro CR, Ojukwu VE, Oniye M, Shaikh WA, Biswas JK et al (2025) Biochar for ameliorating soil fertility and microbial diversity: from production to action of the black gold. iScience 28(1):111524

Jabborova D, Kadirova D, Narimanov A, Wirth S (2021) Beneficial effects of biochar application on lettuce (Lactuca sativa L.) growth, root morphological traits and physiological properties. Ann Phytomed 10(2):93–100

Jabborova D, Annapurna K, Azimov A, Tyagi S, Pengani KR, Sharma P, Sayyed RZ (2022) Co-inoculation of biochar and arbuscular mycorrhizae for growth promotion and nutrient fortification in soybean under drought conditions. Front Plant Sci 13:947547

Janu R, Mrlik V, Ribitsch D, Hofman J, Sedláček P, Bielska L, Soja G (2021) Biochar surface functional groups as affected by biomass feedstock, biochar composition and pyrolysis temperature. Carbon Resour Convers 4:36–46. https://doi.org/10.1016/j.crcon.2021.02.002

Jatuwong K, Aiduang W, Kiatsiriroat T, Kamopas W, Lumyong S (2025) A review of biochar from biomass and its interaction with microbes: enhancing soil quality and crop yield in Brassica cultivation. Life 15(2):284

Jaufmann E, Schmid H, Hülsbergen KJ (2025) Soil carbon accrual and yield response to biochar and compost in a four-year organic field study. Nutr Cycl Agroecosyst 131:1–23

Jiafan LI, Mengyuan SONG, Junhui YIN, Lihong GAO, Yongqiang TIAN (2025) The beneficial effects of biochar on overall soil quality and plant performance are dose-dependent and are closely associated with soil pH. Pedosphere. https://doi.org/10.1016/j.pedsph.2025.05.001

Jin X, Zhou X, Wu F, Xiang W, Pan K (2023) Biochar amendment suppressed Fusarium wilt and altered the rhizosphere microbial composition of tomatoes. Agronomy 13(7):1811

Kafle A, Singh S, Singh M, Bajwa P, Deb S, Simpson C, Ritchie G (2025) Influence of deficit irrigation and biochar amendment on growth, physiology, and yield of cucumber in West Texas. Sci Rep 15(1):9615

Kamal MZU, Sarker U, Roy SK, Alam MS, Azam MG, Miah MY, Hossain N, Ercisli S, Alamri S (2024) Manure-biochar compost mitigates the soil salinity stress in tomato plants by modulating the osmoregulatory mechanism, photosynthetic pigments, and ionic homeostasis. Sci Rep 14(1):21929. https://doi.org/10.1038/s41598-024-73093-5

Kebede T, Berhe DT, Zergaw Y (2023) Effects of biochar and compost application on soil properties and on the growth and yield of hot pepper (Capsicum annuum L.). Appl Environ Soil Sci 2023(1):8546135

Kiani H, Khalesro S, Mokhatssi-Bidgoli A, Sharifi Z (2024) Biochar and conservation tillage affect the agronomic performance and fatty acid composition of Nigella sativa L. under both irrigated and dryland conditions. Sci Rep 14(1):2648

Korai SK, Korai PK, Jaffar MA, Qasim M, Younas MU, Shabaan M et al (2025) Leveraging biochar amendments to enhance food security and plant resilience under climate change. Plants 14(13):1984

Kumar SS, Wani OA, Malik AR, Kumar S, Patel R (2024) Biochar as a new emerging tool to achieve food security and environmental sustainability under current climate change scenario. In: Babu S et al (eds) Agricultural diversification for sustainable food production. Springer Nature Singapore Pte Ltd., pp 61–145. https://doi.org/10.1007/978-981-97-7517-0_3

Kumar Y, Ren W, Tao H, Tao B, Lindsey LE (2025) Impact of biochar amendment on soil microbial biomass carbon enhancement under field experiments: a meta-analysis. Biochar 7(1):2

Layek J, Das A, Mitran T, Nath C, Meena RS, Singh R et al (2022) Cereal + legume intercropping: an option for improving productivity and sustaining soil health. In: Legumes for soil health and sustainable management. Springer, pp 133–172. https://doi.org/10.1007/978-981-19-4277-2_7

Lei Y, Xu L, Wang M, Sun S, Yang Y, Xu C (2024) Effects of biochar application on tomato yield and fruit quality: a meta-analysis. Sustainability 16(15):6397. https://doi.org/10.3390/su16156397

Li Y, Yao N, Liang J, Wang X, Jia Y, Jiang F et al (2022) Optimum biochar application rate for peak economic benefit of sugar beet in Xinjiang. China Agric Water Manage 272:107880. https://doi.org/10.1016/j.agwat.2022.107880

Li X, Wu D, Liu X, Huang Y, Cai A, Xu H, Zhang W (2024) A global dataset of biochar application effects on crop yield, soil properties, and greenhouse gas emissions. Sci Data 11(1):57

Li T, Cui L, Filipović V, Tang C, Lai Y, Wehr B, Dang YP (2025) From soil health to agricultural productivity: the critical role of soil constraint management. Catena 250:108776

López-Cano I, Cayuela ML, Sánchez-García M, Sánchez-Monedero MA (2018) Suitability of different agricultural and urban organic wastes as feedstocks for the production of biochar—Part 2: agronomical evaluation as soil amendment. Sustainability 10(6):2077

Lu M, Powlson DS, Liang Y, Chadwick DR, Long S, Liu D, Chen X (2021) Significant soil degradation is associated with intensive vegetable cropping in a subtropical area: a case study in southwestern China. Soil 7(2):333–346

Luan T, Yan J, Zhao X, Tuoheti T, Xu Y, Gan T et al (2025) Combined effects of biochar and silkworm excrement compost applications on soil properties and vegetable growth. Appl Soil Ecol 210:106067

Luigi M, Manglli A, Dragone I, Antonelli MG, Contarini M, Speranza S, Faggioli F (2022) Effects of biochar on the growth and development of tomato seedlings and on the response of tomato plants to the infection of systemic viral agents. Front Microbiol 13:862075

Luo P, Zhang W, Xiao D, Hu J, Li N, Yang J (2025) Biochar-based fertilizers: advancements, applications, and future directions in sustainable agriculture—a review. Agronomy 15(5):1104

Mahmoud Y, Njenga M, Sundberg C, de Nowina KR (2021) Soils, sinks, and smallholder farmers: examining the benefits of biochar energy transitions in Kenya. Energy Res Soc Sci 75:102033

Malik L, Hussain S, Shahid M, Mahmood F, Ali HM, Malik M et al (2024) Co-applied biochar and drought tolerant PGPRs induced more improvement in soil quality and wheat production than their individual applications under drought conditions. PeerJ 12:e18171

Malone M, Shakya KM (2024) Trace metal contamination in community garden soils across the United States. Sustainability 16(5):1831

Mazibuko DM, Maskey S, Kurashina K, Okazawa H, Oshima H, Kato T, Kikuno H (2025) Effects of biochar on growth, response to water stress, and post-stress recovery in underutilized vegetable hibiscus sabdariffa from Malawi. Crops 5(2):13

Meng F, Wang Y, Wei Y (2025) Advancements in biochar for soil remediation of heavy metals and/or organic pollutants. Materials 18(7):1524

Mohammadi A (2021) A review of the sustainable application of biochar for wastewater treatment and soil improvement. J Clean Prod 295:126346. https://doi.org/10.1016/j.jclepro.2021.126346

Montoya D, Fernández JA, Franco JA, del Carmen Martínez Ballesta, M. (2022) Enriched-biochar application increases broccoli nutritional and phytochemical content without detrimental effect on yield. J Sci Food Agric 102(15):7353–7362

Mugwe J, Mugendi D, Kungu J, Muna MM (2009) Maize yields response to application of organic and inorganic input under on-station and on-farm experiments in central Kenya. Exp Agric 45(1):47–59

Murtaza G, Ditta A, Ullah N, Usman M, Ahmed Z (2022) Biochar for the management of nutrient impoverished and metal contaminated soils: preparation, applications, and prospects. Environ Sci Pollut Res 29(14):20399–20421. https://doi.org/10.1007/s11356-021-18035-9

Ndede EO, Kurebito S, Idowu O, Tokunari T, Jindo K (2022) The potential of biochar to enhance the water retention properties of sandy agricultural soils. Agronomy 12(2):311

Ngome AF, Becker M, Mtei KM, Mussgnug F (2011) Fertility management for maize cultivation in some soils of Western Kenya. Soil Tillage Res 117:69–75

Nguyen CN, Chau HW, Kumar A, Chakraborty A, Muttil N (2024) Biochar amendment in green roof substrate: a comprehensive review of the benefits, performance, and challenges. Appl Sci 14(16):7421

Obadi A, Alharbi A, Alomran A, Alghamdi AG, Louki I, Alkhasha A, Alqardaeai T (2024) Enhancement in tomato yield and quality using biochar amendments in greenhouse under salinity and drought stress. Plants 13(12):1634

Omokaro GO, Kornev KP, Nafula ZS, Chikukula AA, Osayogie OG, Efeni OS (2025) Biochar for sustainable soil management: enhancing soil fertility. Plant Growth and Climate Resilience, Farming System, p 100167

Pattnaik BK, Behera R, Santra SC, Choudhury S, Biswas JK, Hossain A, Moulick D (2025) Potentials of urban waste derived biochar in minimizing heavy metal bioavailability: a techno-economic review. iScience 28(3):111915

Pei J, Liang Y, Xue L, Zamanian K, Sun S, Li W et al (2025) Five years of biochar amendment combined with reduced fertilization and irrigation improved the soil organic carbon composition and structure in a solonchak. Sci Rep 15(1):21823

Phares CA, Amoakwah E, Danquah A, Afrifa A, Beyaw LR, Frimpong KA (2022) Biochar and NPK fertilizer co-applied with plant growth promoting bacteria (PGPB) enhanced maize grain yield and nutrient use efficiency of inorganic fertilizer. J Agric Food Res 10:100434

Picca G, Goñi-Urtiaga A, de Sosa LL, Rodriguez MVC, Navarro IF, Plaza C, Panettieri M (2025) Sustainability in urban agriculture: the role of biochar in enhancing the productive capacity of compost-based growing media for rooftop farming. Urban For Urban Green 107:128774

Pokharel P, Ma Z, Chang SX (2020) Biochar increases soil microbial biomass with changes in extra-and intracellular enzyme activities: a global meta-analysis. Biochar 2(1):65–79

Poveda J, Martínez-Gómez Á, Fenoll C, Escobar C (2021) The use of biochar for plant pathogen control. Phytopathology 111(9):1490–1499

Premarathna K, Rajapaksha AU, Sarkar B, Kwon EE, Bhatnagar A, Ok YS et al (2019) Biochar-based engineered composites for sorptive decontamination of water: a review. Chem Eng J 372:536–550. https://doi.org/10.1016/j.cej.2019.04.097

Ray PK, Bharti P (2023) Biochar: a quality enhancer for fruit crops. In: International year of millets. Times of Agriculture, p 77

Schreinemachers P, Simmons EB, Wopereis MC (2018) Tapping the economic and nutritional power of vegetables. Glob Food Sec 16:36–45

Schröder C, Häfner F, Larsen OC, Krause A (2021) Urban organic waste for urban farming: growing lettuce using vermicompost and thermophilic compost. Agronomy 11(6):1175

Selvarajh G, Ch'ng HY, Zain NM, Wei LS, Liew JY, Azmin SNHM et al (2024) Enriched rice husk biochar superior to commercial biochar in ameliorating ammonia loss from urea fertilizer and improving plant uptake. Heliyon 10(11):e32080

Sharma S, Mukherjee S, Bolan S, de Figueiredo C, Fachini J, Chang SX et al (2025) Biochar as a potential nutrient carrier for agricultural applications. Curr Pollut Rep 11(1):1–36

Sim DHH, Tan IAW, Lim LLP, Yeo FKS, Phornvillay S, Abat M et al (2025) Biochar-based slow/controlled-release fertilizer for sustainable agriculture: recent advances, challenges and future prospects. J Environ Chem Eng 118826:118826

Simiele M, Argentino O, Baronti S, Scippa GS, Chiatante D, Terzaghi M, Montagnoli A (2022) Biochar enhances plant growth, fruit yield, and antioxidant content of cherry tomato (Solanum lycopersicum L.) in a soilless substrate. Agriculture 12(8):1135

Steiner C, Teixeira WG, Lehmann J, Zech W (2016) Long term effects of manure, charcoal and mineral fertilization on crop production and fertility on a highly weathered central Amazonian upland soil. Plant Soil 291(1–2):275–290. https://doi.org/10.1007/s11104-006-9183-2

Sun M, Fan SX, Zhang N (2025) Effects of biochar combined with the application of plant ash and effective microorganisms on the soil in the vegetable facility. Sci Rep 15(1):15824

Thakur N, Kaur J, Kurbah I, Aakriti. (2024) Unlocking the potential of biochar: a comprehensive review of enhancing vegetable production in challenging soils. J Plant Nutr 47(17):2897–2912

United Nations (2024) Global issues, food. https://www.un.org/en/global-issues/food

Wang YC, Ni JJ (2023) Biochar application on heavy metal immobilization in unsaturated soil with vegetation: a review. Int J Geotech Eng 17(3):260–270

Wang J, Xia K, Wai MG, Gao Y, Odinga ES, Ling W et al (2018) Application of biochar to soils may result in plant contamination and human cancer risk due to exposure of polycyclic aromatic hydrocarbons. Environ Int 121:169–177. https://doi.org/10.1016/j.envint.2018.09.010

Wang Z, Li Y-k, Guo W-z, Xu Z, Wang LC et al (2019) Yield, nitrogen use efficiency and economic benefits of biochar additions to Chinese flowering cabbage in Northwest China. Nutr Cycl Agroecosyst 113:337–348. https://doi.org/10.1007/s10705-019-09971-6

Wang C, Luo D, Zhang X, Huang R, Cao Y, Liu G et al (2022) Biochar-based slow-release of fertilizers for sustainable agriculture: a mini review. Environ Sci Ecotechnol 10:100167

Wang J, Zhai B, Shi D, Chen A, Liu C (2024) How does bio-organic fertilizer combined with biochar affect Chinese small cabbage's growth and quality on newly reclaimed land? Plants 13(5):598

Wang T, Cong X, Yu J, Zhao H, Xu L, Pang G, Xu Z (2025) Impact of biochar and deficit irrigation on greenhouse gas emissions reduction in maize-wheat rotation systems. Agric Water Manag 315:109547

Wani OA, Malik AR, Kumar S, Patel R (2023a) Biochar application for sustainable soil management: a review. J Soil Sci Plant Nutr 23(4):5678–5695. https://doi.org/10.1007/s42729-023-01345-6

Wani OA, Malik AR, Kumar S, Patel R (2023b) Enhancing soil carbon sequestration through biochar: challenges and opportunities. Agric Ecosyst Environ 345:108325. https://doi.org/10.1016/j.agee.2023.108325

Wu Q, Zhang J, Liu X, Chang T, Wang Q, Shaghaleh H, Hamoud YA (2023) Effects of biochar and vermicompost on microorganisms and enzymatic activities in greenhouse soil. Front Environ Sci 10:1060277

Xing Y, Wang X, Mustafa A (2025) Exploring the link between soil health and crop productivity. Ecotoxicol Environ Saf 289:117703

Xiu L, Sun Y, Han X (2025) Effects of biochar amendment on potassium supply capacity and potassium accumulation in soybean across diverse soils. Plants 14(13):1959

Xu X, Wang H, Egun IL, Gan T, Huang Z, Liu H et al (2025a) Enhancing vegetable yield and quality with biochar: prospects and challenges. Front Sustain Food Syst 9:1621366

Xu Z, Zhou R, Xu G (2025b) Global analysis on potential effects of biochar on crop yields and soil quality. Soil Ecol Lett 7(1):240267

Yang Y, Chen T, Xiao R, Chen X, Zhang T (2022) A quantitative evaluation of the biochar's influence on plant disease suppress: a global meta-analysis. Biochar 4(1):43

Yang X, Wong SC, Ge L, Chan WP, Lisak G (2025) Enhancing vegetable yield and quality in urban farming under soil and soil-less conditions: the synergistic role of biochar in food security and waste management. J Clean Prod 513:145724

Yilmaz H (2025) Enhancements in morphology, biochemicals, nutrients, and L-Dopa in faba bean through plant growth promoting rhizobacteria and arbuscular mycorrhizal Fungi. Sci Rep 15(1):7390

Yue S, Li W, Qin F, Dong M, Weng G, Ali HM et al (2025) Assessment of biochar filter application in improving chromium stress tolerance and plant physiology in Chinese cabbage (Brassica rapa) under a flow-through water setup. BMC Biotechnol 25(1):74

Zhang M, Liu Y, Wei Q, Liu L, Gu X, Gou J (2022) Biochar-based fertilizer enhances the production capacity and economic benefit of open-field eggplant in the karst region of southwest China. Agriculture 12(9):1388

Zhang K, Khan Z, Khan MN, Luo T, Luo L, Bi J, Hu L (2024a) The application of biochar improves the nutrient supply efficiency of organic fertilizer, sustains soil quality and promotes sustainable crop production. Food Energy Secur 13(1):e520

Zhang P, Liu J, Zhang H, Wang M, Xu J, Yu L, Cai H (2024b) Deficit irrigation interacting with biochar mitigates N2O emissions from farmland in a wheat–maize rotation system. Agric Water Manag 297:108843

Zhang W, Niu W, Luo H (2024c) Effect of biochar amendment on the growth and photosynthetic traits of plants under drought stress: a meta-analysis. Agronomy 14(12):2952

Zheng CG, Pei HH, Zhang YS, Li JX, Liu FW, Qiao XX, Qin JM (2024) Effects of Biochar on Growth and Pollutant Accumulation of Lettuce in Soil Co-contaminated with Tetracycline and Copper. Huan Jing Ke Xue 45(5):3037–3046

Zhou M, Ying S, Chen J, Jiang P, Teng Y (2021) Effects of biochar-based fertilizer on nitrogen use efficiency and nitrogen losses via leaching and ammonia volatilization from an open vegetable field. Environ Sci Pollut Res 28(46):65188–65199

Ziss E, Friesl-Hanl W, Götzinger S, Noller C, Puschenreiter M, Watzinger A, Hood-Nowotny R (2021) Exploring the potential risk of heavy metal pollution of edible cultivated plants in urban gardening contexts using a citizen science approach in the project "heavy metal city-zen". Sustainability 13(15):8626

Zubairu AM, Michéli E, Ocansey CM, Boros N, Rétháti G, Lehoczky É, Gulyás M (2023) Biochar improves soil fertility and crop performance: a case study of Nigeria. Soil Syst 7(4):105

Zulfiqar F, Moosa A, Nazir MM, Ferrante A, Ashraf M, Nafees M et al (2022) Biochar: an emerging recipe for designing sustainable horticulture under climate change scenarios. Front Plant Sci 13:1018646

Biochar and Fruit Crops: Enhancing Yield, Quality and Soil Sustainability in Perennial Horticulture

Vijay Kumar, Vijay Khajuria, Saurav Gupta, Owais Ali Wani, Sabreena Ashraf, and Abid Shabir Bali

Abstract

Biochar is a carbon-rich material produced through the thermal decomposition of organic biomass under limited oxygen conditions. It has emerged as a promising soil amendment in horticulture due to its potential to improve soil fertility, plant growth, fruit yield, and overall soil health. The unique physicochemical properties of biochar, including high porosity, large surface area, and stable carbon content, make it an environmentally sustainable option for enhancing soil structure, water retention, and nutrient availability. In horticultural systems, the application of biochar improves soil physical properties such as aeration and water-holding capacity, which are essential for root development and plant vigor. Biochar also enhances soil chemical characteristics by increasing the soil's cat-

V. Kumar
Rainfed Research Sub-station for Sub-tropical Fruits, Raya, Sher-e-Kashmir University of Agricultural Sciences and Technology, Jammu, Jammu and Kashmir, India

V. Khajuria
Farming System Research Centre, Chatha, Sher-e-Kashmir University of Agricultural Sciences and Technology, Jammu, Jammu and Kashmir, India

S. Gupta
Krishi Vigyan Kendra, Samba, Sher-e-Kashmir University of Agricultural Sciences and Technology, Jammu, Jammu and Kashmir, India

O. A. Wani (✉)
Division of Agronomy, Faculty of Agriculture, SKUAST-Kashmir, Wadura, Sopore, Jammu and Kashmir, India

S. Ashraf
Punjab Agricultural University, Ludhiana, Punjab, India

A. S. Bali
Division of Agronomy, Sher-e-Kashmir University of Agricultural Sciences and Technology of Kashmir, Srinagar, Jammu and Kashmir, India

O. A. Wani et al. (eds.), *Biochar for Sustainable Soil Health and Crop Production*, https://doi.org/10.1007/978-981-92-2230-8_11

ion exchange capacity, thereby improving nutrient retention and reducing nutrient losses through leaching. Furthermore, biochar promotes beneficial soil microbial activity, enhances microbial diversity, and supports nutrient cycling processes that contribute to improved plant growth and productivity. However, the effectiveness of biochar varies depending on feedstock type, production conditions, application rate, soil characteristics, and crop species. Potential limitations associated with biochar application include nutrient immobilization, increased soil pH under alkaline conditions, and the possible presence of harmful compounds depending on production methods. Beyond soil improvement, biochar provides important environmental benefits by sequestering carbon in soil for extended periods, thereby mitigating greenhouse gas emissions and contributing to climate change mitigation. The conversion of crop residues and fruit processing wastes into biochar also supports sustainable waste management and promotes a circular bioeconomy. Biochar-based fertilizers have recently been developed as an innovative approach to enhance nutrient use efficiency and soil fertility. Common biochar feedstocks include fruit peels, shells, woody residues, and other by-products of fruit production systems. Numerous studies have demonstrated the role of biochar in improving soil physicochemical properties, influencing soil microbial communities, and enhancing productivity in fruit orchards. Although the beneficial effects of biochar on soil and plant health are well recognized, important knowledge gaps remain regarding long-term impacts, economic feasibility, and crop-specific management strategies. This chapter reviews recent advances in biochar production from fruit crop residues and evaluates its applications in sustainable fruit production systems, highlighting future research directions and practical considerations for large-scale adoption.

Keywords

Biochar · Amendments · Soil fertility · Fruit productivity · Sustainable horticulture

1 Introduction

Approximately 3.2 billion people living in rural communities of developing countries depend on agriculture for their livelihood and food security (Knickel et al. 2018). Fruit crops play a crucial role in improving human nutrition and enhancing food security because they provide a wide range of essential vitamins, minerals, antioxidants, and bioactive phytochemicals. Fruits and vegetables contribute significantly to the diets of rural populations, particularly women, supplying nearly 38% of vitamin A, 35% of vitamin C, 30% of calcium, and approximately 17% of daily iron requirements (Padhy and Behera 2015). Regular consumption of fruits contributes substantially to human health due to their rich nutritional composition. Apples are an important source of dietary fibre, antioxidants, and vitamin C, supporting digestion, cognitive health, and cholesterol regulation. Mangoes contain high levels of vitamins A and C as well as potassium, which strengthen immune function, improve vision, and may reduce the risk of certain cancers. Citrus fruits are particularly rich in vitamin C, which plays a vital role in immune response and collagen synthesis. Bananas provide potassium, vitamin B6, dietary fibre, and antioxidants that contribute to cardiovascular health, muscle function, and digestive well-being. Fruit fibres are classified into soluble and insoluble fractions, both of which contribute to intestinal health. Fruits such as apples, citrus fruits, and berries contain soluble fibres that form gel-like substances when combined with water, promoting the growth of beneficial gut microbiota (Rist et al. 2013). These microbial processes produce short-chain fatty acids (SCFAs), which help regulate immune responses. Vitamin C rich fruits such as kiwi and strawberry strengthen intestinal lining integrity and protect tissues against oxidative damage caused by reactive oxygen species (Miles and Calder 2021). Additionally, fruits such as mango and avocado provide vitamin E, which protects digestive system cell membranes from oxidative stress (Vincente et al. 2014). Cherries and grapes contain antioxidants including carotenoids and anthocyanins that reduce oxidative stress, repair cellular damage, and suppress inflammation. Consequently, fruit consumption lowers the risk of chronic diseases and promotes a healthy lifestyle.

India possesses diverse agro-climatic conditions that support large-scale horticultural production and ranks second globally in fruit and vegetable production. During 2022–23, India produced approximately 108.34 million tonnes of fruits and 212.91 million tonnes of vegetables, contributing nearly 10% of global production in each category (FAO Statistics 2023). Globally, fresh fruit production increased from 576.65 million metric tonnes to approximately 933.1 million metric tonnes by 2022 (OECD/FAO 2023). India is the world's leading producer of banana and mango, with domestic markets serving as the primary destination for most fruit and vegetable production. Nearly three-quarters of total production is consumed domestically, whereas only about 1% is exported and around 2% undergoes processing. Post-harvest losses and waste account for nearly 20% of total production (National Academy of Agricultural Sciences 2019). Despite their nutritional, economic, and cultural importance, fruit and vegetable production faces several environmental challenges, including declining soil fertility, drought stress, nutrient deficiencies, and climate variability. Climate change has emerged as a major threat affecting fruit crop productivity and sustainability. Fluctuations in temperature, variations in light intensity, and nutrient imbalances present significant challenges for orchard management. Both abiotic and biotic stresses influence fruit yield and quality. Biotic stresses caused by insects, herbivores, nematodes, bacteria, fungi, and weeds can significantly damage fruit crops, altering their sensory and nutritional characteristics (Umar et al. 2022). Among environmental factors, temperature is one of the most critical determinants of plant growth and development. It influences morphological, physiological and biochemical processes, including photosynthesis, respiration, transpiration, flowering, fruit set and seed development (Taiz et al. 2018). Declining soil fertility further reduces fruit productivity by lowering yields, deteriorating fruit quality, and increasing susceptibility to pests and diseases. Soil degradation results in nutrient depletion, loss of organic matter, and poor soil structure, ultimately affecting plant health and orchard sustainability.

Water scarcity and drought stress represent additional constraints to fruit production, as large areas worldwide are increasingly affected by changing climatic conditions that lead to yield losses (Rachappanavar et al. 2022). Since plant roots directly interact with soil to absorb water and nutrients, maintaining soil health is essential for sustainable fruit production systems. In this context, biochar has gained considerable attention as a sustainable soil amendment capable of improving soil quality, enhancing agricultural productivity, and mitigating climate change impacts. Addressing climate change, improving soil fertility, managing agricultural waste, conserving water resources and developing renewable energy sources are among the major challenges of modern agriculture. Rapid urbanization and population growth have further intensified pressure on soil resources, leading to declining soil health worldwide. Therefore, innovative and sustainable soil management strategies such as biochar application are increasingly being explored to ensure long-term horticultural productivity and environmental sustainability.

Biochar has emerged as a sustainable alternative to conventional soil amendments such as farmyard manure, compost, and synthetic fertilizers due to its long-term stability, environmental benefits, and positive effects on soil health. Unlike traditional amendments that decompose rapidly, biochar persists in soil for extended periods, improving soil quality while simultaneously contributing to greenhouse gas mitigation. Its multifunctional properties make it highly suitable for modern sustainable agricultural systems. Biochar improves soil physical properties by enhancing soil structure, increasing water-holding capacity, improving aeration, and supporting soil aggregation. These improvements promote the development of a stable soil microbial ecosystem, which is essential for maintaining long-term soil health and productivity (Umar et al. 2022). Compared with compost and synthetic fertilizers, biochar exhibits a greater ability to retain nitrogen and other essential nutrients, thereby reducing nutrient losses through leaching (Gao et al. 2023). In addition to physical improvements, biochar significantly enhances soil chemical properties. It increases cation exchange capacity (CEC), stabilizes soil pH, and improves soil porosity and texture, facilitating the retention of both cations and anions within the soil matrix. Soil pH strongly influences microbial abundance and activity, and biochar application can create favorable conditions for microbial functioning and nutrient cycling. Because biochar can be produced from agricultural and organic waste materials, it represents a cost-effective soil amendment that also provides economic incentives for farmers through waste valorization.

The physicochemical characteristics of biochar largely determine its field performance in fruit production systems. Therefore, careful characterization of feedstock type, production conditions, and biochar properties is essential before its application in orchard soils. Proper soil analysis and testing help determine compatibility between biochar and soil conditions, enabling researchers and growers to optimize application strategies for maintaining ideal soil environments and promoting healthy plant growth. Application of biochar in orchard systems has been associated with improvements in fruit quality attributes, including fruit size, color, firmness, and shelf life. Enhanced nutrient availability, improved soil moisture retention, and suppression of soil-borne diseases collectively contribute to the production of high-quality fruits (Mahmud et al. 2019). For example, citrus orchard soils amended with biochar exhibit improved water retention and reduced nutrient leaching, allowing greater water availability during dry periods (Jeffery et al. 2011). Abiotic stresses include extreme temperatures, drought, salinity, and heavy metal contamination which pose significant challenges to plant growth, development, and productivity. Climate change is increasing the frequency and severity of these stresses, threatening global food security. Biochar has shown considerable potential in enhancing plant tolerance to abiotic stress conditions. Its high porosity and large surface area improve soil moisture dynamics, thereby helping plants cope with water stress (Kuzyakov et al. 2014).

Another important environmental benefit of biochar is its capacity for carbon sequestration. During pyrolysis, biomass carbon is converted into a stable form that can remain stored in soil for decades or even centuries. This process removes carbon dioxide from the atmosphere and reduces greenhouse gas emissions, making biochar an effective tool for climate change mitigation (Wu et al. 2019). Soil salinity represents a major constraint to horticultural productivity, particularly in arid and semi-arid regions where improper irrigation and climate variability exacerbate salt accumulation. Saline soils create drought-like conditions by limiting water uptake and causing ion toxicity in plants. Biochar application helps reduce sodium availability in the soil solution, thereby limiting the uptake of harmful ions by plant roots and improving plant performance under saline conditions. Beyond agriculture, biochar also serves as a renewable energy resource. It can be utilized in biomass combustion and gasification processes for the generation of heat and electricity (Azeta et al. 2021). Additionally, converting agricultural residues into biochar contributes to sustainable waste management by reducing landfill disposal and methane emissions while transforming waste biomass into a valuable resource. Overall, biochar represents an integrated solution linking soil health improvement, climate change mitigation, renewable energy generation, and circular bioeconomy development. This chapter therefore synthesizes current scientific knowledge on the role of biochar in improving soil quality and productivity in fruit orchard systems.

2 Mechanism of Biochar Benefits in Fruit Crops

2.1 Soil Physical Improvements

Application of biochar to soil modifies several physical characteristics of the soil porous medium, including water retention capacity, bulk density, pore distribution, and soil aggregation. These improvements largely result from the highly porous and sponge-like structure of biochar particles, which enhance soil aeration and moisture storage within orchard soils. In mango orchards, biochar application has been reported to reduce irrigation requirements and alleviate plant stress during dry periods (Mahmud et al. 2019). According to Zanutel et al. (2021), biochar influences soil physical properties through multiple mechanisms. First, incorporation of biochar alters soil pore-size distribution due to its intrinsic structural properties

(Hyvaluoma et al. 2018). Second, biochar interacts with existing soil structure by occupying mineral voids, particularly in coarse-textured soils, thereby improving soil packing and water retention (Edeh et al. 2020). Third, biochar particles may replace portions of mineral or organic matter and act as binding agents that promote aggregation, similar to the effects of organic amendments (Burgeon et al. 2021; Du et al. 2017).

Water movement within soil is strongly influenced by changes in saturated and unsaturated hydraulic conductivity. Modification of the rhizosphere through biochar incorporation improves root-zone structure and enhances water transport efficiency (Kroener et al. 2018). Soil strength represents another critical physical parameter affecting sustainable fruit production because excessive compaction restricts root penetration and nutrient uptake. Soil penetration resistance is therefore widely used as an indicator of root growth limitations in orchard systems. Sharma et al. (2021) reported that application of biochar at 2 t ha^{-1} significantly increased soil aggregate stability by approximately 8.1% in subtropical soils. This improvement is largely attributed to enhanced microbial activity stimulated by biochar, leading to the production of microbial binding substances that strengthen soil aggregates (Han et al. 2020). In erosion-prone mudstone soils, biochar application improved aggregate stability and increased macropore formation, thereby reducing erosion susceptibility and enhancing soil permeability (Hseu et al. 2014). Overall, biochar-mediated improvements in soil structure enhance nutrient availability, regulate soil physicochemical and biological processes, and promote efficient nutrient cycling within the root zone (Guo et al. 2022; Li et al. 2024).

2.2 Soil Chemical and Nutrient Dynamics

Long-term intensive cultivation in fruit orchards often results in deterioration of soil physicochemical and biological properties. Chemical degradation commonly manifests as soil acidification, nutrient leaching, and depletion of organic matter, which are major concerns in perennial orchard systems such as apple, citrus, mango, and banana plantations. Under these conditions, biochar has gained attention as a sustainable soil amendment capable of restoring soil fertility and improving long-term nutrient management. Biochar is a carbon-rich material produced through pyrolysis, a thermochemical process in which biomass is decomposed under limited oxygen conditions. Owing to its porous structure and abundance of surface functional groups, biochar has a strong capacity to retain nutrients and regulate soil acidity, making it particularly beneficial for acidic and low-fertility orchard soils. Long-term field experiments in citrus orchards have demonstrated that repeated biochar application significantly alters soil chemical properties. Zhang et al. (2022) reported that 4 years of biochar amendment increased soil pH by approximately one unit, effectively neutralizing soil acidity and creating favorable growth conditions for citrus trees. The same study observed a 120.77% increase in cation exchange capacity along with a 16.21% rise in soil organic matter content, greatly enhancing the soil's capacity to retain essential nutrients. In nutrient-depleted orchard soils, nutrients are often either immobilized in unavailable forms or lost through leaching. Biochar improves nutrient availability by modifying both soil physical and chemical environments, thereby enhancing orchard sustainability and fruit quality. Application of calcium-modified biochar in citrus orchards significantly increased available phosphorus levels due to both the inherent phosphorus content of biochar and improved phosphorus solubilization under optimized soil pH conditions (Bai et al. 2024).

Similarly, incorporation of biochar into papaya cultivation systems enhanced plant uptake of nitrogen, phosphorus, potassium, calcium, and magnesium, resulting in higher fruit yield, increased fruit size, and improved total soluble solids (Sanchez-Monedero et al. 2019). Enhanced nutrient availability, improved water retention, and suppression of soil-borne diseases collectively contribute to improved tree health and productivity (Liu et al. 2012). The effectiveness of biochar largely depends on feedstock type and pyrolysis conditions, particularly temperature, which influence surface chemistry, nutrient composition, and stability. Numerous studies

Table 1 Features of biochar types used in orchard systems

Fruit crops	Feed stock	Soil type	Pyrolysis temperature (°C)	Effect of soil properties and nutrient content	References
Grapevine	Grapevine-pruning residues	Acid soil	400, 500, and 600	Affected EC, ash, and TC content and specific surface area	Andelini et al. (2023)
Apple	Apple branches	Alkaline soil	450	Increased soil water content and soil pH	Li et al. (2024)
Citrus cv. "Shatangju"	Citrus branches and grass	Acid red soil	500–550	Improved soil moisture content, porosity and available macro and micro nutrients	Guo et al. (2022)
Grapevine	Grapevine pruning residues	Sandy loam	500	Increasing soil pH, CEC, soil water content and decrease bulk density	Beatrice et al. (2024)
Mango	Poultry manure	Saline soil	355	Increases in ash content percent	Iqbal et al. (2021)
Banana	Wood sawdust	Saline-sodic soil	Gradual pyrolysis is done in a kiln with a retention time of 2 h	Increase leaf nutrient content i.e. NPK and reduction of Na and EC	Ogiala (2018)
Grapevine (muscadine)	Pinewood biochar	Sandy soil	400	Enhanced soil electrical conductivity, pH, buffering capacity, alkali hydrolysed N, available P, available K	Chang et al. (2021)

across fruit crops have demonstrated the effectiveness of biochar as a soil amendment, and representative examples are summarized in Table 1.

3 Yield and Quality Enhancement

Numerous biological mechanisms operate within the rhizosphere that regulate nutrient availability to plants, and soil microorganisms associated with plant roots play a central role in these processes. Alterations in soil properties following biochar application can significantly influence microbial community composition, which ultimately governs soil functionality and productivity. Soil microbial populations may reach densities of up to one billion cells per gram of soil, highlighting their fundamental importance in soil ecological processes and agricultural sustainability (Fierer 2017). Soil microorganisms contribute to nutrient mineralization, pesticide degradation, organic matter turnover, and suppression of pathogenic organisms through competitive exclusion. Therefore, changes in microbial community structure directly influence nutrient cycling and plant health. To better understand plant nutrient acquisition mediated by soil microorganisms, Shen et al. (2016a) employed a root-exclusion experimental system using ^{15}N-labelled KNO_3 and Ca_3PO_4 combined with diffusion barriers. This approach enabled differentiation between direct root uptake and nutrient acquisition mediated by microbial communities and arbuscular mycorrhizal fungi (AMF). Results demonstrated that biochar application enhanced AMF abundance and stimulated root colonization. When biochar was applied in combination with AMF inoculation, phosphorus availability increased significantly, while plant tolerance to moisture stress improved (Li and Cai 2021; Shen et al. 2016a, b). Biochar-amended soils also showed selective pathogen suppression under continuous cropping systems. For instance, reductions in pathogenic fungal genera such as *Ceratobasidium* and *Monosporascus* were observed, while overall fungal diversity increased, helping restore rhizosphere ecological balance (Shen et al. 2016a, b).

Although biochar may suppress specific pathogens such as *Fusarium*, maintaining overall microbial diversity remains essential for sustaining healthy orchard ecosystems. Improved soil nutrient status and moderated soil pH following biochar application promote the proliferation of beneficial fungal communities, which show negative correlations with pathogens such as *Phytophthora capsici* and associated disease incidence. Several studies have further demonstrated that combining biochar with biocontrol fungi, including *Trichoderma* and *Paecilomyces*, enhances biological disease suppression in soils (Elmer 2016; Zhang et al. 2019). Long-term field experiments indicate that biochar amendments progressively improve soil structure, fertility, and biological activity over time, providing sustained benefits within the root zone and contributing to resilient orchard production systems (Wan et al. 2023).

3.1 Yield Effects

Biochar application has demonstrated significant potential to enhance fruit yield through improvements in soil health, nutrient availability, and plant physiological performance. In apple orchards, biochar influences fruit set, fruit retention, fruit size, and overall tree vigor, ultimately leading to increased productivity and reduced fruit drop. Biochar has gained global scientific attention as an innovative soil conditioner and has become a rapidly expanding research area in horticulture and environmental management (Zhang et al. 2012). Studies conducted in China emphasize the importance of soil improvement strategies for sustaining productivity in aging apple orchards, particularly those exceeding 20 years of cultivation (Li et al. 2024). Appropriate biochar application improves soil moisture retention, enhances nutrient availability, and promotes root development, facilitating efficient nutrient uptake and improved fruit yield and quality.

Citrus orchards have similarly benefited from biochar amendments through enhanced soil fertility and improved nutrient dynamics. Biochar application has been associated with reduced incidence of root diseases such as *Phytophthora*-induced root rot, resulting in healthier plants and increased fruit production (He et al. 2016). Increased yields observed in biochar-amended orchards are largely attributed to improved water retention, enhanced nutrient availability, and suppression of soil-borne pathogens (Liu et al. 2012). Banana (*Musa* spp.), one of the most important tropical and subtropical fruit crops worldwide, also shows strong positive responses to biochar application. Biochar improves plant water relations and nutrient utilization efficiency, thereby enhancing productivity (Genesio et al. 2015). Application of rice-husk biochar at 10 Mg ha^{-1} significantly improved soil fertility and fruit yield in banana plantations (Mankasingh et al. 2011). Similarly, wood sawdust biochar applied to 'Grande Naine' banana orchards in Egypt increased yields, particularly under saline–sodic soil conditions (Ogiala 2018). Improvements in yield and related parameters under biochar application at 20 Mg ha^{-1} in saline soils are illustrated in Fig. 1.

3.2 Quality Improvements

In addition to yield enhancement, biochar application has demonstrated substantial improvements in fruit quality attributes across various horticultural crops. Long-term field studies in mature apple orchards located in the Loess Plateau of China showed that biochar improved external fruit characteristics such as size, shape, coloration, and weight, as well as internal quality parameters including sugar content, flavor balance, and nutritional composition (Li et al. 2024). Biochar acts as a nutrient reservoir, adsorbing nutrients and releasing them gradually according to plant demand, thereby improving nutrient use efficiency (Van Zwieten et al. 2010).

Banana	Control	Biochar (20 Mg ha^{-1})
Cluster weight (kg)	1.70	1.94
Finger weight (g)	96.135	106.6
Pulp weight (g)	56.98	71
yield (ton/ha)	72.005	82.45
TSS (^{0}Brix)	18.445	21.61
Total sugars (%)	15.985	17.56

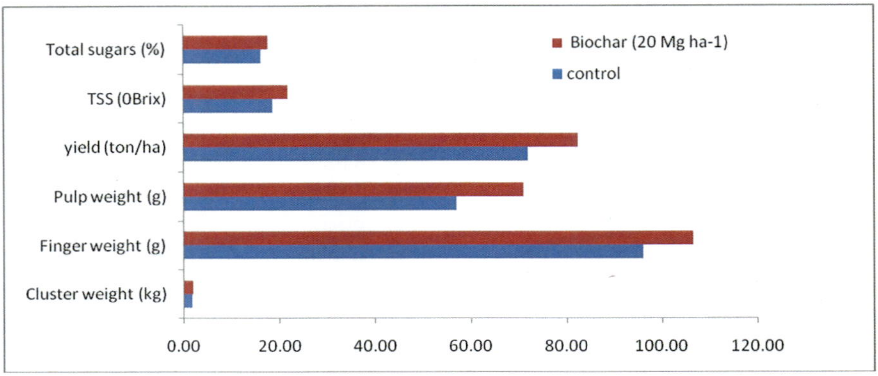

Fig. 1 Fruit yield and quality changes with biochar application

Enhanced nutrient availability strengthens tree growth and physiological performance, leading to superior fruit development and quality. Improvements reported in apple orchards include increased fruit size, enhanced coloration, improved firmness, and higher soluble sugar content (Yin et al. 2020), all of which contribute to greater market value and consumer acceptance. Biochar amendments have also improved quality characteristics in citrus production systems. Studies indicate that biochar application increases fruit size, enhances peel color, improves flavor, and elevates vitamin C content (Shen et al. 2016a, b). Additionally, biochar modifies soil properties and microbial communities across soil depths, creating favorable conditions for sustained fruit quality improvement (Zhang et al. 2021). Reported benefits include improved peel texture, higher soluble solids-to-acidity ratios, increased firmness, extended shelf life, and enhanced visual appeal. The magnitude of yield and quality improvements associated with biochar application depends on feedstock type, pyrolysis conditions, soil characteristics, and crop species. Representative studies demonstrating the effectiveness of biochar amendments in fruit crops are summarized in Table 2.

Table 2 Effects of biochar on yield and quality parameters in selected fruit crops

Name of fruit crops	Feed stock	Soil type	Effect of yield and quality attributes	References
Citrus cv. "Shatangju"	Grasses	Acid red soil	Increased titratable acid, vitamin C content, total sugar, reducing sugar and juice yield by 5.86–22.35%	Guo et al. (2022)
Banana	Wood sawdust	Saline-sodic soil	Enhanced in fruit quality (T.S.S, total sugars, starch) and fruit yield by 12–24%	Ogiala (2018)
Apple	Apple branches	Alkaline soil	Increase in vertical and transverse diameter of fruits, fruit shape index and weight up to 9.1–26.0%	Li et al. (2024)
Mango	Poultry manure	Saline soil	Improved in fruit yield and quality attributes	Iqbal et al. (2021)
Kinnow	Foliar application of nano-biochar suspension (NBS)	Clay loam	Improved fruit setting, flower retention fruit drop and fruit retention and fruit yield	Shani et al. (2023)

4 Case Studies and Experimental Evidence

4.1 Apple Orchards (Temperate Zones)

Apples are among the most widely cultivated fruit crops worldwide and play a significant role not only in human nutrition but also in multiple processing industries. Increasing global demand for apples necessitates sustainable strategies to maintain productivity, particularly in aging orchard systems. A case study conducted in the Loess Plateau region of China evaluated the effects of biochar application on yield and fruit quality in old apple orchards (Li et al. 2024). The Loess Plateau represents one of the largest apple-producing regions globally, where approximately 85% of orchards are more than 20 years old. Aging orchards commonly face challenges such as poor soil water retention, declining tree vigor, reduced productivity, and deterioration of fruit quality. Therefore, identifying effective soil management strategies is essential for sustaining long-term orchard productivity.

In this study, six biochar application rates (2, 4, 6, 8, 10, and 12 kg tree^{-1}) were evaluated. Results demonstrated that biochar significantly enhanced soil water-holding capacity and improved overall tree performance. Biochar application also moderated excessively high soil pH conditions that are unfavorable for apple growth and fruit development. Optimal biochar application improved fruit characteristics by increasing fruit size, enhancing texture, improving sweetness, and achieving a better balance between soluble sugars and organic acids. Consequently, appropriate biochar application contributed to both higher yields and superior fruit quality in mature apple orchards.

4.2 Citrus and Mango Orchards (Tropical/Subtropical)

Citrus species have shown substantial positive responses to biochar application through improved nutrient availability, enhanced flowering, greater fruit retention, and improved fruit quality. A field investigation conducted in an acid red soil "Shatangju" (*Citrus reticulata* cv.) orchard in China during 2020–2021 evaluated the effects of biochar on soil physicochemical properties, rhizosphere microbial diversity, leaf nutrient status, and fruit quality (Guo et al. 2022). All biochar treatments resulted in improved fruit quality compared with the control, along with enhanced fruit coloration. Application of biochar at 3.0 kg plant^{-1} produced the most favorable results. Pooled data from both experimental years revealed significant improvements in several quality parameters, including:

- Increase in soluble solids to titratable acidity ratio (SSC/TA) by 19.9%
- Increase in vitamin C content by 15.7%
- Increase in total sugar content by 12.3%
- Increase in reducing sugars by 15.2%
- Improvement in edible fruit proportion by 3.3%
- Enhancement of juice yield by 7.4%
- Slight improvement in fruit water content

These findings indicate that appropriate biochar application significantly improves foliar nutrient status and overall fruit quality attributes in citrus orchards.

Fruit quality in mango strongly influences storage life, marketability, and economic returns at both national and international levels. A study evaluating the effects of acidified biochar on the 'Sufaid Chaunsa' mango cultivar grown under alkaline soil conditions investigated biochar application rates of 0, 5, 10, 20, and 40 Mg ha^{-1} (Iqbal et al. 2021).

Results showed that application rates of 20 and 40 Mg ha^{-1} significantly improved fruit retention, total soluble solids, sugar content, and ash content compared with untreated controls. A substantial increase in fruit weight and total yield per plant was recorded, with 40 Mg ha^{-1} identified as the most effective treatment. These findings demonstrate that acidified biochar can effectively improve both productivity and fruit quality in mango orchards established on alkaline soils. An experiment conducted in a farmer-managed orchard at Kot Addu, Punjab, Pakistan evaluated the effect of nano-biochar suspension (NBS) applied as a foliar spray on Kinnow mandarin (*Citrus reticulata* L.) (Shani et al. 2023). Four treatment concentrations—0% (control), 1%, 3%, and 5% NBS were applied during the flowering stage. Application of 3% and 5% NBS significantly enhanced flowering intensity, fruit set, fruit retention, and overall yield performance. Biochar treatments increased fruit size, fruit weight, and juice volume while reducing premature fruit drop. Improvements in total soluble solids were also observed compared with untreated plants. The 5% NBS treatment produced the highest fruit yield and superior juice quality, indicating the potential of nano-biochar technology as an emerging tool for citrus productivity enhancement.

4.3 Banana and Grape Vineyards

Banana (*Musa* spp.) is one of the most important tropical and subtropical fruit crops worldwide. A 2-year study (2013–2014) was conducted to evaluate the influence of wood sawdust biochar (WB) on the performance of the 'Grande Naine' banana cultivar grown in saline sodic soils at a private farm in Kafr Yaaqob village, Egypt (Ogiala 2018). Biochar was applied at rates of 0, 5, 10, and 20 Mg ha^{-1}. The best results were obtained with the application of 20 Mg ha^{-1} sawdust biochar. Increasing biochar application rates significantly improved vegetative growth, yield, and fruit quality parameters. All fruit quality attributes include number of hands per bunch, finger length and diameter, pulp weight, peel weight, total soluble solids (TSS), total sugars, and starch content showed marked improvement with increasing biochar application. Additionally, leaf mineral nutrient contents increased, indicating enhanced nutrient uptake efficiency following biochar amendment. These findings demonstrate that biochar application can effectively improve banana productivity and fruit quality under saline sodic soil conditions.

Grapevine (*Vitis vinifera* L.) is among the most widely cultivated fruit crops globally, covering approximately 7.4 million hectares worldwide. A long-term study conducted in a vineyard in Tuscany, Italy, evaluated the effects of biochar application on grapevine fine and coarse root dynamics and associated soil characteristics within the 0–20 cm soil layer (Beatrice et al. 2024). Grapevines established in 2009 and 2010 received one or two biochar applications and were evaluated over two growing seasons for root production, turnover rate, standing biomass, root length, and specific root length. Long-term biochar application significantly increased soil pH, nutrient availability, and water-holding capacity. These improvements were associated with reduced production of fine and coarse roots, reflected in a decrease in total root network size but accompanied by higher root efficiency and turnover rates. The results suggest that biochar amendments can enhance soil quality, thereby reducing carbon investment in root production while maintaining grapevine growth and productivity. Andelini et al. (2023) further investigated biochar produced from grapevine pruning residues of the local grape variety 'Istrian Malvasia' (*Vitis vinifera* L.), focusing on the influence of rootstock type and pyrolysis temperature. Biochar was produced at peak temperatures of 400, 500, and 600 °C. Higher pyrolysis temperatures resulted in lower biochar yield but significantly increased electrical conductivity (EC), ash content, and total carbon (TC). Rootstock type also influenced EC, ash content, and specific surface area of the resulting biochar. The study indicated that more vigorous rootstocks improved biochar quality characteristics, and biochar derived from grapevine pruning waste particularly at 500–600 °C represents a valuable resource for soil improvement through sustainable biomass valorization.

5 Biochar Application Strategies in Orchards

In many established orchards, trench or ditch methods are commonly used to apply biochar in zones where active tree roots develop, often in combination with mulching or deep soil placement around mature trees. Soil surrounding the trees is typically excavated in circular basins or trenches along the tree row or around the trunk area. These trenches are then filled with biochar, either mixed with topsoil or replaced with amended soil containing biochar. The principal approaches for biochar incorporation into orchard soils include surface mixing, deep placement, and surface mulching applications (Verheijen et al. 2010). Biochar application significantly improves soil physical and chemical properties, including enhanced water-holding capacity, improved soil permeability, and increased crop productivity. The particle size distribution of biochar varies considerably depending on feedstock type and pyrolysis conditions. Most biochar materials possess low structural strength and may break into smaller particles during handling and soil incorporation. For fruit crops such as mango, banana, and citrus, biochar application rates generally range between 5 and 20 t ha^{-1}, although optimal rates depend on soil type, crop species, and biochar availability. Biochar is frequently combined with complementary soil amendments such as compost, farmyard manure (FYM), or biofertilizers to maximize agronomic benefits. In perennial orchard systems, reapplication every 2–3 years is often recommended to maintain long-term soil improvement effects, as suggested by the National Bamboo Mission and related research initiatives.

Biochar should ideally be placed within the root activity zone, commonly referred to as the drip line, which represents the ground area beneath the outer canopy where roots actively absorb water and nutrients. Ideally, biochar should be incorporated throughout the entire drip-line area rather than confined to the initial planting pit; however, this may not always be practically feasible. Various application strategies can therefore be employed during tree planting or in established orchards, including subsurface banding, localized soil incorporation, or surface broadcasting. Care must be taken during application to avoid root injury and soil compaction. Seedlings raised in growing media amended with biochar may exhibit early growth advantages; however, as plants mature, roots often extend beyond the treated zone, emphasizing the importance of wider soil incorporation. Biochar applications have also shown promising results in landscaping, turfgrass establishment, and urban green spaces, demonstrating its broader horticultural utility. Biochar may be blended with sand, topsoil, compost, or turfgrass substrates before field application to ensure uniform distribution. Additionally, biochar can be formulated through pelleting, prilling, or co-composting with organic amendments such as manure or compost. Different biochar formulations are suited to specific application methods, and finely ground biochar may be advantageous for certain horticultural uses.

Several experimental studies have demonstrated the positive effects of biochar across horticultural crops. Application of biochar-based fertilizers within the root zone increased aboveground plant biomass by 56%, total leaf area by 222%, and

Table 3 Integrated biochar management in fruit orchards

Fruit crops	Biomass sources	Pyrolysis temperature (°C)	Application methods	Impacts
Apple	Apple branches	450	Biochar mixed with the upper 30 cm soil	Increased fruit vertical and transverse diameters, fruit shape index and fruit weight by 9.1–26.0%
Citrus cv. "Shatangju"	Citrus branches and grass	500–550	Biochar soil were mixed	Increased titratable acid, vitamin C content total sugar reducing sugar juice yield by 5.86–22.35%
Mango	Poultry manure	355	Biochar mixed with soil	Improved in fruit yield, fruit retention, fruit weight and quality i.e. soluble sugar, acidity, ash content and total soluble solids
Banana	Wood sawdust	Slow pyrolysis in a kiln with a retention time of 2 h	Biochar mix with surface soil	Enhanced in fruit quality (T.S.S, total sugars, starch) and fruit yield by 12–24%

chlorophyll content index by 140%. Leaf phosphorus concentration increased by 53%, improving the nitrogen-to-phosphorus ratio and alleviating phosphorus deficiency (Meyerzu Drewer et al. 2022). Similarly, application of *Eucalyptus citriodora* Hook-derived biochar at 2.5 g kg^{-1} soil enhanced basil growth by approximately 10% under slightly alkaline soil conditions (Pandey et al. 2016). Biochar produced from *Eichhornia* biomass applied at 20 kg ha^{-1} significantly increased soil enzymatic activity, with dehydrogenase activity increasing by 21% and acid and alkaline phosphatase activities increasing by 32% and 22.8%, respectively (Guenet et al. 2021). In saline–sodic soils, sawdust biochar applied at 20 Mg ha^{-1} significantly improved pseudostem length, girth, leaf area, bunch weight, and finger weight in 'Grande Naine' banana orchards (Ogiala 2018). Careful selection of biomass feedstock, optimization of pyrolysis conditions, and appropriate soil incorporation techniques allow biochar applications to be tailored to different orchard systems and environmental conditions, thereby maximizing improvements in soil health, plant growth, and fruit productivity (Table 3).

6 Integration with Climate Resilient Horticulture

Climate change poses a significant threat to horticultural production by increasing temperatures, causing irregular rainfall patterns, and intensifying extreme weather events that adversely affect crop productivity, fruit quality, and long-term sustainability of horticultural systems. The integration of biochar into climate-resilient horticulture involves the use of biochar, a carbon-rich material produced through biomass pyrolysis to improve soil health and enhance crop tolerance to climate-induced stresses. Biochar improves soil structure, enhances water-holding capacity, increases soil carbon storage, and improves nutrient availability to plants. These improvements enable crops to better withstand adverse environmental conditions and contribute to sustainable agricultural production. When combined with other sustainable management practices such as cover cropping, conservation tillage, crop rotation, crop diversification, organic farming approaches, climate-resilient cultivar selection, efficient pest management, adaptive local farming strategies, and farmer knowledge sharing biochar significantly enhances soil quality and ecological stability. The integration of cover crops with biochar application protects soil structure, reduces erosion, enhances nutrient cycling, and improves soil moisture retention. Conservation tillage systems, including no-till farming, further reduce soil disturbance and erosion while maintaining biochar stability within the soil profile (Ayompe et al. 2021). Perennial cropping systems and diversification strategies contribute to long-term soil fertility by supplying continuous organic matter inputs and improving nutrient retention.

Efficient water harvesting and irrigation management practices become more effective in biochar-amended soils due to enhanced water storage and reduced moisture loss, thereby stabilizing fruit yield during drought conditions. Incorporation of biochar into horticultural production systems also increases the efficiency of organic inputs, promoting sustainable soil fertility and improving resilience against environmental stresses (Hussain et al. 2021). Planting climate-resilient crop varieties in biochar-amended soils ensures improved access to nutrients and moisture even under unfavorable climatic conditions. Integrated Pest Management (IPM) strategies further benefit from biochar application through enhanced microbial activity and increased populations of beneficial organisms that act as natural biological control agents. Community-based adaptation approaches encourage the production of biochar from locally available biomass resources and its integration into regional horticultural systems. Education and knowledge dissemination play a critical role in enabling fruit growers to make informed decisions regarding biochar adoption and climate-adaptive management practices (Veni et al. 2020). Biochar is produced by heating organic biomass under limited oxygen conditions through a process known as pyrolysis. It has emerged as a promising sustainable technology for improving soil fertility, enhancing plant growth, and mitigating climate change impacts. Application of biochar to soil contributes to carbon sequestration, reduces greenhouse gas emissions, improves soil functionality, and supports bioenergy production systems (Marzeddu et al. 2021). Biochar possesses considerable carbon storage potential and can sequester more than 920 kg of CO_2 per tonne of agricultural biomass residue for extended periods, potentially lasting hundreds to thousands of years (Yang et al. 2021). The extent of carbon stabilization depends on biochar structure and its interactions with soil minerals and microbial communities (Luo et al. 2023). Recent studies have demonstrated that biochar application can reduce emissions of major greenhouse gases such as nitrous oxide (N_2O) and methane (CH_4), both of which significantly contribute to global warming (Lyu et al. 2022). For example, biochar produced from paper mill residues, biosolids, and poultry manure has shown substantial reductions in N_2O emissions in acidic soils. However, the mitigation efficiency of biochar varies depending on soil moisture conditions, biomass feedstock type, and pyrolysis temperature, all of which influence greenhouse gas dynamics (Purakayastha et al. 2016).

Biochar also exhibits strong adsorption capacity for gaseous pollutants such as hydrogen sulphide (H_2S). Adsorption efficiency is influenced by factors including biochar moisture content (>85% v/w), alkaline pH (>8.0), surface area, and the presence of functional groups such as hydroxyl (−OH) and carboxyl (−COOH). Under moist and aerobic conditions, H_2S interacts with alkaline biochar surfaces to form sulphate compounds such as (K, Na_2SO_4), which may subsequently become available to plants as sulphate ions (SO_4^{2-}) (Oni et al. 2019). Overall, biochar provides substantial environmental benefits by promoting carbon sequestration, improving soil health, reducing pollution, and lowering atmospheric concentrations of harmful gases. Nevertheless, careful management of biochar production and application is essential to minimize potential environmental risks associated with pyrolysis emissions and biomass sourcing. When responsibly produced and strategically applied, biochar represents a powerful tool for improving soil productivity while contributing to climate change mitigation and sustainable horticultural development.

7 Challenges and Considerations

The major challenges associated with biochar application in orchard systems include the high cost and labour requirements for large-scale production, variability in biochar properties, and the need for site-specific and crop-specific application strategies. The composition and effectiveness of biochar vary considerably depending on the type of feedstock used and the production conditions, particularly pyrolysis temperature. Different feedstocks such as wood residues, animal manure, and crop wastes produce biochars with distinct physical and chemical characteristics, which influence their ability to improve soil health, enhance water retention, and promote carbon sequestration. Moreover, pyrolysis temperature significantly affects biochar properties such as surface area, porosity, nutrient availability, and structural stability, ultimately determining its agronomic performance. Because soil characteristics, climatic conditions, and crop requirements vary widely across orchard systems, biochar application requires carefully tailored management practices. Soil texture, pH, organic matter content, nutrient status, and environmental conditions strongly influence biochar responses in agricultural soils. Therefore, a uniform or generalized recommendation is ineffective, and site-specific as well as crop-specific guidelines are essential to maximize benefits and ensure sustainable use. Management of crop residues, particularly orchard biomass waste, also remains a major concern in many agricultural regions worldwide. A significant proportion of crop residues are still burned in open fields, contributing to environmental pollution. Converting these residues into biochar represents a sustainable waste management strategy that reduces residue burning while providing a valuable soil amendment. Limited farmer awareness and insufficient extension services constitute additional barriers to the adoption of biochar technology in horticulture. Common challenges faced by farming communities include the limited availability of biochar production units, lack of technical training, incompatibility with traditional cooking practices, poor technological development, and low durability of existing pyrolysis systems. These constraints increase adoption costs and perceived risks for farmers. Consequently, greater investment is required to develop farmer-friendly biochar production technologies that are efficient, durable, affordable, and suitable for on-farm use, including systems safe for household environments.

In many rural areas, biomass resources such as dry leaves, wood residues, corn cobs, and coconut shells serve as primary household energy sources. Integrating biochar production with household energy systems offers a practical and attractive solution. Multifunctional pyrolysis units capable of simultaneously producing biochar and providing energy for cooking or heating can reduce opportunity costs and encourage wider adoption among smallholder farmers. Such integrated systems allow farmers to generate biochar locally for use in nearby plantations while meeting household energy needs. Beyond soil improvement, biochar production from agricultural waste has significant potential to contribute to the global bioeconomy. Farmers may benefit economically through improved crop productivity and potential commercialization of biochar products. However, large-scale adoption requires further research to evaluate economic feasibility, environmental safety, and possible risks such as emissions during the pyrolysis process. Long-term, multi-location, and multi-crop field experiments are needed to assess biochar performance under diverse soil types and climatic conditions. Future research should also focus on developing environmentally safe, low-cost, and efficient biochar production technologies capable of generating standardized biochar-based fertilizers. Establishing quality standards, certification systems, and scientifically validated field application guidelines will be essential for promoting responsible and effective biochar use. Such coordinated efforts will enable growers and stakeholders to utilize biochar more efficiently within sustainable horticultural production systems and integrated nutrient management programs.

8 Conclusion

Biochar is a carbon-rich material produced through the process of pyrolysis and has gained significant attention as a sustainable soil amendment in horticultural systems. Numerous studies have demonstrated that biochar provides multiple agronomic benefits, including enhanced plant growth, increased crop yield, improved fruit quality, and overall productivity enhancement. These positive effects are primarily attributed to its ability to improve soil health, enhance soil fertility, and mitigate the adverse impacts of changing climatic conditions. The effectiveness of biochar largely depends on the appropriate selection of feedstock materials and careful control of pyrolysis conditions, which determine its physical, chemical, and biological properties. Biochar contributes to climate change mitigation by promoting carbon sequestration through the stabilization of carbon in soils and by reducing greenhouse gas emissions. Its high adsorption capacity improves nutrient retention and minimizes nutrient losses through leaching, making essential nutrients more available to plants over longer periods. Consequently, biochar application enhances nutrient use efficiency, leading to improved fruit quality, higher yields, and better overall plant performance. In addition, biochar helps alleviate chemical constraints in fruit cultivation systems. Through its adsorption–desorption mechanisms, biochar can regulate soil pH, reduce sodium toxicity, and improve calcium balance, thereby creating a more favorable soil environment for plant growth and development. Owing to its nutrient content and slow-release characteristics, biochar also serves as a sustainable alternative to synthetic fertilizers, supporting long-term soil productivity while reducing environmental risks associated with excessive chemical fertilizer use. Future research should focus on understanding crop-specific responses to biochar application across different soil types and climatic conditions. Since individual fruit crops may respond differently, determining optimal application rates, timing, and methods is essential for maximizing benefits. Continued investigation into biochar-based management strategies will provide a strong scientific foundation for sustainable fruit production systems and facilitate the wider adoption of biochar as an environmentally friendly technology for improving orchard productivity and resilience.

References

Andelini M, Major N, Išic N, Kovacevic TK, Ban D, Palcic I, & Radunic M (2023). Goreta Ban, S. Sugar and Organic Acid Content Is Dependent on Tomato (Solanum Lycoperiscum L.) Peel Color. Horticulturae 9, 313

Ayompe LM, Schaafsma M, Egoh BN (2021) Towards sustainable palm oil production: the positive and negative impacts on ecosystem services and human wellbeing. J Clean Prod 278:123914

Azeta O, Ayeni AO, Agboola O, Elehinafe FB (2021) A review on the sustainable energy generation from the pyrolysis of coconut biomass. Sci Afr 13:e00909

Bai Y, Huang R, Li S, Li X, Fan Q, Liu S, Hu L (2024) Potential of calcium-modified biochar for soil nutrient and carbon sequestration in citrus orchards. Agriculture 14(12):2222. https://doi.org/10.3390/agriculture14122222

Beatrice P, Dalle Fratte M, Baronti S, Miali A, Genesio L, Vaccari FP, Cerabolini BEL, Montagnoli A (2024) The long-term effect of biochar application to *Vitis vinifera* L. reduces fibrous and pioneer root production and increases their turnover rate in the upper soil layers. Front Plant Sci 15:1384065. https://doi.org/10.3389/fpls.1384065

Burgeon V, Fouche J, Leifeld J, Chenu C, Cornelis JT (2021) Organo-mineral associations largely contribute to the stabilization of century-old pyrogenic organic matter in cropland soils. Geoderma 388:114841. https://doi.org/10.1016/j.geoderma.2020.114841

Chang Y, Rossi L, Zotarelli L, Gao B, Shahid MA, Sarkhosh A (2021) Biochar improves soil physical characteristics and strengthens root architecture in Muscadine grape (Vitis rotundifolia L.). Chem Biol Technol Agric 8:7. https://doi.org/10.1186/s40538-020-00204-5

Du ZL, Zhao JK, Wang YD, Zhang QZ (2017) Biochar addition drives soil aggregation and carbon sequestration in aggregate fractions from an intensive agricultural system. J Soils Sediments 17:581–589. https://doi.org/10.1007/s11368-015-1349-2

Edeh IG, Masek O, Buss W (2020) A meta-analysis on biochar's effects on soil water properties—new insights and future research challenges. Sci Total Environ 714:136857. https://doi.org/10.1016/j.scitotenv.2020.136857

Elmer WH (2016) Effect of leaf mold mulch, biochar, and earthworms on mycorrhizal colonization and yield of asparagus affected by Fusarium crown and root rot. Plant Dis 100:2507–2512. https://doi.org/10.1094/PDIS-10-15-1196-RE

Fierer N (2017) Embracing the unknown: disentangling the complexities of the soil microbiome. Nature reviews microbiology, 15(10), 579–590

FAO (2023) World food and agriculture – statistical yearbook. FAO, Rome. https://doi.org/10.4060/cc8166en

Gao S, Harrison BP, Thao T, Gonzales ML, An D, Ghezzehei TA et al (2023) Biochar co-compost improves nitrogen retention and reduces carbon emissions in a winter wheat cropping system. GCB Bioenergy 15:462–477. https://doi.org/10.1111/gcbb.13028

Genesio L, Miglietta F, Baronti S, Vaccari FP (2015) Biochar increases vineyard productivity without affecting grape quality: results from a four years field experiment in Tuscany. Agric Ecosyst Environ 201:20–25

Guenet B, Gabrielle B, Chenu C, Arrouays D, Balesdent J, Bernoux M, Bruni E, Caliman JP, Cardinael R, Chen S, Ciais P, Desbois D, Fouche J, Frank S, Henault C, Lugato E, Naipal V, Nesme T, Obersteiner M, Pellerin S, Powlson DS, Rasse DP, Rees F, Soussana JF, Su Y, Tian H, Valin H, Zhou F (2021) Can N_2O emissions offset the benefits from soil organic carbon storage? Glob Chang Biol 27:237–256. https://doi.org/10.1111/gcb.15342

Guo YJ, Qiu BL, Khan Z, Jiang H, Ji QH, Fan QZ, Khan MM (2022) The potential for biochar application in "Shatangju" (Citrus reticulate cv.) orchard on acid red soil: biochar prepared from its organic waste in an orchard. Front Plant Sci 13:1001740. https://doi.org/10.3389/fpls.2022.1001740

Han L, Sun K, Yang Y, Xia X, Li F, Yang Z, Xing B (2020) Biochar's stability and effect on the content, composition and turnover of soil organic carbon. Geoderma 364:114184

He W, Xu G, Zhang X, He Y (2016) Effect of biochar on the control efficacy of citrus root rot caused by Phytophthora parasitica. J Plant Pathol 98(2):299–305

Hseu ZY, Jien SH, Chien WH, Liou RC (2014) Impacts of biochar on physical properties and erosion potential of a mudstone slopeland soil. ScientificWorldJournal 2014:602197. https://doi.org/10.1155/2014/602197

Hussain S, Hussain S, Guo R, Sarwar M, Ren X, Krstic D, Aslam Z, Zulifqar U, Rauf A, Hano C, El-Esawi MA (2021) Carbon sequestration to avoid soil degradation: a review on the role of conservation tillage. Plants 10(10):2001

Hyvaluoma J, Kuljun S, Hannula M, Wikberg H, Källi A, Ras K (2018) Quantitative characterization of pore structure of several biochars with 3D imaging. Environ Sci Pollut Res Int 25:25648–25658. https://doi.org/10.1007/s11356-017-8823-x

Iqbal J, Kiran S, Hussain S, Iqbal RK, Ghafoor U, Younis U, Zarei T, Naz M, Germi SG, Danish S, Ansari MJ, Datta R (2021) Acidified biochar confers improvement in quality and yield attributes of Sufaid Chaunsa mango in saline soil. Horticulturae 7:418. https://doi.org/10.3390/horticulturae7110418

Jeffery S, Verheijen FG, van der Velde M, Bastos AC (2011) A quantitative review of the effects of biochar application to soils on crop productivity using meta analysis. Agric Eco Environ 144(1):175–187

Knickel K, Redman M, Darnhofer I, Ashkenazy A, Chebach TC, Šūmane S, Tisenkopfs T, Zemeckis R, Atkociuniene V, Rivera M (2018) Between aspirations and reality: making farming, food systems and rural areas more resilient, sustainable and equitable. J Rural Stud 59:197–210

Kroener E, Holz M, Zarebanadkouki M, Ahmed M, Carminati A (2018) Effects of mucilage on rhizosphere hydraulic functions depend on soil particle size. Vadose Zone J 17:170056. https://doi.org/10.2136/vzj2017.03.0056

Kuzyakov Y, Bogomolova I, Glaser B (2014) Biochar stability in soil: decomposition during eight years and transformation as assessed by compound-specific 14C analysis. Soil Biol Biochem 70:229–236

Li M, Cai L (2021) Biochar and arbuscular mycorrhizal fungi play different roles in enabling maize to uptake phosphorus. Sustainability 13:1–15. https://doi.org/10.3390/su13063244

Li W, Gao J, Zhou S, Zhou F (2024) Effect of biochar on apple yield and quality in aged apple orchards on the loess plateau (China). Agronomy 14:1125. https://doi.org/10.3390/agronomy14061125

Liu Z, Han G, Liu J, Wang D, Chen Z (2012) Effects of biochar amendment on soil aggregates and hydraulic properties. J Soil Sci Plant Nutr 12(3):459–470

Luo L, Wang J, Lv J, Liu Z, Sun T, Yang YI, Zhu YG (2023) Carbon sequestration strategies in soil using biochar: advances, challenges, and opportunities. Environ Sci Technol 57:11357–11372. https://doi.org/10.1021/acs.est.3c02620

Lyu H, Zhang H, Chu M, Zhang C, Tang J, Chang SX, Masek O, Ok YS (2022) Biochar affects greenhouse gas emissions in various environments: a critical review. Land Degrad Dev 33(17):3327–3342. https://doi.org/10.1002/ldr.4405

Mahmud S, Islam MT, Rahman MM, Rahman MM, Akhtar MF (2019) Effects of biochar on soil properties and growth of fruit trees. In: Rahman SE (ed) Advanced agro engineering technologies for rural business development. IGI Global, pp 101–118

Mankasingh U, Choi PC, Ragnarsdottir V (2011) Biochar application in a tropical, agricultural region: a plot scale study in Tamil Nadu, India. Appl Geochem 26:218–221

Marzeddu S, Cappelli A, Ambrosio A, Décima MA, Viotti P, Boni MR (2021) A life cycle assessment of an energy-biochar chain involving a gasifcation plant in Italy. Land (Basel) 10:1256. https://doi.org/10.3390/land10111256

Meyerzu Drewer J, Köster M, Abdulai I, Rotter RP, Hagemann N, Schmidt HP (2022) Impact of different methods of root-zone application of biochar-based fertilizers on young cocoa plants: insights from a pot-trial. Horticulturae 8:328. https://doi.org/10.3390/horticulturae8040328

Miles EA, Calder PC (2021) Effects of citrus fruit juices and their bioactive components on inflammation and immunity: a narrative review. Front Immunol 12:712608. https://doi.org/10.3389/fimmu.2021.712608

National Academy of Agricultural Sciences (2019) Saving the harvest: reducing the food loss and waste. National Academy of Agricultural Sciences, New Delhi. Available online: http://naasindia.org/documents/Saving%20the%20Harvest.pdf

OECD-FAO (2023) "OECD-FAO agricultural outlook", OECD agriculture statistics (database). https://doi.org/10.1787/agr-outl-data-en

Ogiala AMA (2018) Impact of biochar on vegetative parameters, leaf mineral content, yield and fruit quality of Grande Naine banana in saline-sodic soil. Egypt J Hortic 45(2):315–322

Oni BA, Oziegbe O, Olawole OO (2019) Signifcance of biochar application to the environment and economy. Ann Agric Sci 64:222–236. https://doi.org/10.1016/j.aoas.2019.12.006

Padhy C, Behera S (2015) Role of horticulture in human nutrition: an analytical review. Int J Eng Technol Manag Appl Sci 3(6):167–176

Pandey V, Patel A, Patra DD (2016) Biochar ameliorates crop productivity, soil fertility, essential oil yield and aroma profiling in basil (*Ocimum basilicum* L.). Ecol Eng 90:361–366. https://doi.org/10.1016/j.ecoleng.2016.01.020

Purakayastha TJ, Das KC, Gaskin J, Harris K, Smith JL, Kumari S (2016) Effect of pyrolysis temperatures on stability and priming effects of C_3 and C_4 biochars applied to two different soils. Soil Tillage Res 155:107–115. https://doi.org/10.1016/j.still.2015.07.011

Rachappanavar V, Padiyal A, Sharma JK, Gupta SK (2022) Plant hormone mediated stress regulation responses in fruit crops – a review. Sci Hortic 304:111302. https://doi.org/10.1016/j.scienta.2022.111302

Rist VTS, Weiss E, Eklund M, Mosenthin R (2013) Impact of dietary protein on microbiota composition and activity in the gastrointestinal tract of piglets in relation to gut health: a review. Animal 7(7):1067–1078. https://doi.org/10.1017/S1751731113000062

Sanchez-Monedero MA, Cayuela ML, Roig A, Jindo K (2019) Biochar increases papaya yield and enhances fruit quality in tropical soils. J Agric Food Chem 67(1):397–404

Shani MY, Ahmed SR, Ashraf MY, Khan Z, Cocozza C, De Mastro F, Gul N, Pervaiz S, Abbas S, Nawaz H, Brunetti G (2023) Nano-biochar suspension mediated alterations in yield and juice quality of Kinnow (*Citrus reticulata* L.). Horticulturae 9:521. https://doi.org/10.3390/horticulturae9050521

Sharma P, Abrol V, Sharma V, Chaddha S, Srinivasa Rao C, Ganie AQ, Ingo Hefft D, El-Sheikh MA, Mansoor S (2021) Effectiveness of biochar and compost on improving soil hydro-physical properties, crop yield and monetary returns in inceptisol subtropics. Saudi J Biol Sci 28:7539–7549

Shen H, Huang Q, Chen Y, Xiao L (2016a) Biochar composting positively affects the availability of nutrients in a soilless cultivation system. J Plant Nutr Soil Sci 179(2):241–248

Shen Q, Hedley M, Camps Arbestain M, Kirschbaum MUF (2016b) Can biochar increase the bioavailability of phosphorus? J Soil Sci Plant Nutr 16:268–286. https://doi.org/10.4067/S0718-95162016005000022

Taiz L, Zeiger E, Møller IM, Murphy AS (2018) Fundamentals of plant physiology. Oxford University Press, Oxford. ISBN 978-0-1605-7559-8

Umar OB, Ranti LA, Abdulbaki AS, Bola AL, Abdulhamid AK, Biola MR, Victor KO (2022) Stresses in plants: biotic and abiotic. IntechOpen, London

Van Zwieten L, Kimber S, Morris S, Chan KY, Downie A, Rust J, Joseph S, Cowie A (2010) Effects of biochar from slow pyrolysis of papermill waste on agronomic performance and soil fertility. Plant Soil 327:235–246

Veni VG, Srinivasarao C, Reddy KS, Sharma KL, Rai A (2020) Soil health and climate change. In: Climate change and soil interactions. Elsevier, pp 751–767

Verheijen F, Jeffery S, Bastos AC, van der Velde M, Diafas I (2010) Biochar application to soils: a critical scientific review of effects on soil properties, processes and functions. Europe Commission

Vincente AR, Manganaris GA, Ortiz CM, Sozzi GO, Crisosto CH (2014) Nutritional quality of fruits and vegetables. In: Postharvest handling. Academic Press, pp 69–122

Wan H, Liu X, Shi Q, Chen Y, Jiang M, Zhang J, Cui B, Hou J, Wei Z, Hossain MA, Liu F (2023) Biochar amendment alters root morphology of maize plant: its implications in enhancing nutrient uptake and shoot growth under reduced irrigation regimes. Front Plant Sci 14:1122742

Wu P, Ata-Ul-Karim ST, Singh BP, Wang H, Wu T, Liu C, Fang G, Zhou D, Wang Y, Chen W (2019) A scientometric review of biochar research in the past 20 years (1998–2018). Biochar 1:23–43

Yang Q, Masek O, Zhao L, Nan H, Yu S, Yin J, Li Z, Cao X (2021) Country-level potential of carbon sequestration and environmental benefits by utilizing crop residues for biochar implementation. Appl Energy 282:116275. https://doi.org/10.1016/j.apenergy.116275

Yin C, Mueth NA, Hulbert SH, Bezrukova MV, Gao F (2020) Biochar improves apple fruit quality via in luencin growth development and sugar metabolism. Plant Soil 449(1–2):463–478

Zanutel M, Garré S, Bielders C (2021) Long-term effect of biochar on physical properties of agricultural soils with different textures at pre-industrial charcoal kiln sites in Wallonia (Belgium). Eur J Soil Sci 73:e13157. https://doi.org/10.1111/ejss.13157

Zhang A, Liu Y, Pan G, Hussain Q, Li L, Zheng J, Zhang X (2012) Effect of biochar amendment on maize yield and greenhouse gas emissions from a soil organic carbon poor calcareous loamy soil from Central China Plain. Plant Soil 351:263–275

Zhang M, Zhang L, Riaz M, Xia H, Jiang C (2021) Biochar amendment improved fruit quality and soil properties and microbial communities at different depths in citrus production. J Clean Prod 292:126062. https://doi.org/10.1016/j.jclepro.2021.126062

Zhang J, Zhou S, Sun H, Lü F & He P (2019) Three-year rice grain yield responses to coastal mudflat soil properties amended with straw biochar. J. Environ. Manage, 239, 23–29

Zhang M, Riaz M, Xia H, Li Y, Wang X, Jiang C (2022) Four-year biochar study: positive response of acidic soil microenvironment and citrus growth to biochar under potassium deficiency conditions. Sci Total Environ 813:152515. https://doi.org/10.1016/j.scitotenv.2021.152515

Kolibri 360 GmbH
Ferdinand-Jühlke-Straße 7
99095 Erfurt, Deutschland
produktsicherheit@kolibri360.de